LC-NMR

Chromatographic Science Series

A Series of Textbooks and Reference Books

Editor:
Nelu Grinberg

Founding Editor:
Jack Cazes

The *Chromatographic Science Series* offers an in-depth treatment of the latest developments and applications in the field of separation sciences. The series enjoys a broad international readership in part due to the accomplished list of authors and editors that have contributed their expertise to series books covering both classic and cutting-edge topics.

Thin Layer Chromatography in Drug Analysis
Edited by Łukasz Komsta, Monika Waksmundzka-Hajnos, and Joseph Sherma

Pharmaceutical Industry Practices on Genotoxic Impurities
Edited by Heewon Lee

Advanced Separations by Specialized Sorbents
Edited by Ecaterina Stela Dragan

High Performance Liquid Chromatography in Pesticide Residue Analysis
Edited by Tomasz Tuzimski and Joseph Sherma

Planar Chromatography-Mass Spectrometry
Edited by Teresa Kowalska, Mieczysław Sajewicz, and Joseph Sherma

Chemometrics in Chromatography
Edited by Łukasz Komsta, Yvan Vander Heyden, and Joseph Sherma

Chromatographic Techniques in the Forensic Analysis of Designer Drugs
Edited by Teresa Kowalska, Mieczyslaw Sajewicz, and Joseph Sherma

Determination of Target Xenobiotics and Unknown Compounds Residue in Food, Environmental, and Biological Samples
Edited by Tomasz Tuzimski and Joseph Sherma

LC-NMR: Expanding the Limits of Structure Elucidation, Second Edition
Nina C. Gonnella

For more information about this series, please visit: https://www.crcpress.com/Chromatographic-Science-Series/book-series/CRCCHROMASCI

LC-NMR

Expanding the Limits of Structure Elucidation

Second Edition

Nina C. Gonnella

CRC Press
Taylor & Francis Group
Boca Raton London New York

CRC Press is an imprint of the
Taylor & Francis Group, an **Informa** business

CRC Press
Taylor & Francis Group
6000 Broken Sound Parkway NW, Suite 300
Boca Raton, FL 33487-2742

First issued in paperback 2021

© 2020 by Taylor & Francis Group, LLC
CRC Press is an imprint of Taylor & Francis Group, an Informa business

No claim to original U.S. Government works

ISBN 13: 978-1-03-223774-9 (pbk)
ISBN 13: 978-1-138-49340-7 (hbk)

DOI: 10.1201/9781351023740

Dedication

I wish to dedicate this book to the memory of my mother, Antoinette Gonnella, who taught me to never let others place limitations on who you are or what you can become.

Contents

Foreword

LC-NMR (liquid chromatography-nuclear magnetic resonance) is not a new technique. As is the case with many of the most commonly used techniques, it has been around for decades. This hyphenated technique gives researchers the power to employ one of the most information-rich spectroscopy methods directly after separation by liquid chromatography. It sounds simple, but several details need to be considered. This book does a nice job of helping the reader wade through those details.

The mindset for LC-NMR is a bit different than what a typical chromatographer is used to. When tackling a minor impurity, it may be necessary to push the limits of column loading. Peak shape and resolution are of course important, but the most import aspect is getting enough materials into the flow cell to collect the desired data. With UV and MS detection, narrow peaks are preferable and necessary to achieve a good signal-to-noise ratio. In contrast, the volume of an NMR flow cell can be up to 250 μL. This corresponds to 15 s at 1 mL/min, so sensitivity may not be lost due to poor peak shape. Baseline resolution may also not be required, as NMR can often distinguish between different species.

The final analysis is done on NMR. The questions needed to be answered will undoubtedly determine the amount of analyte needed. The so-called LOD (limits of detection) is not a precise value. It can span orders of magnitude dependent on what information is needed. Perhaps a quick one-dimensional (1D) proton spectrum from an on-flow experiment will confirm your suspicions. However, you will need a different strategy if a full dataset including HMBC (heteronuclear multiple bond correlation) and 1D carbon data sets are required for a minor impurity. In the end, the quantity needed is no different from what would be required when using a standard NMR tube. However, this amount is several orders of magnitude higher than what a typical LC-MS run would require. Conversely, the technique is nondestructive and sample recovery is possible in most cases. The advent of the cryogenically cooled NMR probe head has drastically improved the detection capabilities and even more so with the microcryoprobe. Combining this technology with isolation techniques, such as solid-phase extraction, has expanded the capability to exciting new levels.

Right from the introduction, the author is providing practical information covering the nuances of data collection ranging from the deuterium lock to chemical shift tracking. After a good introduction to NMR theory, the key subjects are explored. It starts off in an appropriate area, i.e., separation methods. Here, the author explores topics such as method development and detector types as well as includes a nice discussion on what needs to be considered for compatibility with NMR.

The chromatography instrumentation already available or what might be made available should also be considered. A number of isolation technologies are examined. Someone new to the field may not realize the different roles these technologies play. Researchers often think of liquid chromatography-nuclear magnetic resonance (LC-NMR) only in terms of on-flow, a secondary detector to something like UV. While this may be the case, other modalities exist which can and should be exploited. Which would you choose: stopped flow, loop collection, solid phase extraction? These all have their role and the author puts these into perspective with ample references. In addition, nonchromatographic flow NMR along with the emerging field of capillary electrophoresis NMR is also explored.

The NMR hardware configuration also comes into play. The magnetic field strength might be important, but perhaps a more important consideration would be the NMR probe head. Other options besides a flow probe exist. Is a cryogenically cooled probe available or even necessary? Perhaps a microcoil probe would be a better fit. The author gives a comprehensive survey of probe types and configurations allowing the reader to make an informed decision.

As mentioned earlier, the ultimate goal is to collect the NMR data. The author begins a chapter focusing on NMR experiments with solvent suppression. Solvent suppression will undoubtedly be an important part of your data collection. Although some deuterated solvents will most likely be used, it is also very likely that your peaks of interest will be orders of magnitude lower than the solvent signals. A few pulse sequences are discussed including the popular WET (water suppression enhanced through T_1 effects) sequence. In addition to 1D solvent suppression, the author goes on to discuss the typical experiments used for small-molecule structure elucidation. Here you will find a plethora of useful tips for running these experiments in general as well as specific to LC-NMR.

In this second edition, the author delves into two new topics while expanding on another. The already diverse chapter discussing several practical applications ranging from unstable products to impurity analyses has now been expanded to include formulation adducts. The author describes a first-hand account during the production of a pharmaceutical product.

The first new chapter deals with the extremely important topic of quantitative NMR. HPLC using UV detection and LC-MS is widely used for quantitative analysis. However, it is well known that these methods strongly rely on a multi-point calibration curve. This is due to the varying detectors' responses of these methods. NMR has the advantage of being a primary method. The area of the signal obtained is a direct measurement of the number of nuclei giving rise to that signal. Because of this, multi-point calibration and a cabinet full of reference standards are not necessary. However, quantitative NMR does require some forethought. A standard for instrument calibration is required. The author discusses this along with other important aspects such as instrument setup, data processing, and quantitation limits. Examples from different research fields are presented, making this an excellent primer for quantitative NMR for everyday use and not specific to LC-NMR.

In recent years, the application of computational methods to structural elucidation has become common. Also, added to the second addition is a discussion of computational tools for calculating chemical shifts using quantum chemistry and density functional theory. Here, the author presents a comparison between different functionals and basis sets. Application examples from the literature are presented for varying structural questions such as regioisomers and tautomers. Finally, a functional and basis set are recommended after evaluation, using several pharmaceutically relevant compounds.

For those looking at LC-NMR from either side, a wealth of information is presented. The NMR spectroscopist will benefit from the in-depth information on what is required to interface with liquid chromatography. The details provided on the isolation technologies along with other caveats such as shimming and solvent suppression will be appreciated. The chromatographer will benefit by learning what is needed from their end in contrast to what is typically used for something like LC-MS. This book will be an excellent resource for those venturing into LC-NMR as well as an invaluable reference to those already familiar.

Dr. Robert Krull
Bruker Bio-Spin Corp.

Preface

Since the first publication of this book, the application of LC-SPE-NMR and microcryoprobe technology has become a routine practice in addressing the scientific challenges found in isolating and structurally characterizing complex molecules. The types of molecules characterized by LC-SPE-NMR include very low levels of degradants, reactants, and impurities that may be present in APIs and formulations. Identification of such molecules is needed to satisfy the requirements of regulatory agencies. This is because these compounds can have potential toxic or mutagenic properties when present at ~0.1% or greater; hence, characterization is critical for the assessment of drug safety and efficacy. In addition to degradants and impurities, LC-SPE-NMR plays a critical role in the isolation and spectroscopic characterization of metabolites. With LC-SPE-NMR, unambiguous structural characterization can be achieved, which is not possible with mass spectrometry alone. A key structure elucidation application of an unusual formulation adduct has been added to Chapter 7, which demonstrates the strengths of LC-SPE-NMR methodology when faced with a major structural challenge.

Although NMR spectroscopy is a strong and effective technology for structure elucidation, it is still inherently an insensitive technique. Obtaining sufficient quantities of material for NMR characterization can sometimes be difficult even when employing LC-SPE methods. Often, when studying biosynthesized hydrolysis products of APIs, identification of the structural modifications can be elusive. This may be due to the limited quantities of isolated material as well as potential instability of the material. In addition, structural characterization of isolated metabolites from DMPK studies or process impurity compounds may produce materials that need to be recovered for use in follow-up toxicology studies. In such cases, it becomes imperative to be able to identify and subsequently measure the purity and the quantity of material that are isolated, without compromising the purity or chemical integrity of the sample.

LC-SPE-NMR combined with technological advances in probe design has expanded structure elucidation applications to include metabonomics and metabolomics studies as well as characterization of compound mixtures formed *in situ* (i.e., compounds sensitive to light, oxygen, volatility, time, and so on), natural products, and compound libraries. Nonetheless, structure elucidation of such compounds or complex mixtures can yield multiple structural hypotheses or ambiguous results. Hence, despite the many impressive advances that have been made, there are still limitations to full structure characterization using conventional NMR experiments. The challenges typically arise from severe signal overlap, insufficient nuclei in the molecule needed to trace spectral peak connectivities, absence of information due to low concentrations, and stereochemistry uncertainties when investigating diastereomers with flexible chiral centers.

The expansion of this book was undertaken to augment and enhance the capability of LC-SPE-NMR by the addition of chapters covering qNMR (quantitative NMR) and QM/DFT (quantum mechanics/density functional theory) chemical shift prediction. The first newly added chapter (Chapter 9) presents an overview of qNMR methods that allow measurements of purity and microgram quantities of solubilized isolated materials to be achieved. A discussion on the use of external qNMR references is included, which is required for quantification and recovery of microgram amounts of solubilized materials. Besides ^{1}H qNMR, the chapter also covers the use of ^{31}P and ^{19}F nuclei that can simplify the qNMR method, especially when dealing with complex mixtures of proton-rich organic compounds. Applications of qNMR methods to studies related to natural products, biometabolites, pharmaceutical, and industrial products are incorporated in this chapter.

The second new chapter (Chapter 10) covers fast, accurate, and automated *ab initio* NMR chemical shift prediction programs that we have developed to enhance and accelerate the structure elucidation process. The programs, named H*i*PAS/D*i*CE, use QM/DFT and probability theory, and are routinely applied to elucidate challenging molecular structures. These programs have been extremely powerful in solving problems of regiochemistry, preferred tautomers, impurities, metabolites, decomposition products, assignment ambiguities, and stereochemistry of diastereomers. Overall, many aspects of the technologies presented herein are now a regular part of our structure elucidation practice.

The development and implementation of these technologies would not have been possible without the contributions of creative and dedicated colleagues, especially Keith Fandrick, Paul-James Jones, Om Chaudhary, C. Avery Sader, and Dongyue Xin. Special thanks are extended to Xiaorong He and Keith Horspool for their support and encouragement in the completion of this work. I also wish to thank Aaron Teitelbaum and Rob Krull for providing critical review of portions of this manuscript. Lastly, I would like to extend my deep appreciation to Nelu Grinberg for his friendship and moral support in the pursuit of this work.

Finally I wish to extend my heartfelt love and gratitude to my family, especially my husband Jerry, my sisters Patricia and Maria, and my mother Antoinette, for their continued understanding, encouragement, and support throughout this process.

Author

Nina C. Gonnella, Ph.D., is a Senior Associate Director at Boehringer Ingelheim (BI), where she heads a molecular structure and solid form informatics group, encompassing NMR Spectroscopy, Mass Spectrometry, Single-Crystal X-ray, and Computational Chemistry. She has extensive experience leading Pharmaceutical Research and Development groups spanning structural characterization/ligand screening and *in vitro* and *in vivo* biological studies.

Nina received her Ph.D. in Synthetic Organic Chemistry from the University of Pennsylvania, and subsequently held postdoctoral positions at California Institute of Technology and Columbia University. Prior to joining BI, Nina was group leader/manager at Novartis, where she established and led an NMR group in areas that included small-molecule characterization, protein-ligand structure elucidation, and *in vivo* drug metabolism. She has advanced discovery research through application of structure-based drug design and ligand-based screening technologies to identify and optimize new lead series in both kinase and protease programs. Nina also held a position as adjunct professor at the University of Medicine & Dentistry of NJ (UMDNJ) (Graduate Program). She subsequently joined AbbVie as Group Leader of an NMR-LC-MS group with focus on protein-ligand binding studies. At BI, Nina has advanced BI projects through the development of innovative solutions. She championed solid-state NMR as a company-wide Center of Expertise and initiated the acquisition of small-molecule single-crystal X-ray for global R&D. She introduced "state-of-the-art" performance in integrating LC-NMR and microcryoprobe NMR technology to enable elucidation of microscale degradation products and impurities. Nina initiated and led an international team in the development of powerful commercial ready *in silico* structure elucidation programs that use quantum chemistry, density functional theory, and probability theory to solve challenging chemical structures. Nina co-founded, organized, and chaired a new Gordon Research Conference in "Molecular Structure Elucidation." She has served on scientific review boards, published over 95 peer-reviewed journal articles, authored book chapters, taught courses, and presented numerous national and international lectures.

1 Introduction to LC-NMR

1.1 HISTORICAL REVIEW

Nuclear magnetic resonance (NMR) is a powerful technology that has been extensively used for the structural elucidation and characterization of organic, inorganic, and biological molecules. This technology is over 65 years old; however, during this time frame numerous applications have emerged spanning nearly all scientific disciplines ranging from nuclear physics to medicine.

The origins of NMR began with the simultaneous discovery of nuclear resonance in 1945. The discoveries were made by physicists in the laboratories of Edward Purcell at Harvard and Felix Bloch at Stanford using parafilm and water, respectively [1, 2]. Bloch and Purcell were awarded the Nobel Prize in Physics in 1952 for their discovery. Although the initial goal of the physicists was to determine the magnetic moments of all known elements in the periodic table, their studies revealed unexpected findings regarding the magnetic properties of nuclei and their local environment. These early investigators found that the effects of electrons surrounding the nucleus of a resonating atom introduced slight changes in the precessional frequency of nuclei. This phenomenon gave rise to the concept of the chemical shift, which relates the electronic environment around the nuclei in a molecule to distinct resonance frequencies that are diagnostic of the molecular structure. Although the discovery of NMR occurred in 1945, it was not until 1951 that the ability of the chemical shift to provide structural information at the molecular level was demonstrated. The hypothesis was initially made by an organic chemist, S. Dharmatti, and it was Dharmatti and coworkers who subsequently demonstrated the capability of the NMR technology in structure elucidation using the organic molecule ethanol [3]. The importance of this discovery became apparent and the technology was rapidly embraced by organic chemists who recognized the power of NMR in establishing the chemical structure. However, in the early 1950s, the technology was still in its infancy and significant development soon followed. The initial challenge was to build homogeneous magnets with resolution better than 0.1 ppm and line widths at 0.5 Hz. This was achieved in the mid-1950s, resulting in subsequent development and implementation of new techniques and processes, such as computer-assisted time averaging, to enhance sensitivity for chemical structure elucidation [4].

Despite the introduction of homogeneous magnets and processes such as time averaging [5], NMR still suffered from poor sensitivity. One approach in improving sensitivity was to increase the magnetic field strength. This led to the development of superconducting solenoid magnets with circulating current in the coils that were cooled to 4 K with liquid helium [6]. The initial commercial magnets were available at 200 MHz. Today, magnets of approximately 1 GHz and beyond are being built and may be purchased. The introduction of Fourier transformation (FT) provided two orders of magnitude improvement in signal-to-noise (S/N) ratio and enabled simultaneous recording of all spectral resonances [7]. Decoupling of abundant nuclei, such as protons, enabled isotopically dilute nuclei such as ^{13}C to be more readily observed [8]. In 1971, a new era in NMR spectroscopy was created when Jeener introduced the concept of two-dimensional (2D) FT NMR spectroscopy [9]. Inspired by the lecture of Jeener, Richard Ernst and coworkers showed that 2D FT NMR spectroscopy could be successfully applied to chemical research. Expansion of multidimensional NMR was soon followed by the development of powerful experiments to enable structure elucidation of complex molecules. Structures of organic, inorganic, and biological macromolecules such as proteins could be determined with these multidimensional techniques with accuracies rivaling those of single-crystal X-ray. Richard Ernst received the Nobel Prize in Chemistry for his contributions to NMR in 1991.

The development of NMR technology for structure elucidation employed the use of cylindrical NMR tubes. These tubes spanned a range of diameters from 3 mm up to 25 mm, with 5 mm being the most widely used. This configuration allowed chemists to dissolve a compound of interest in deuterated solvent (for 5-mm tubes, this was typically about 0.5–0.7 mL solvent) and easily recover the sample when sufficiently volatile solvent was used. Since NMR is a nondestructive technology, the sample could be used and reused indefinitely as long as the compound remained stable in the NMR tube. Application of pulse FT acquisition afforded a S/N ratio increase through the addition of free-induction decays (FIDs) (see Chapter 2), thereby enabling meaningful spectral data to be acquired on dilute sample concentrations or chemically dilute nuclei. Hence, the sample would reside in the magnet for as long as may be required to collect a spectrum. Although experiments with flow NMR were reported as early as 1951 [10], use of the tube configuration would remain as the generally accepted practice until the integration of liquid chromatography (LC) and NMR spectroscopy.

While LC dates back to 1903 [11], the invention of high-performance liquid chromatography (HPLC), comprising columns packed with small particles and high pressure pumps, did not occur until the late 1960s. Extensive development of this technology involved incorporation of in-line detectors and automatic injectors, all under computer control. In fact, highly sensitive detectors have extended the limits of detection to femtogram levels. These advancements led to a myriad of applications [11], and HPLC became widely used as an indispensable technology throughout analytical laboratories in both industry and academia. The capabilities of HPLC continue to improve with advancements in automation, speed, efficiency, and sensitivity. However, although HPLC has advanced significantly, one major limitation is the lack of a universal detector. In addition, separation of complex mixtures can still be challenging, requiring extensive method development.

The first integration of an HPLC system with an NMR spectrometer was published by Watanabe and Niki in 1978 [12]. At that time, there had not yet been a paper published about the direct coupling of HPLC with NMR, although its potential ability in qualitative analysis was well recognized. However, a major issue that had to be considered was the inherent low sensitivity of NMR. In addition, at that time, the solvents used for NMR measurements were limited to a few possibilities, such as deuterated chloroform or carbon tetrachloride. This limitation had to be tempered with HPLC considerations regarding the selection of the optimum solvents needed to achieve successful separation of analytes. Since optimal requirements for solvents in using NMR and HPLC may not be compatible, solvent selection was a notable challenge in interfacing the two technologies. For sensitivity issues, the development of FT NMR was recognized as a means to overcome such difficulties, particularly with respect to the samples exhibiting low solubility in an LC-NMR-compatible solvent. In addition, the difficulty with the solvent compatibility could be overcome by the improvements in hardware and software of both technologies.

In considering the direct coupling of LC and NMR, the resolution of NMR was acknowledged to be just as important as the sensitivity. Although sample spinning is not possible in the LC-NMR configuration, it was generally recognized that good resolution still needed to be achieved. For HPLC, the separation and column efficiency were of critical importance. To minimize the dilution of solute that would be transferred to the LC-NMR probe, it was necessary to utilize an HPLC system with high column efficiency. Regarding the important role that solvent plays with both NMR and HPLC, there was a deliberate requirement to use solvents that did not contain hydrogen atoms (^1H). Another recognized complication was the potential presence of impurities in the solvents that could compromise spectral quality due to the low content of the isolated solute. Hence the success of the first LC-NMR study depended upon a careful balance of solvent distribution and purity.

The first direct coupling of HPLC to an NMR spectrometer employed the use of a JEOL FT-NMR FX-60. The NMR probe was modified to be more sensitive than a standard probe because of the requirement of its exclusive use for observing proton (^{1}H) signals. The flow cell consisted of a thin-wall Teflon tube of 1.4-mm inner diameter (ID) that penetrated the NMR probe and provided a flow-through structure. The effective length and volume of the LC-NMR probe were about 1 cm and 15 μL, respectively. The flow cell was connected to the detector of HPLC with thick-wall Teflon tubing of 0.5-mm ID and 165 cm total length. Because this connecting tubing contributed to most of the broadening after the HPLC detector, it was made as short as possible. A schematic drawing of the LC-NMR system is shown in Figure 1.1. A series of valves was used to control the sample flow and resolution to the probe. Valve 1 was turned at a certain delay after the maximum of the chromatographic peak monitored by the HPLC detector and the eluent was held in the LC-NMR probe during measurement of NMR spectrum. The spectral resolution was achieved by introducing acetone-d$_6$ into the probe through valve 2, to allow the signal to remain as durable as possible. This was because the HPLC solvent did not contain a deuterium signal; hence, it was necessary to introduce deuterated acetone through valve 2 to serve as the magnetic field lock. The investigators observed that when multiple transient acquisitions were required, the use of an external lock method resulted in compromised resolution.

The HPLC system used in this first reported LC-NMR study was composed of Milton Roy pump and Rheodyne universal injector with a loop of 20 μL. A silica column (30-cm × 4-mm ID) was used along with carbon tetrachloride or tetrachloroethylene as the mobile phase. Effluents were monitored by the dielectric constant detector that was built in the investigators' laboratory.

To determine how much delay should be applied before turning valve 1 after the maximum detector peak appeared, the response of NMR versus the elapsed time was measured.

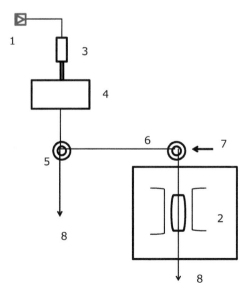

FIGURE 1.1 Schematic diagram of LC-NMR direct coupling. (1) HPLC column inlet, (2) FT-NMR, (3) column, (4) dielectric constant detector, (5) three-way valve 1, (6) three-way valve 2, (7) injection port for resolution, and (8) drain.

The profile of the detector peaks was used to estimate the volume of the connecting tube and the flow rate of mobile phase.

To test the LC-NMR system, 20 µL of sample solution containing three isomers of dimethylphenol were injected into an ETH-silica column (ETH-silica gel supplied by Toyo Soda Co. Inc.) using tetrachloroethylene as mobile phase. Stopping the eluents in the LC-NMR probe at 52 s after the maxima of each chromatographic peak, an NMR spectrum for each of the eluents was measured. The flow rate used was 0.77 mL/min. Examination of the spectra showed impurities in the mobile phase (Figure 1.2). These spectra provided information sufficient to identify the individual solutes. When an impurity was observed producing a single resonance, selective irradiation was used to suppress the impurity and enhance the spectral quality. The results obtained in this work demonstrated the first application of LC-NMR with adequate sensitivity and resolution.

The HPLC used in this work was described as normal phase chromatography of restricted applicability. However, these investigators proposed that for future applications it would be preferable to use reversed-phase ion-exchange chromatography. In addition, the importance of excluding the overwhelmingly large signal(s) due to solvent in NMR spectrum (particularly if reverse-phase chromatography is used) was recognized as a critical challenge.

Hence it was acknowledged at the earliest stages of LC-NMR development that the direct coupling of FT-NMR to LC required an effective selective-irradiation method that could suppress more than one solvent signal at the same time. Finally, apart from modifications to NMR hardware, the refinement of HPLC with serial linking of two columns was acknowledged as a viable separation strategy. Two successive columns – the reversed-phase column for separation and the normal-phase column to enable conversion to the solvent suitable for NMR measurement – were also proposed.

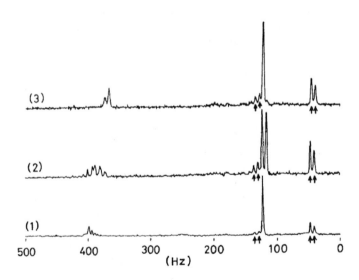

FIGURE 1.2   LC-NMR injection of solution containing 2,6-dimethylphenol (1.2%), 2,3-dimethylphenol (2%), and 3,5-dimethylphenol (2%) in tetrachloroethylene. The column used was ETH-silica, 4-mm × 30-cm; mobile phase: tetrachloroethylene, 0.77 mL/min. (1) 2,6-dimethylphenol, elution time 7 min. (2) 2,3-dimethylphenol, elution time 20.2 min. (3) 3,5-dimethylphenol, elution time 22.7 min. Signals from impurities in solvent are indicated with up arrows. (The figure is modified, with permission, from Watanabe and Niki [12].)

One year following the report by Watanabe and Niki [12], Bayer and coworkers [13] reported studies on the chromatographic separation and NMR evaluation of an equimolar mixture (50 μmol) of compounds ionol, anisole, and salol. Both stop-flow and continuous-flow experiments were performed. These investigators used a specially constructed flow-probe design that was interfaced with a conventional Bruker WH90 spectrometer.

To design an appropriate system for flow NMR, a number of problems needed to be addressed that do not occur in a static tube environment. The first challenge involved the fact that the intensity of an NMR signal is dependent on the flow rate of the solvent. Hence when the flow rate is low, there is a steady increase in magnetization of nuclei, and the signal amplitude increases. With high flow rates the signal amplitude decreases. When a flow-rate range of 0.5–2 mL/min is compatible with the HPLC separation process, not much change in the signal amplitude is expected. However, the half-width amplitude of the signal, and thus the overall resolution of the spectrum, can change with a large detector volume. The half-width amplitude is dependent on the viscosity of the solvent, the radio-frequency field strength, the nonhomogeneity of the magnetic field, and above all on the time the sample dwells in the detector cell. Thus, the line-broadening depends on the flow-rate/detector volume ratio. In an HPLC-NMR combination, the flow rate is related to the separation process and, therefore, the detector volume is related to the elution volume of a peak.

To enable optimal performance, the following conditions for the construction of the flow cell had to be established:

(1) The cell must guarantee a sufficient particle concentration for the NMR measurement. For dilute samples, this would mean a relatively large volume. In addition, the pulse FT technique must be used in order to overcome the problems of low concentration and to accumulate the signals during the passage of the peaks through the NMR flow cell.
(2) The detector volume should not exceed the volume of a separation peak. Fortunately, the sample volume is typically larger than that of the normal detectors. The cells should accommodate continuous-flow measurement as well as the stop-flow technique. No turbulent flow should occur in the connections between the HPLC column and the NMR flow cell that would compromise sample integrity.
(3) The geometry of the flow cell must enable an optimum synchronization with the measuring coil. This requires straight and parallel cell walls in the actual measuring range.
(4) An intermingling of the separated peaks should not occur in the flow cell. This means that the geometry of the cell and the connections should be constructed in such a way that no turbulent flow arises.

Applying the aforementioned considerations to optimize the performance of the insert coil, a flow cell was constructed having a detector volume of approximately 0.416 mL and an elution volume of 0.416 mL. A flow rate of 1 mL/min was used. The residence time in the detector flow cell was approximately 25 s. The initial flow cell design was found to cause peak mixing when the PTFE (polytetrafluoroethylene) capillary connection tubing (ID, 0.23 mm) was directly connected to the cell (ID, 7 mm). Using a dye system, the flow profiles of various cell configurations were studied with the goal of optimizing the flow cell geometry, and an optimal geometric arrangement was realized with sample entering through the bottom of the flow cell and exiting from the top. This configuration has been the basis for modern flow-cell design.

The chromatography used in this study employed normal-phase columns and carbon tetrachloride as solvent. Because deuterated solvents required for field-frequency stabilization were not used in the chromatography, the problem of introducing a deuterium signal was addressed by employing an external lock arrangement. Thus field-frequency stabilization was achieved with an ampoule containing D_2O, which was placed near the actual NMR cell. The use of conventional, nondeuterated solvents is possible with this arrangement.

One of the observations of the LC-NMR study was that the resolution of the NMR spectra in the LC-NMR system was poorer than for the conventional NMR system, which makes the measurement of small coupling constants difficult. This difference was initially attributed to nonrotation of the cell. A comparison of resolution in the stop-flow versus the continuous-flow modes was also reported. Without the flow of the solvent (i.e., stop-flow measurement), a resolution of 2.5 Hz was obtained. Upon changing the measuring technique from stop flow to continuous flow at a flow rate of 1.5 mL/min, the observed resolution was about 3 Hz. Hence it was determined that no substantial changes occur between the measurements in the stop-flow and continuous-flow method.

The early investigators concluded that a major advantage of the constructed flow cell was its compatibility with conventional NMR systems, which in principle enables the coupling of the HPLC instrument and the NMR spectrometer without any nonroutine modifications to the NMR magnet and console [13]. The examples presented demonstrated that large amounts of information can be obtained from the coupling of HPLC separation with NMR detection. Of the three substantial pieces of information from the ^1H spectrum (ppm values, integration ratio, and coupling constant), the first two were easily obtained. Only the coupling constant was limited to >3 Hz. The detection limit was found to be in the micromolar range. The solvent limitation was noted as a remaining challenge.

The initial LC-NMR reports led to many similar investigations, and the technology expanded rapidly [14]. Because LC-NMR required the use of an NMR probe configured for the observation of a sample flowing through tubing, this configuration led to further optimization of the design of the NMR flow probe, which is a probe designed to acquire NMR data on a sample introduced to a probe inside a magnet via capillary tubing.

In 1980, continuous-flow LC-NMR was used in a practical application where the investigators were analyzing the mixtures of several jet fuel samples [15]. Deuterated chloroform and Freon-113 and normal-phase columns were the common conditions for LC-NMR used in these studies [16–19], limiting the application of this technique.

In 1982, Bayer et al. [20] reported the direct coupling of reverse-phase HPLC and high-field NMR spectroscopy. This was the first time when superconducting magnets were used as opposed to iron magnets. NMR resolution was improved from 1–3 Hz to 0.5 Hz, and the NMR sensitivity was increased by a factor of 10. The experimental detection limit with the continuous-flow collection modes was found to be in the lower micromolar range. For the stop-flow collection mode, the detection limit was in the higher nanomolar range. Using either the continuous- or stop-flow procedure allowed meaningful NMR data to be obtained without the use of deuterated solvents.

A study was reported using two NMR instruments of different magnetic field strengths, a Bruker WM 250 and Bruker WM 400, operating at 5.9 and 9.4 T, respectively. Solvent line suppression was not necessary when using the 250-MHz instrument, which employed a 16-bit analog to digital converter (ADC). With the 400-MHz instrument, solvent line suppression was necessary. This was achieved using either homonuclear decoupling in the case of solvents yielding only one signal, or homonuclear-gated decoupling for solvents with several signals.

Because superconducting magnets were used, a deuterium lock for field frequency stabilization was not required, since the field drift during data acquisition was negligible. The solvent resonance signals of the nondeuterated solvents, however, were sometimes problematic. Depending upon the chromatographic parameters required for the particular separation, the ratio of the solvent to sample signal was between 1000:1 and 50,000:1. Although the dynamic range of the computer system for the 250-MHz spectrometer was adequate, the 400-MHz instrument had a 12-bit ADC; hence, the dynamic range was insufficient for the determination of very weak signals in the presence of very intense ones. The result of this condition was that suppression of solvent peaks had to be achieved by a factor of at least 16.

Since chloroform-d was relatively inexpensive and amenable to recovery and repurification by distillation, it could be used for normal-phase chromatographic applications. Typically 30 mL of deuteriochloroform was used in each experiment. For reversed-phase chromatography, however, deuterated solvents were and still are considerably more expensive, so, nondeuterated solvents were used.

A further problem arose when using a gradient between two solvents (e.g., acetonitrile/water). Since the difference between the two solvent signals changes according to the composition of the gradient, the ability to observe the analyte signals that resonate near solvent signals depended largely on the gradient.

Reversed-phase separations were carried out in acetonitrile as the mobile phase. The separation of diethyl and diisononyl phthalate (10 μmol) in acetonitrile was clearly observed in the UV chromatograms and the NMR spectra. The NMR data consisted of 29 ^{1}H NMR spectra, each spectrum being the sum of 36 scans that were acquired within 0.49 min. The resonance of the relatively large acetonitrile signal spanned a 0.5-ppm range. Nonetheless, analyte resonances lying within the aliphatic resonance region could be detected and assigned.

The reports of reverse-phase chromatography using an acetonitrile-water gradient were quite encouraging [20]. The example of a drug separation showed that the NMR resolution of the stop-flow spectra was equal to that of the conventionally recorded spectra. This experiment showed that the LC-NMR probe head does not lead to loss of spectral information in relation to the conventional tube-based probe. Over the range of concentration of acetonitrile in water used, the resulting solvent window was sufficient to allow the classification of the drugs L-ephedrine hydrochloride and propyphenazone at 4 μmol concentration.

Separation was also reported using an aqueous buffer system as solvent. The ^{1}H NMR spectra of a separation of L-lysine and L-alanine (5 μmol) in a KH_2PO_4/H_3PO_4 buffer were obtained. The ^{1}H NMR signal of the phosphate buffer was located near the water resonance. Due to the high viscosity of the solvent, the resolution of the ^{1}H NMR spectra was poor. To acquire sufficient signal, a stop-flow technique had to be employed.

Although there were still issues with solvent suppression and dynamic range, the promise HPLC/NMR as a rapid and nondestructive technology in the isolation and spectroscopic characterization of organic molecules approaching complex higher molecular weight was evident.

Despite the advances made in the early 1980s, there were still clear challenges that had to be addressed. In particular, the use of reversed-phase columns in LC-NMR complicated the NMR analysis because:

(1) The use of more than one protonated solvent will very likely interfere with observation of the sample.
(2) The change in solvent resonances during the course of the chromatographic run when using solvent gradients will affect sample homogeneity and line broadening.
(3) Small analyte signals relative to those of the solvent will cause ADC overflow and pose serious problems in observing signals of interest.

In 1995, Smallcombe et al. [21] overcame some of these problems by developing a robust solvent suppression technique, which greatly improved the quality of the spectra obtained by continuous-flow or stop-flow experiments. The optimization of the WET (water suppression enhanced through T_1 effects) solvent-suppression technique enabled the generation of high-quality data for 1D continuous-flow and stop-flow experiments and 2D experiments for the stop-flow mode, such as WETTOCSY, WET-COSY, WET-NOESY, and others [21]. For more details on solvent suppression, see Chapter 6 on NMR Experiments.

1.2   FLOW AND NMR

Following the initial pioneering work in LC-NMR, several variations of flow NMR were developed, including "stopped-flow NMR" (stopping a sample inside the NMR flow cell) [22–37] "continuous-flow" NMR (pumping a sample continuously into a flow cell as it elutes from a column) [38, 39], or "flow-injection or direct-injection" NMR (pumping a sample into a flow cell without a chromatographic step) [40]. The stopped-flow mode of sample delivery to the NMR probe may be used with NMR systems both integrated and nonintegrated with a chromatographic unit. For cases where the NMR is interfaced with an HPLC unit, a selected compound peak that elutes from the column may be sent to an NMR flow cell and parked there for an indefinite period of time. In such cases, the chromatography would be stopped while the NMR data collection was taking place. This technique is ideal for stable samples that require long data acquisition for structure elucidation. It is not ideal if subsequent peaks remain on the column and need to be collected for NMR evaluation. Rapid injection is used to acquire NMR data on unstable or rapidly reacting samples. With tube-based NMR (where the sample is placed in a glass tube that is lowered into the NMR probe in the magnet), there is often a lag time between "preparing" the sample and acquiring the NMR data. Under such conditions, this time may be too long to allow a transient species to be observed. To address this issue, it was necessary to devise a system that would reduce sample transport time and allow observation of a transient species as it occurs inside the NMR probe. This becomes particularly important when monitoring chemical reaction intermediates and product formation. Reports where two reactive species were mixed inside the NMR flow cell have been published by various investigators. One method involved placing the first reactant in an NMR tube in the magnet, and then injecting a second reactant into the tube using a syringe located outside the magnet. A second method was to flow two reactants into a mixer located outside the magnet and then flow the resulting mixture (through tubing) into the NMR flow probe [23–26, 41–55].

Because the continuous-flow experiment is best suited for the detection and characterization of multiple component systems involving known or similar structures, this experiment has undergone significant evaluation and optimization. The technique is heavily dependent on the flow rate and retention volume to obtain sufficient resolution. This is because, in a continuous-flow experiment, the sample resides in the flow cell for a specified period of time. The residence time, τ, depends upon the volume in the flow cell and the flow rate.

$$\text{Residence time } \tau = \frac{\text{Detection volume}}{\text{Flow rate}} \tag{1.1}$$

Table 1.1 outlines the relationship of the cell volume and flow rate on the residence time and linewidth [14].

As the residence time τ in the flow cell increases, the linewidth decreases. Residence time is proportional to the effective lifetime of particular spin states. The effective relaxation rates ($1/T$) are increased by $1/\tau$ (Eq. 1.2). This effect is the same for longitudinal relaxation T_1 and transverse relaxation T_2.

$$1/T_{1/2\text{flow}} = 1/T_{1/2\text{static}} + 1/\tau \tag{1.2}$$

For a particular flow-cell volume, as the flow rate increases, the decrease in T_1 can be compensated by increasing the repetition rate. For T_2, however, the shorter transverse relaxation

TABLE 1.1

Continuous-Flow NMR Spectroscopy: Residence Time and Line Broadening as a Function of Cell Volume and Flow Rate

	Dwell Time τ(s)		Line Broadening $1/\tau$ (Hz)	
Volume (µL)	Flow Rate (0.5 mL/min)	Flow Rate (1.0 mL/min)	Flow Rate (0.5 mL/min)	Flow Rate (1.0 mL/min)
44.0	5.28	2.64	0.19	0.38
60.0	7.20	3.60	0.14	0.28
120.0	14.40	7.20	0.07	0.14
200.0	24.00	12.00	0.04	0.08
Volume (µL)	Flow Rate (0.01 µL/s)	Flow Rate (0.02 µL/s)	Flow Rate (0.01 µL/s)	Flow Rate (0.02 µL/s)
0.005	0.05	0.25	2.00	4.00
0.050	5.00	2.50	0.20	0.40
0.5	50.00	2500	0.02	0.04

Adapted from Reference [14].

results in broadening of the signal linewidth. The relationship between linewidth LW and $T_{2\text{flow}}$ is proportionally reciprocal. Hence as the flow increases in a fixed volume, the linewidth increases. Therefore, careful calculation of flow rate and detection volume needs to be performed when initiating a continuous-flow experiment.

From its first introduction in 1978 [12], the coupling of separation systems, such as HPLC, with NMR spectroscopy is now at the forefront of emerging and cutting-edge technologies. Over the past several decades, LC-NMR has matured into a viable analytical tool for structure identification and characterization [20]. Combined improvements in magnet design, probe technology, solvent-suppression techniques [21], and automation have contributed greatly to its scope and utility in the pharmaceutical industry. The advances in automation and sensitivity have pushed the limits of detection to enable unprecedented mass sensitivity for NMR detection. In addition, the modes of isolation that were developed have been integrated with the NMR technology in an automated fashion to facilitate the process. These collection modes include continuous-flow detection, stop flow, solid-phase extraction (SPE), loop collection, and capillary electrophoresis [20, 21, 56]. Combinations of LC-NMR with detectors, such as PDA and MS, have resulted in powerful isolation and structure-elucidation systems.

Overall, LC-NMR has evolved into a robust technology that is now considered to be nearly routine [14].

This is in part due to the advancements made to the individual components of the hyphenated system. Notably, HPLC has become a robust technology through the development of versatile instrumentation and advances in column solid support. In addition, sophisticated advances in NMR spectroscopic instrumentation along with advances in multidimensional NMR and solvent-suppression techniques have vastly expanded the range and breadth of NMR capabilities. This has been realized particularly with respect to the more challenging issues regarding sensitivity and limits of detection. These advances have had an acute impact on the hyphenated chromatographic-NMR capabilities.

Numerous reviews on chromatographic-NMR hyphenated systems have been written, some of which have provided condensed detailed descriptions of the LC-NMR technology with an historical perspective [14, 56–60]. This book will provide general guidance in setting up an LC-NMR system, an introduction to NMR theory, discussion of LC methods that are compatible with NMR, an update on recent advances in system performance (both hardware and software), and an overview of NMR experiments and techniques used in structure elucidation, and will also focus on current developments in chromatographic-NMR integration with particular emphasis on applications and utility in the pharmaceutical industry. Finally, novel applications and emerging technologies that challenge detection limitations will be discussed.

1.3 SETTING UP THE LC-NMR SYSTEM

The installation of the LC-NMR equipment requires skill sets in both HPLC and NMR technologies. Ideally, a collaboration between persons skilled in NMR with knowledge of HPLC and persons skilled in HPLC with knowledge of NMR is desired. If one only has knowledge in NMR, it is possible that parts of the LC system could be damaged or solvents may become polluted. The values for the UV absorption pump pressures and retention time need to be calibrated by an experienced operator. Likewise for NMR, the homogeneity of the flow probe needs to be adjusted and the pulse widths and power levels need to be calibrated; otherwise, spectral integrity will be compromised or serious damage to the system can occur. Often installation of LC-NMR systems is performed by a vendor's service engineer; however, proper use and maintenance of the system requires individual(s) with knowledge of both technologies. Below is some practical guidance in setting up an LC-NMR system.

When setting up an LC-NMR system, the coupling of different kinds of chromatographic separation systems to a flow probe of an NMR spectrometer needs to be established. With the currently available LC-NMR probes, it is possible to use standard analytical scale chromatography.

Separation techniques such as RP (reversed phase), NP (normal phase)-HPLC (high-performance liquid chromatography), GPC (gel permeation chromatography), and SEC (size-exclusion chromatography) may be integrated with an LC-NMR system. With special equipment, SFC (supercritical fluid chromatography) may also be used.

LC-NMR experiments can be performed in four main modes.

- On-flow mode
- Direct stopped-flow mode
- Loop-storage mode
- Solid-phase extraction (cartridge storage) mode.

Several levels of automation are possible with modern NMR spectrometers. The NMR experiment can be run by manually adjusting parameters and individual functions such as peak selection. Individual automated programs can also be employed or the experiment can be run in a fully automated mode.

1.3.1 CONTINUOUS-FLOW MODE

When configuring a continuous-flow mode, the outlet of the LC detector must be connected directly to the NMR probe. While the peaks are eluting, NMR spectra are continuously acquired. The chromatographic system is used to move the compounds/peaks through the NMR cell.

The basic equipment added to an NMR spectrometer for a continuous-flow NMR experiment includes the following:

- An LC-NMR flow probe
- A stable HPLC system

1.3.2   DIRECT STOPPED-FLOW MODE

With a direct stopped-flow mode, the outlet of the LC detector is connected directly to the NMR probe. An LC detector (typically UV) is used to detect peaks eluting from the column. When a peak is detected, the flow continues until the peak arrives in the NMR cell. Once the sample is in the NMR flow cell, the chromatography pump and gradient stops and the NMR experiments are acquired. After the NMR experiments are completed, the chromatography continues until the next peak is found. This process can be repeated for several peaks within one chromatogram.

The basic equipment added to an NMR spectrometer for a stopped-flow NMR experiment includes:

- Controlling station with chromatography software
- HPLC system controlled by chromatography software
- LC-NMR flow probe
- Peak router (sampling unit for multiple collections)

1.3.3   STORAGE MODE

The outlet of the LC detector may be connected to sample loops or cartridges. An LC detector (typically UV) is used to detect peaks eluting from the column. A detected peak is moved into one of the sample loops without interrupting the chromatography. When the chromatography is completed, the HPLC pump can be used to transfer the peaks from the loops into the NMR probe.

The basic equipment added to an NMR spectrometer for storage collection in NMR flow experiments includes:

- LC-NMR probe
- Pump under control of chromatography software for transfer
- An HPLC system
- Peak router (sampling unit for multiple collections)
- Controlling station with chromatography software.

1.3.4   CALIBRATING THE SYSTEM

When setting up an LC-NMR system, the following materials are recommended:

- L acetonitrile. Use premium quality solvents for NMR
- L-D_2O 99.8% or better
- 0.5 mL of formic acid (preferably deuterated DCOOD) or acetic acid-d_4.
- HPLC syringe $\approx$ 100 μL
- Installation column "125 $\times$ 4 mm" and RP column with C18 material dimension 4.0 ID $\times$ 125 mm length
- The LC-NMR starter kit containing suitable solvents, syringe, and injector loop
- Stopwatch with display of seconds.

A mixture of four p-hydroxy-benzoic-acid methyl, ethyl, and propyl-esters (ratio 1:1:1:1) at a concentration of 32 mmol/μL may be used for testing LC-NMR systems. This mixture is appropriate to demonstrate stop flow in the NMR probe, sampling in loops for transfer to the flow probe, and trapping (single and multiple) on cartridges for transfer to the NMR probe. Another useful chemical for calibrating flow times is tetramethylbenzidine (TMB). These compounds are commercially available.

In preparation for testing the system, a number of conditions need to be met. First, the capillary and data connections for all instruments must be installed. Solvents need to be connected to the pump. Additives such as formic acid (preferably) or acetic acid in the D_2O should be used. (If no mass spectrometry system is interfaced with the unit, then tri-fluoro acetic acid may also be used.)

The HPLC solvent systems usually contain four channels, which may be appropriated as follows:

- Channel A: D_2O 0.1 % acid (or H_2O for systems with SPE)
- Channel B: acetonitrile + 0.1% acid
- Channel C: empty or methanol
- Channel D: H_2O + 0.1% of acid (only for SPE)

All solvents should be properly degassed for approximately 30 min. The hardware setup should contain the appropriate settings according to the installation for pumps, detectors, and connection ports.

Accurate descriptions of the connected solvents need to be recorded and saved along with the column description. For most chromatography systems, the pump pressure is monitored digitally.

In setting up an LC-NMR system, the "system parameters" need to be carefully calibrated. These parameters are the time/volumes between certain positions in the system. Such parameters are required to position the sample during an LC-NMR measurement. The volumes in an LC-NMR system are determined when the system is installed. They will not change as long as the flow rate of the pump is stable and reliable, and the critical compounds and capillaries in the system are not changed.

Parameters can be stored in hardware setup files. The times are typically entered for a flow rate of 1 mL/min. During the LC-NMR analysis, the times would be recalculated for the actual flow rate used in the experiment.

Standard system parameters are shown in Table 1.2.

The precision values describe the level of exactness that may be achieved for each of the parameters. The values given are representative of a 120-μL flow-probe system. The transfer volume parameters for SPE are based upon the use of a 60-μL flow probe.

It is generally advised that once all system parameters are determined, a backup copy be generated along with a printout of the hardware setup and kept in a safe place.

Flow rates determined at 1 mL/min may be used reliably for 0.5–1.5 mL/min. However, if a selected flow rate is always used, it would be best to optimize for that flow rate. Depending

TABLE 1.2

Standard System Timing Parameters and Precision Estimates

Collection mode	From	To	Default at 1 mL/min	Precision
LC-NMR (stop flow)	UV cell	NMR cell	20–30 s	1 s
LC-loop (sampling)	UV cell	Sample loop	10–20 s	1 s
Loop-NMR (transfer)	Loop	NMR cell	20–30 s	4 s
LC-SPE	UV cell	Cartridge	—	—
Divert to waste	Loop	Waste valve	5 s	1 s
Transfer: volume	SPE cartridge	NMR cell	300 μL	10 μL

on the probe or length of capillary tubing, the absolute values for time and volume may differ considerably.

In LC-NMR the "stop-flow time" is the delay that is needed to move the sample from the UV detector to the NMR cell. This timing has to be determined for the LC-NMR interfaced with the sampling unit. The NMR instrument is necessary for this determination.

To determine the stop-flow time, a very coarse chromatographic separation of 100 μg TMB may be performed under isocratic conditions. The peak is transferred to the NMR probe but the flow is not stopped. During the chromatogram, an NMR continuous-flow experiment is carried out. The peak is visible in the UV and some seconds later in the NMR spectrum. The time between the appearance of the UV peak and the maximum NMR spectral signal needs to be determined. Because the time resolution of the NMR experiment is known, the delay between the two events can be calculated and corresponds to the stop-flow delay.

A typical experiment would involve injecting 20 μL of TMB sample and setting the NMR program for continuous acquisition. The TMB compound should elute within 3–5 min under standard conditions. The UV signal should be monitored relative to the maximum NMR signal. This process should be repeated two to three times to produce a result with an error of ± 1 s. If the flow rate is changed, the process must be repeated. This process may be optimized through repetition with stopping the sample in the magnet and acquiring an NMR spectrum. The signal intensity can be measured and the process repeated using different stop-flow times to optimize the NMR signal.

It should be noted that TMB will produce a decomposition product that has a strong UV absorption. Even small quantities can produce a significant UV peak that may be more intense than the TMB peak, yet no NMR signal will be observed. Hence when using this standard, it is important to ensure that TMB is the UV signal used and not the impurity peak.

Setup of an LC-NMR system also requires that a significant part of the mobile phase be sent to waste. This may include mobile phase containing unwanted compounds or solvent residing in the transfer capillaries. Prior to the transfer of a sample to the probe, it is necessary that the transfer capillaries are emptied of solvent. The solvent should be diverted into the waste to prevent it from reaching the NMR probe and diluting the sample.

To determine the time required to divert to waste Crystal Violet (a commercially available dye, which has an intense blue color) can be injected. The color is transported to the storage loop without using a column. This procedure should be repeated at least four times. Next, the loop content containing the blue dye is transferred toward the NMR, and when the color is observed at the outlet of the storage loop the time is noted. This recorded time (i.e., when the first traces of blue color reach the outlet), is the "divert to waste" time.

This procedure requires that one replaces the chromatography column by a union with a column-switching valve. The first step is to disconnect the waste capillary and connect a short 5 cm × 0.5 mm ID capillary. The hardware with the previously determined loop-sampling time entered may be used. Enter a time for "loop to NMR," which is longer than the expected time (approximately 40 s). Enter a "divert to waste" time that is slightly shorter than the value previously selected (approximately 35 s). One may then wash five loops with an appropriate solvent composition and make five injections of 20-μL Crystal Violet. When the previously injected peak is stored into a loop, then the next sample may be injected. A wash flow of 1.0 mL/min and the transfer flow rate to 0.25 mL/min can then be set. The NMR is not needed in this process.

Once the loops are filled the transfer may begin. The end of the short capillary needs to be carefully monitored for traces of blue color. This may be achieved by placing a white towel at the capillary outlet. When a blue color is observed, the time should be recorded. This time should subsequently be entered into the hardware setup parameters. Enter the determined timing as divert to waste and then recalculate the delay for a flow rate of 1 mL/min.

1.3.5   Preparation of a Solid-Phase Extraction System

Because SPE plays such a critical role in the isolation and identification of trace amounts of compound, determination of the parameters governing the SPE transfer volume into NMR flow probe becomes critical. In SPE-NMR, the "transfer volume" is the volume that is required to transfer a sample from a SPE cartridge to the NMR flow cell.

Specific NMR experiments are not required for this process. The process begins by drying the NMR probe and a cartridge with dry nitrogen gas. A cartridge transfer is done but the solvent (deuterated acetonitrile) amount is chosen so that the flow cell is *not* reached by the solvent front. Additional amounts of solvent are transferred toward the probe until the flow cell is completely filled. The correct filling of the NMR cell is checked by the shimming of the solvent signal. The sum of the initially used volume and the additionally delivered volume is the transfer volume.

In preparation of the SPE system (for example, a Prospekt-2 SPE system may be used, as this system is ideal for isolating micro quantities of compound), the following parameters must be set:

- Cartridge drying: 1 min
- Transfer volume: 1000 µL
- Transfer flow rate: 250 µL/min
- Wash and dry NMR probe: drying time 1 min, volume 100 µL

A new (unused) cartridge must be selected prior to starting the transfer. This will ensure that the whole system is flushed with acetonitrile, the cartridge is cleaned from any other solvent and the probe is completely filled with deuterated solvent. A reference spectrum for the probe completely filled with acetonitrile-d$_3$ should then be obtained. Switching the lock off and opening the lock display window will allow monitoring of the entry and exit of the sample into and out of the probe. For determination of the required volume, a cartridge drying ~2 min and a transfer flow rate of 250 µL/min should be set.

The NMR probe should be washed and dried using a drying time of 1 min with a volume (acetonitrile) about 2.5 times the active volume of the flow cell: for a 30-µL probe, use ~75 µL solvent, for a 60-µL probe, use ~150 µL solvent, and for a 120-µL probe, use ~300 µL solvent. Afterward a "transfer volume" may be selected, which is lower than the expected volume; for example, with a 30-µL probe, use 150 µL solvent, with a 60-µL probe, use 200 µL solvent, and with a 120-µL probe, use 300 µL solvent. A new dried cartridge should then be selected and the transfer with deuterated solvent initiated. When the transfer is completed, no deuterium signal should be visible in the lock window. Solvent should be dispensed in steps of 10 µL until the signal of the deuterated solvent appears in the lock window. When a signal appears in the lock window, the solvent has reached the active volume of the probe. Approximately 1.5 times the active volume of the flow cell is required to get a reasonable filling of the probe (i.e., 45 µL for a 30-µL probe).

When the signal appears in the probe, the lock may be engaged and one may continue to dispense 10 µL portions. After each step only the Z1 shim should be adjusted. Acquire a spectrum and compare it with the previously acquired reference spectrum. When the probe is filled, the spectrum will achieve the quality of the reference spectrum and the transfer volume will be determined. At this point one may sum the volume dispensed during the initial transfer and the additional volume from the direct control plus an additional 10 µL.

It is also possible to divert the SPE elution volume into capillary tubes instead of sending the sample elution directly to the flow probe. In this case, the transfer volume is the volume that is required to transfer a sample from an SPE cartridge to a capillary tube. When the sample is transferred into a sample tube, the determination and usage is slightly different. When determining the volume for transfer to a tube, the NMR probe is not required.

To perform the calibration of the transfer volume, the capillaries and cartridges should be dried with nitrogen gas. A cartridge transfer is done but the solvent amount is chosen so that the tip of the outlet capillary is not reached by the solvent front. Additional amounts of solvent can subsequently be transferred until the first drop of solvent becomes visible. The sum of the initially used volume and the additionally delivered volume is the volume required to reach the needle tip. This volume is the "transfer volume" in SPE-NMR. The amount of liquid required in the NMR tube is defined as excess volume in the transfer settings and can be adjusted for the type of NMR tubes used.

In the hardware setup, the system should be configured to divert the sample from the flow probe. When using a Prospekt-2 SPE system, the transfer settings should include the following parameters and steps:

- Cartridge drying: 10–30 min
- Transfer flow rate: 250 µL/min
- Wash and dry NMR probe: drying time 1 min, volume 150 µL
- An NMR capillary tube or a white paper towel should be placed below the outlet capillary
- No excess volume should be used.

A new (unused) cartridge may be selected and the transfer initiated. When the transfer is finished, there should be no solvent visible at the outlet capillary. If solvent is visible, repeat the procedure with reduced transfer volume.

Set the flow rate to 250 µL/min and dispense solvent in steps of 5 µL until the first drop of solvent is observed at the outlet of the capillary. The sum of the volume dispensed during the initial transfer and the additional volume from the direct control comprise the "transfer volume" (SPE–NMR) in the hardware setup.

A needle may then be inserted into a capillary and liquid may be dispensed until the tubes are filled to the desired level. The additional volume should be noted and used as *excess volume* in the transfer settings. For a 1.7-mm capillary tube, the excess volume is 30 µL. The total excess and transfer volume may be ~130 µL. In the hardware setup parameters, the determined elution volume may be entered. It is important to keep a record of this parameter for future reference.

To ensure accurate determination of solvent, it is critical to dry the cartridges for at least 10 min before the calibration. If this is not done, residual solvent can affect the observed transfer volume. For a 2-mm cartridge, the minimum volume which ensures that the whole sample is eluted from cartridge and capillaries is about 50 µL. This value must be increased if the sample shows strong affinity to the cartridge material.

1.3.6   ADDING MORE SAMPLE TO A CARTRIDGE

Before transferring a sample onto a SPE cartridge, one must first equilibrate the cartridge. Cartridge equilibration may be achieved with 2000 µL conditioning solvent and 500 µL equilibration solvent. This is required before sample collection onto a particular cartridge may occur. However, once a cartridge has sample stored, additional sample cannot be added unless the cartridge is once again conditioned. The problem is that in conditioning the cartridge, sample that is already stored on the cartridge will be washed out, defeating the purpose of enabling addition of more sample. To circumvent this problem so that the sample already on the cartridge is not washed off the cartridge, the conditioning volume may be set to 1 μL and equilibration to 100 μL. The cartridge will then become ready for sample collection and more material may be added without washing away what was previously acquired.

When sample collection is completed, the cartridge must be dried prior to elution into the detection chamber. The drying process is achieved with a stream of nitrogen gas. A drying time of 30 min has been found to be adequate for most samples. Once the loaded cartridge is dried, the sample is ready to be eluted from the cartridge.

Before sending a sample to the probe or to a capillary tube it is important to check the system for air bubbles. This may be achieved with an aspirate and dispense routine and visually inspecting the tubing for the presence of bubbles. To perform the aspirate/dispense routine, there must be a direct control of the system. Choosing the appropriate syringe pump (one that is normally used to dispense deuterated acetonitrile to the cartridge), the unit is switched to dispense mode and 500 μL of solvent is taken up. The valve may then be switched to waste and 500 μL of solvent is dispensed. This process should be repeated until the flow in the tubing is no longer interrupted by air bubbles. Failure to ensure the lines are free of air bubbles can result in trapping bubbles in the flow cell. This condition will prevent optimized shimming of the sample and spectral quality will be seriously diminished.

1.4   SOLVENT REQUIREMENTS IN LC-NMR

Solvents that meet the high requirements of HPLC and proton NMR are needed when performing LC-NMR experiments. But the high throughput of solvents in HPLC does not allow the usage of expensive deuterated solvents as usually required for NMR. Acetonitrile is often contaminated with impurities containing protons from compounds, such as propionitrile, which does not affect UV detection but can complicate NMR spectra. Therefore, a dramatic decrease of proton containing contaminants for LC-NMR solvents is essential.

When considering the selection of LC-NMR solvents, one must determine whether deuterated or nondeuterated solvents are required. For nondeuterated chromatography solvents (ideal for SPE studies), solvents such as acetonitrile LC-NMR CHROMASOLV® >99.9%, and water LC-MS CHROMASOLV® are the solvents of choice. The acetonitrile product was developed with superior specifications required for LC-NMR. There are no signals in the ^{1}H-NMR spectrum that exceed the size of methyl signals (more than 5% at 3.78 ppm of 0.0006% trimethoxybenzene) at typical LC-NMR conditions.

Deuterated solvents may also be used in chromatography; however, these materials can be cost prohibitive for routine chromatography trials. The deuterated solvents most frequently used but are not limited to acetonitrile-d_3 99.8%, methanol-d_4 99.8%, and deuterium oxide 99.9%.

Use and handling of NMR solvents require care, since most deuterated NMR solvents readily absorb moisture. One may minimize the chance of water contamination by using dried NMR tubes and handling NMR solvents in a dry atmosphere.

To obtain a nearly moisture-free surface, it is necessary to dry glassware at ~150 °C for 24 hours and cool under an inert atmosphere. For NMR sample tubes, care must be taken when heating not to warp the concentricity of the tube so they are usually not recommended. Another technique is to rinse the NMR tube with dry deuterated solvent prior to preparing the sample. This allows for a complete exchange of protons from any residual moisture on the glass surface. In certain less demanding applications, a nitrogen blanket over the solvent reservoir may be adequate for sample preparation.

With a number of applications (i.e., trace analysis studies), it is necessary to avoid sources of impurities and chemical residues. To avoid such impurities, always use clean and dry glassware. Use a vortex mixer instead of shaking the tube contents. Shaking a tube or vial can introduce contaminants from the tube cap. Avoid equipment that can introduce residual chemical vapor, which can be a source of impurities, for example, residual acetone in pipette bulbs is a common source of unwanted solvent signal.

Volatile solvents may be removed readily. Protonated solvent residue can be removed by co-evaporation.

Adding a small quantity of the desired deuterated solvent, followed by brief high vacuum drying (5–10 min), prior to preparation of the NMR sample serves to remove unwanted solvents.

For solvents such as chloroform-d, benzene-d_6, and toluene-d_8, residual water can be removed azeotropically.

Most deuterated solvents are highly hygroscopic, absorbing moisture from the surface of glass bottles, pipettes, and so on. For compounds that are sensitive to water, avoiding water contamination requires special care in sample preparation. All glassware that will be in contact with the solvent can be dried at 150 °C for 24 hours, and then allowed to cool under a dry inert atmosphere. The glassware and NMR tubes can also be transferred to a glove box, in which the sample can be prepared under a dry inert atmosphere. A portable means of producing a semi-dry atmosphere for samples in which water contamination is not critical is to use an inverted plastic bag or funnel, and flow dry nitrogen over the bottle and NMR tube when transferring the solvent.

Deuterated solvents are available in a variety of sizes and package types. Solvents are available in gram units, as well as in ampoules. Ampoules are primarily recommended for final sample preparation where high solvent integrity is required.

1.5 SOLVENT SUPPRESSION AND REFERENCING

In reverse-phase LC-NMR studies, the solvent resonances are typically large enough that both the CH_3CN and the water resonances need to be suppressed. Even D_2O will absorb enough H_2O to form a considerable amount of partially deuterated water (HOD) during LC-NMR, unless significant drying steps are performed. As mentioned previously, a robust means of solvent suppression is to use a two-frequency shaped (WET [21]) pulse in which the transmitter is set on one resonance and the other resonance is irradiated by the SLP (shifted laminar pulse; phase-ramped pulse) technique [61]. This requires the user to decide which of these two resonances to keep the transmitter on, either the CH_3CN or water. As these two options are not interchangeable, the choice will severely impact the resulting data.

The reason setting the solvent resonance is so important that solvent resonances will move due to solvent-composition changes. When this happens the frequencies used for solvent suppression need to be re-optimized. The rate at which re-optimization needs to happen depends upon the rate of change of solvent composition. In continuous-flow NMR, the solvent composition is influenced by the flow rate and the HPLC method, which are both under user control. Re-optimization is influenced by how much each resonance moves in response to changes in solvent composition (which is not under user control). To address this problem, a scanning technique was developed [62] to automatically re-optimize these frequencies. The scanning technique, which has been termed Scout Scan, takes a small-tip-angle ^1H spectrum without using solvent suppression, analyzes that spectrum to both set the transmitter on the desired resonance and calculate the resulting frequency offset(s) for the solvent signal(s) to be suppressed, creates a shaped pulse that excites all these resonances [63, 64]. It then resets the parameters to do a signal-averaged solvent-suppressed ^1H spectrum and starts acquisition. The entire Scout Scan process requires only a few seconds and the frequency at which this happens is determined by the number of transients used per spectrum (a common default is 16). This scanning method enables the number of resonances to be selected for suppression, whether to use ^{13}C-satellite suppression, and which resonance to keep at a constant chemical shift. To achieve optimal results, it is necessary to reference the multiple spectra (increments) within a pseudo-2D dataset. The normal way to do this is to actively maintain one resonance at a constant chemical shift. The Scout Scan technique does this by actively adjusting the transmitter (the center of the spectrum) for every spectrum onto one of the tall resonances. For a typical LC-NMR solvent system this would be either the CH_3CN or the water resonance. With such a solvent system, normally data would be acquired by actively maintaining the transmitter on the CH_3CN resonance, and referencing the CH_3CN resonance in every increment of the pseudo-2D experiment to 1.93 ppm. If one is using D_2O instead of H_2O, the HOD signal will be much smaller than the CH_3CN signal due to the predominance of D_2O over H_2O. Under such circumstances, the water resonance will move several ppm relative to the CH_3CN resonance during the 0–100% CH_3CN ramp. During this process, however, the frequency of the CH_3CN resonance remains unchanged. This enables the acetonitrile solvent resonance to be used as a reference compound.

A careful evaluation of the relative movement of the ^{1}H resonances from different compounds was performed by spiking the mobile phase with a variety of compounds whose relative ^{1}H chemical-shift movements could be monitored during the 0–100% solvent-gradient ramp [65]. Selection of the mobile-phase additives required that they be soluble in both acetonitrile and water, and be able to elute from the chromatographic column. In addition, sharp ^{1}H resonances were desired, so measurement of lineshape distortions could be made. The cocktail used contained DSS (sodium 2,2-dimethyl-2-silapentane-5-sulfonate), CH_3CN, CH_3OH, sucrose, H_2O, CH_2Cl_2, $CHCl_3$, and sodium formate. Their chemical shifts were determined in 50:50 CH_3CN:D_2O, with CH_3CN set as the reference. Some of these additives were partly retained on the chromatography column under 100% D_2O conditions, hence the HPLC method was started at 50:50 CH_3CN:D_2O (to equilibrate the system) then changed to 100% D_2O for 5 min (between $t = 5$ and $t = 10$ min) before starting the solvent-gradient ramp at $t = 10$ min. This allowed reproducible amounts of additives to be retained on the HPLC column for each run, and also allowed each LC-NMR dataset to have multiple "control" spectra at the beginning of each run where all of the ^{1}H resonances were visible (for lineshape verification).

If a solvent change from 50:50 CH_3CN:D_2O to 100% D_2O is applied, the spectra will degenerate because such a sudden radical solvent change moves all the resonances during one single increment while the NMR spectrometer continues to signal average. In addition, the lineshape of every resonance will become grossly distorted due to the magnetic-susceptibility inhomogeneities in the NMR flow cell caused by incomplete mixing of the incoming solvent. Such effects demonstrate the limitations in how rapidly the solvent composition can be changed during an LC-NMR experiment. Typically, the solvent composition should only be changed by 1–2% per minute or else the spectral data quality will severely suffer.

When acquiring reversed-phase LC-NMR data, the spectroscopist needs to determine whether to maintain the transmitter on the CH_3CN or the H_2O resonance. To illustrate the ramifications of nonoptimal transmitter selection, a careful study was conducted [65]. In this study, the first experiment involved actively maintaining the transmitter on the CH_3CN resonance for each increment (and referencing the CH_3CN resonance to 1.95 ppm). In a second experiment, the transmitter was actively maintained on the HOD resonance (and referencing the HOD resonance to 4.19 ppm). The HOD signal was referenced to 4.19 ppm, since this placed the CH_3CN resonance in 50:50 CH_3CN:D_2O at 1.95 ppm. This referencing enabled comparisons between experimental datasets.

The first observation extracted from these studies showed that the chemical shifts arising from all the additives were generally constant when CH_3CN is kept on resonance. The HOD signal was the only signal that moved significantly. In contrast, when the HOD signal was kept on resonance, the resonances of all the additives moved extensively. This study demonstrated that holding the HOD resonance constant results in vast distortions in chemical shifts of the analytes.

1.6 THE DEUTERIUM LOCK

When using the $CH_3CN:D_2O$ solvent system, the only source of deuterium for a ^2H lock is D_2O. Hence, if the ^2H lock is used under such conditions (without any form of Scout Scan-type correction), spectral distortions will occur. This is because the only available ^2H signal in the solvent system is D_2O (or HOD) whose frequency is highly susceptible to shifting depending on experimental conditions. The ^2H lock holds the ^2H resonance of D_2O/HOD at a fixed frequency, which then holds the ^1H resonance of HOD/H_2O at a fixed frequency (since they track together). This is why the use of the ^2H lock alone (without a Scout Scan resetting of the transmitter frequency) will produce unacceptable spectra. The ^2H-lock problem is even more pronounced when one considers that the tugging of the ^2H lock (on D_2O/HOD) that happens during a signal-averaged increment manifests itself as a broadening of all the other resonances within each spectrum (increment). Hence expansion of an otherwise sharp singlet is broadened by an inconsistent D_2O resonance that compromises the ^2H lock stability. It was found that when the ^1H HOD resonance moves most rapidly, resonances from an analyte such as DSS suffer more severe linebroadeing from a "locked" spectrum as compared to the "unlocked" spectrum [65].

1.7 THE SOLVENT-GRADIENT RAMP

When the solvent-gradient ramp becomes steep in time or volume, the solvent mixture in the NMR flow cell becomes more inhomogeneous and magnetic susceptibility effects result. Magnetic susceptibility may be viewed as a measure of how well a particular solution can accommodate magnetic field lines. Whenever there is a steep transition from one solution composition to another, the density of field lines will change at the interface and this will introduce magnetic field heterogeneity in the sample. Shimming to compensate for the heterogeneities is compromised due to the severity of the discontinuity across the sample; hence this situation broadens the NMR resonances. The heterogeneity of the sample may be attributed either to incomplete mixing of the contents in the flow cell, or to a steep ramp in solvent composition in the NMR flow cell. While smaller NMR flow cells tolerate steeper solvent ramps, they also decrease the inherent NMR sensitivity. During steep solvent-gradient ramps, the potential for inappropriate chemical-shift change induced by the ^2H lock becomes greater, because the lock is tracking a D_2O/HOD resonance that moves and broadens at a fast rate similar to the solvent composition. Datasets were reported with 0–100% CH_3CN ramps (all flowing at 1 mL/min) that occurred over 100 min (1%/min), 50 min (2%/min), 25 min (4%/min), and 12.5 min (8%/min). A narrow DSS linewidth was observed at low flow rates; however, the DSS linewidth was shown to become broader and less uniform as the solvent-gradient ramp becomes faster. This linebroadening is caused by the steepness of the solvent gradient and cannot be removed by shimming the sample.

In summary, linebroadening caused by the steepness of the solvent-concentration ramp is influenced by both the volume- and time-steepness of the ramp. If the ramp is steep in volume, but slow in time, then the effect of poor mixing within the NMR flow cell can be reduced. Studies conducted showed that if the flow rate was slow enough, it may compensate for a solvent-gradient ramp that is steep in volume [65]. Similar but less dramatic effects were also observed with shallower solvent ramps. A comparison of data acquired with 0–100% ramps over either 100 min at 1.0 mL/min (1%/min) or 800 min at 0.5 mL/min (0.125%/min) showed that the DSS linewidth remained narrower for the slower, shallower-ramp run. The S/N ratio was also higher for the slower-ramp dataset because the slower pumping speed allowed more scans per increment to be used with no penalty.

1.8   DIFFUSION

The lineshape of a compound's resonances may also be affected by diffusion properties. When more time for active diffusion within the NMR flow cell is available, this condition would render the magnetic susceptibility homogeneous over time. Inhibiting fresh solvent from entering the flow cell avoids creating the inhomogeneity. The effects of diffusion were evaluated by manually stopping the HPLC pump 22 min into a run, while the NMR spectrometer continued to acquire data [65]. This allowed monitoring of the effects of active diffusion upon the NMR lineshapes. A study using DSS showed the lineshape recovers from a 20-Hz-wide lump the moment the pump was stopped (at 23 min into the run), to a 4.6-Hz-wide split singlet within 90 s of stopping the pump, to a 3.4-Hz-wide split singlet within 2.5 min, to a 2.1-Hz singlet within 5 min. The DSS linewidth was 0.86 Hz at the start of the experiment in 50:50 $CH_3CN:D_2O$, where it was optimally shimmed.

1.9   SHIMMING

Although the diffusion effect will improve lineshape, the DSS lineshape will not recover back to the initial optimal shimming value obtained at the beginning of the experiment. This nicely illustrates an aspect of shimming integral to all LC-NMR experiments that use a solvent gradient. The best lineshapes are always obtained when the same solvent composition is used to both shim the probe and acquire the data. As the solvent composition used for data acquisition changes further away from the solvent composition used for shimming, the lineshape will increasingly degrade. When the solvent composition is returned to that used for shimming, the lineshape will recover (due to the fixed geometry of both the flow cell and the solvent (provided no meniscus or air bubbles are present in the flow cell). Hence it is recommended that a spectroscopist shim on a solvent composition that is in the mid-point of the solvent-gradient ramp (when conducting continuous-flow solvent-gradient experiments).

In contrast to tube-based probes that must be reshimmed on every sample tube, a flow probe that has been optimally shimmed on a 50:50 solvent mixture should remain viable for several days unless the solvent composition is changed. Hence the NMR lineshape will recover to original values as soon as the solvent composition is restored to the initial values.

Studies have been conducted that show there is an upper limit to the steepness of the solvent gradients that can (or should) be used in LC-NMR. This limit is influenced by flow rate, LC-column size, NMR-flowcell size, and the magnetic susceptibilities of the solvents being mixed. A limit of 1%/mL and 1%/min was reasonable here, for a 4.6-mm LC column in a 60-μL NMR flow cell. There have been other reports that solvent gradients can cause NMR line broadenings, but with less quantification of the limit or under different conditions [66, 67]. Possible means of circumventing this problem have included stopped-flow LC-NMR, SPE-NMR column trapping, and inserting a delay before NMR acquisition [67–70]. A mobile-phase compensation method has also been proposed as a way to actually increase steepness of the solvent gradient [65, 71].

1.10   ACQUISITION PARAMETERS

The effect of the balance between the number of transients and the number of increments during continuous-flow solvent-gradient in LC-NMR experiments needs to be appropriately planned. In a given amount of experiment time, there are trade-offs that must be considered.

More transients will increase S/N ratio for each increment; however, more increments will increase the "chromatographic resolution" of the NMR experiment. During a continuous-flow solvent-gradient experiment, as the number of transients (per increment) increases, the risk increases that a changing lock signal will broaden the NMR linewidths. This is because there are fewer Scout Scans for a given change in the solvent gradient. Hence data acquired with less scans per increment will have narrower linewidths but a lower S/N ratio than data acquired with more scans per increment [65].

1.11 CHEMICAL-SHIFT TRACKING

A question on whether to use an organic resonance or the water resonance as a fixed-position chemical-shift reference during a solvent-gradient experiment was answered previously. The drastically more-constant chemical shifts made it clear that maintaining the transmitter on the CH_3CN resonance gives much better NMR data than keeping water constant, despite the fact that the latter method has been routinely reported in the literature [72–74]. The latter method is easy to run – either by locking on the D_2O/HOD ^2H resonance or by actively tracking the HOD/H_2O ^1H resonance. However, it can produce misleading ^1H chemical-shift scales.

Since an organic resonance is recommended as a fixed-position chemical-shift reference, one may ask which organic resonance should be used. Although conventional solution-state NMR has defined that a dilute solution of TMS (tetramethylsilane) in $CDCl_3$, with the ^1H resonance of the TMS set to 0.0 ppm, is the primary chemical-shift reference standard, most LC-NMR experiments use reversed-phase (aqueous) HPLC conditions, which precludes the use of TMS. Although DSS and TSP (trimethylsilyl propanoic acid) may be considered, they are not usually added to NMR samples, and their chemical shifts are known to be dependant upon solvent composition and pH.

The most common HPLC and LC-NMR mobile phase is acetonitrile:water, hence studies were conducted to determine the ^1H chemical shift of CH_3CN over the vast range of possible solvent compositions [65]. By comparing the ^1H chemical shift of CH_3CN to the ^1H chemical shifts of a number of other (potential) secondary standards during the solvent-gradient experiment, the ^1H chemical-shift movements were determined. Chemical-shift movements were tracked over the entire 0–100% CH_3CN ramp, for data acquired by keeping the ^1H resonance of the CH_3CN fixed. Protons on the methyl carbon of acetonitrile may be set to a defined value (for example, 1.93 ppm).

To determine which NMR resonance had the most stable chemical shift during application of a solvent gradient, the cocktail of DSS (sodium 2,2-dimethyl-2-silapentane-5-sulfonate), CH_3CN, CH_3OH, sucrose, H_2O, CH_2Cl_2, $CHCl_3$, and sodium formate was used. Immediately water was eliminated. It should be noted that no resonance is perfectly well behaved. The DSS singlet was found to track with the CH_2Cl_2 resonance as well as the CH_3 of the CH_3OH resonance. The DSS methylene at 0.5 ppm tracked fairly well with the CH_3OH and the anomeric sucrose signals. The CH_3OH signal somewhat tracked the sucrose anomeric signal, which tracked a few of the other sucrose resonances, but many of the other sucrose resonances track in the other direction. The $CHCl_3$ signal tracked well with the CH_3CN signal. The formate signal tracked somewhat with the CH_3CN signal until about 85% CH_3CN when the formate signal then moved drastically. The $CHCl_3$ and formate signals were tracked in one direction compared to CH_3CN (as did the DSS downfield methylene and most of the sucrose signals), whereas the DSS singlet, DSS 0.5 ppm, CH_3OH, CH_2Cl_2, and sucrose anomeric signals were tracked in the other direction. Overall, the majority of signals drifted well under 0.2 ppm in contrast to the 2.4 ppm changes seen for water, which supported keeping the CH_3CN signal at a fixed frequency as the correct way to acquire the data. The movement of the CH_3CN resonance relative to all the organic (non-water) signals is close to the mathematical average of all the other relative chemical-shift movements [75].

Although the investigators were limited to compounds that were not significantly retained by the HPLC column and were soluble over the entire 0–100% CH_3CN solvent-gradient ramp, the results showed that the ^1H resonance of CH_3CN was consistent, and can serve as a suitable secondary chemical-shift standard. An alternate LC-NMR study also concluded that, when all the chemical shifts in the sample move as a function of solvent composition, then it is important to select a reference compound whose chemical shift moves most like the solutes of interest [76]. Hence it was shown that maintaining the CH_3CN resonance at a fixed frequency in conjunction with the Scout Scan technique provides accurate chemical shift referencing.

1.12   OTHER CONSIDERATIONS

Because the ^2H lock can degrade lineshapes when solvent gradients are employed during the NMR acquisition (as can happen during continuous-flow LC-NMR experiments), the choice of whether or not to use a ^2H lock on water (D_2O) needs to be evaluated. The situation would be expected to be different for continuous-flow versus stopped-flow LC-NMR experiments. The ^2H lock on D_2O/HOD can be an asset during stopped-flow experiments, as long as the chemical-shift scale and possibly the transmitter are set appropriately, because the ^2H lock can compensate for magnetic drift. It can also be an asset for isocratic (constant-solvent) in contiuous-flow experiments. But for continuous-flow solvent-gradient experiments, the ^2H lock on D_2O/HOD must be used very carefully, or even turned off. The steeper the solvent gradient, the more important it is to turn off the ^2H lock. To allow good lineshapes to be maintained during continuous-flow LC-NMR runs, the data show that the solvent-gradient ramps should be limited depending on NMR flow-cell volume and solvent mixtures. Faster solvent ramps will result in degraded lineshapes that cannot be corrected, and which will reduce overall S/N ratio. If a steep solvent ramp (by volume) must be used, these data show that slowing down the flow rate will help reduce the degradation of the lineshape. However, it was reported that this approach sometimes does not work, especially when concentrated solutes elute, and that diffusion does not always happen in the NMR flow cell, presumably due to stable density gradients. It has also been reported that the 1%/min limit is influenced by the choice of solvents used in the mobile phase; the rate limit is projected to be determined by the difference in magnetic susceptibilities between the two solvents.

With respect to shimming it is important to recognize that different methods can produce different results in LC-NMR. When the mobile phase is a combination of CH_3CN and D_2O, then shimming on the ^2H signal using either the lock signal or a ^2H pulsed-field-gradient (PFG) map – may result in selection of the water (HOD/D_2O) signal, which is notoriously sensitive to broadening from both temperature gradients and solvent-composition gradients (as noted above). If either temperature- or solvent-gradients exist, and if the ^2H signal is used for shimming, then the uniquely broadened water resonance is made narrower using any of the shim gradients (for example, temperature gradients are often linearly axially dependant, like the Z1 shim, so the temperature-gradient-induced linebroadening can then be offset by missetting Z1). This results in the "removed resultant linebroadening" that was only on the HOD/H_2O resonance being unfortunately transferred to every other resonance in the ^1H spectrum. When this happens, a ^1H spectrum with narrow organic resonances and a broad water resonance is converted into an (undesirable) spectrum with broad organic resonances and a narrow water resonance. Similar effects can sometimes be observed when the water resonance is broadened by a concentration gradient. This can occur because the concentration gradient has a linear component caused by the solvent flowing unidirectionally into the NMR flow cell.

In contrast, when ^1H PFG shimming is used on $CH_3CN:D_2O$, the ^1H signal is dominated by the ^1H resonance of CH_3CN, which is a more well-behaved resonance, and this results in the narrowest possible ^1H linewidths for all of the organic resonances (but it may leave the HOD/H_2O signal broad or split). This is also true in ^1H spectral shimming, (depending upon which ^1H resonance is used), and is true to a lesser extent in ^1H FID shimming, depending upon how much of the FID is made up of the HOD/H_2O signal.

In continuous-flow LC-NMR experiments, the user must strike a balance between how many transients per spectrum to use versus how many spectra (increments) can be obtained. Unlike the conventional 2D NMR spectra, which typically have no short-term limit to the duration of the experiment, continuous-flow LC-NMR experiments have a fixed duration, which is limited by the longest retention time (of the most-retained chromatographic peak). The pseudo-2D nature of LC-NMR experiments means that there is a maximum number of "transient increments" that can be obtained during this time. The repetition rate of an on-flow NMR experiment does not have the same dependence upon T_1 as in a conventional 2D spectrum, because the flowing mobile phase can effectively shorten T_1 [77]. In addition, each solute has only a finite residence time in the NMR flow cell which depends upon the flow rate of the mobile phase and the bandwidth of the chromatographic peak. This condition sets an upper limit on the NMR S/N ratio that can be obtained at a given flow rate. Given these boundaries, if the number of transients per increment is large, the maximum possible NMR S/N ratio will be obtained; however, the chromatographic resolution will be compromised. If a solvent gradient is being used, the data will be susceptible to NMR line broadening caused by chemical-shift movements arising from solvent changes during the signal averaging of a single increment. The NMR line broadenings arise from an inhomogeneous magnetic susceptibility in the flow cell, and are caused by uneven mixing of the solvents. Conversely, if the number of transients per increment is small, a lower NMR S/N ratio will result, but the chromatographic resolution will be maximized and the NMR line broadenings caused by solvent gradients will be reduced. Therefore, as the chemical shift of a resonance moves faster because of the solvent-composition ramp, the number of transients per increment should be made smaller, so as to allow narrow linewidths to be observed. This requires that higher concentrations of solute be captured in the flow cell to enable adequate signal to be observed. Hence the importance of selecting a good balance between the number of transients and the number of increments in LC-NMR is critical to a successful continuous-flow experiment. The balance depends upon the solvent-gradient rate, the magnitude of chemical-shift changes (as a function of solvent composition), sample concentration, and the desired narrowness of linewidths. This balance is less critical if the NMR resonances are naturally broad or if solvent gradients are not used (i.e., if an isocratic LC method is used), and it is a nonissue if the NMR data are acquired as a stopped-flow LC-NMR experiment [65].

1.13   SOURCES OF ERROR

It is important to identify the sources of error for measuring chemical shifts in continuous-flow solvent-gradient LC-NMR. It has been shown previously that water moves enormously during a 0–100% gradient, so this solvent should never be used as a reference signal. It has been demonstrated that because the D_2O resonance moves, it alters the ^2H lock during signal averaging whenever D_2O is used as a lock signal. This then moves and broadens the observed ^1H resonances. The literature contains many examples of solute resonances moving – sometimes in different directions – during application of a solvent gradient, and that effect has frequently been observed. This event complicates both the reporting of the chemical shifts, and the ability of others in repeating the work, because the solvent composition in the NMR flow cell may not be accurately known when the chemical shifts are measured. Also because a solvent gradient changes the proportion of solvents in the mobile phase, this condition can generate different amounts of radiation-damping-induced NMR line broadening in the solvent resonances (as a function of water concentration), and can create a variable uncertainty in the Scout Scan determination of the solvent-resonance frequency.

The phenomenon known as radiation damping occurs because the strong transverse magnetization of the water signal induces a rotating electromotive force in the RF coil strong enough to act significantly back on the sample. Hence the induced currents in the NMR coil generate magnetic field feedback to the sample. These magnetic fields in turn broaden the resonance line. The width of the water line is a function of the strength of the water signal, which depends on the amount of water, as well as on other factors such as coil size and probe tuning. The effect becomes greater with increasing field strength. Radiation damping can also affect the symmetry and phase of the peak. At 500 MHz, a proton NMR spectrum of a sample of 90% H_2O/10% D_2O will have a linewidth at half height of about 35–40 Hz. A sample of 0.1% H_2O/99.9% D_2O will have a linewidth at half height of about 0.5–1 Hz in a well-shimmed magnet. Although the small tip angles used in the Scout Scan method will not eliminate the radiation-damping effect, this is not an insurmountable problem.

The radiation-damping-induced NMR line broadenings are a bigger problem in cryogenically cooled probes in higher field magnets, where the Q of the probe and the signal response are highest. Next, concentration-dependent frequency shifts of the solvent resonances due to flux density changes are readily observable on solvent signals [78–80]. This can cause the solvent signal's frequency to move. These shifts may be ameliorated by progressively moving postprocessing DSP (digital signal processing) notch filters off-resonance" with increasing solvent concentration. Finally, it has been reported that a misset lock phase can induce phase distortions in solute resonances whenever PFG-solvent-suppression sequences like WET [21] are used. This effect can be a sensitive test for how accurately the lock phase has been adjusted, but could also contribute to errors in chemical-shift measurement [65].

1.14 SUMMARY

NMR data have shown that using the water resonance to reference, shim, or lock the NMR spectra in LC-NMR experiments will not produce optimal results for continuous-flow solvent-gradient experiments, despite the reports in the literature claiming success with this approach [72–74]. This was shown to be true regardless of whether the water resonance is used as a ^2H lock (for D_2O) or as an internal standard (for H_2O or HOD). Using the ^1H resonance of CH_3CN as a secondary reference and a signal for shimming (and effectively as a lock signal) consistently gave rise to higher integrity NMR data. Software tools for referencing on the CH_3CN resonance are automated and easy to use on modern spectrometers and can be set for a variety of solvent conditions. When comparing CH_3CN as a secondary chemical-shift reference to other available compounds, it has been found to be the most robust option available for reverse-phase LC-NMR studies.

These conclusions were deemed applicable to a vast majority of LC-NMR experiments such as LC-NMR-MS, LC-PDA-NMR-MS, LC-MS-NMR-CD, CapLC-NMR, and LC-SPE-NMR [81]. The approach applies to all LC-NMR experiments that use solvent gradients and are acquired in the continuous-flow mode. Some of these conclusions would apply to isocratic continuous-flow experiments, and some apply to solvent-gradient stopped-flow experiments. These concepts for experimental setup also apply to other flow NMR methods such as flow-injection-analysis NMR (FIA-NMR) [82] and direct-injection NMR (DINMR) techniques [62]. This approach even applies to conventional 5-mm tube-based experiments where the samples have temperature gradients or use solvent mixtures that may not be sufficiently well mixed [65].

REFERENCES

1. Purcell, E.M., Torrey, H.C., and Pound, R.V. 1946. Resonance absorption by nuclear magnetic moments in a solid. *Phys. Rev.* 69:37–38.
2. Bloch, F., Hansen, W., and Packard, M.E. 1946. The nuclear induction experiment. *Phys. Rev.* 70:474–485.
3. Arnold, J.T., Dharmatti, S.S., and Packard, M.E. 1951. Chemical effects on nuclear induction signals from organic compounds. *J. Chem. Phys.* 19:507.
4. Anderson, W.A. 1956. Nuclear magnetic resonance spectra of some hydrocarbons. *Phys. Rev.* 102:151–167.
5. Klein, M.P., and Barton, G.W. 1963. Enhancement of signal-to-noise ratio by continuous averaging: application to magnetic resonance. *Rev. Sci. Instr.* 34:754–759.
6. Nelson, F.A., and Weaver, H.E. 1964. Nuclear magnetic resonance spectroscopy in superconducting magnetic fields. *Science.* 146:223–232.
7. Ernst, R.R., and Anderson, W.A. 1966. Application of Fourier transform spectroscopy to magnetic resonance. *Rev. Sci. Instr.* 37:93–102.
8. Ernst, R.R. 1966. Nuclear magnetic double resonance with an incoherent radio-frequency field. *J. Chem. Phys.* 45:3845–3861.
9. Jeener, J. 1971. Pulse-pair techniques in high-resolution NMR. Paper presented at Ampere International Summer School, Basko Polje, Yugoslavia.
10. Suryan, G. 1951. Nuclear resonance in flowing liquids. *Proc. Indian Acad. Sci., Sect A.* 33:107–111.
11. Ettre, L.S. 2002. *Milestones in the Evolution of Chromatography*, New York: ChromSource, Wiley.
12. Watanabe, N., and Niki, E. 1978. Direct-coupling of FT-NMR to high-performance liquid chromatography. *Proc. Jpn. Acad. Ser. B.* 54:194–199.
13. Bayer, E., Albert, K., Nieder, M., Grom, E., and Keller, T. 1979. On-line coupling of high-performance liquid chromatography and nuclear magnetic resonance. *J. Chromatogr.* 186:497–507.
14. Albert, K. 2002. *LC NMR and Related Techniques*, West Sussex, England: John Wiley & Sons.
15. Haw, J.F., Glass, T.E., Hausler, D.W., Motell, E., and Dorn, H.C. 1980. Direct coupling of a liquid chromatograph to a continuous-flow hydrogen nuclear magnetic resonance detector for analysis of petroleum and synthetic fuels. *Anal. Chem.* 52:1135–1140.
16. Haw, J.F., Glass, T.E., and Dorn, H.C. 1981. Continuous flow high field nuclear magnetic resonance detector for liquid chromatographic analysis of fuel samples. *Anal. Chem.* 53:2327–2332.
17. Haw, J.F., Glass, T.E., and Dorn, H.C. 1981. Analysis of coal conversion recycle solvents by liquid chromatography with nuclear magnetic resonance detection. *Anal. Chem.* 53:2332–2336.
18. Haw, J.F., Glass, T.E., and Dorn, H.C. 1982. Conditions for quantitative flow FT-^{1}H NMR measurements under repetitive pulse conditions. *J. Magn. Reson.* 49(1):22–31.
19. Haw, J.F., Glass, T.E., and Dorn, H.C. 1983. Liquid chromatography-proton nuclear-magnetic-resonance spectrometry average-composition analysis of fuels. *Anal. Chem.* 55:22–29.
20. Bayer, E., Albert, K., Nleder, M., Grom, E., Wolff, G., and Rindlisbacher, M. 1982. On-line coupling of liquid chromatography and high-field nuclear magnetic resonance spectrometry. *Anal. Chem.* 54:1747–1750.
21. Smallcombe, S.H., Patt, S.L., and Keifer, P.A. 1995. WET solvent suppression and its applications to LC-NMR and high-resolution NMR spectroscopy. *J. Magn. Reson. A.* 117:295–303.
22. Sudmeier, J.L., and Pesek, J.J. 1971. Fast kinetics by stopped-flow chlorine-35 nuclear magnetic resonance: reactions of mercury(II)-bovine serum albumin with various ligands. *Inorg. Chem.* 10 (4):860–863.
23. Grimaldi, J., Baldo, J., McMurray, C., and Sykes, B.D. 1972. Stopped-flow nuclear magnetic resonance spectroscopy. *J. Am. Chem. Soc.* 94(22):7641–7645.
24. Couch, D.A., Howarth, O.W., and Moore, P. 1975. Kinetic studies by stopped-flow pulse Fourier transform nuclear magnetic resonance. *J. Phys. E: Sci. Instrum.* 8:831–833.
25. Grimaldi, J.J., and Sykes, B.D. 1975. Stopped flow Fourier transform nuclear magnetic resonance spectroscopy: application to the alpha-chymotrypsin-catalyzed hydrolysis of tert-butyl-L-phenylalanine. *J. Am. Chem. Soc.* 97(2):273–276.
26. Fyfe, C.A., Sanford, W.E., and Yannoni, C.S. 1976. Electronic isotopic substitution in flow and stopped flow nuclear magnetic resonance studies of chemical reactions. *J. Am. Chem. Soc.* 98 (22):7101–7102.

27. Kuehne, R.O., Schaffhauser, T., Wokaun, A., and Ernst, R.R. 1979. Study of transient chemical reactions by NMR: fast stopped-flow Fourier transform experiments. *J. Magn. Reson.* 35(1):39–67.

28. Frieden, C., Hoeltzli, S.D., and Ropson, I.J. 1993. NMR and protein folding: equilibrium and stopped-flow studies. *Protein Science.* 2(12):2007–2014.

29. Hoeltzli, S.D., Ropson, I.J., and Frieden, C. 1994. In *Techniques in Protein Chemistry V*, ed. J. W. Crabb, 455, San Diego, CA: Academic Press.

30. Hoeltzli, S.D., and Frieden, C. 1995. Stopped-flow NMR spectroscopy: real-time unfolding studies of 6–19F-tryptophan-labeled *Escherichia coli* dihydrofolate reductase. *Proc. Natl. Acad. Sci. USA.* 92:9318–9322.

31. Hoeltzli, S.D., and Frieden, C. 1996. Real-time refolding studies of 6–19F-tryptophan-labeled *E. coli* dihydrofolate reductase using stopped-flow NMR spectroscopy. *Biochemistry.* 35:16843–16851.

32. Hoeltzli, S.D., and Frieden, C. 1998. Refolding of 6–19F-tryptophan-labeled *E. coli* dihydrofolate reductase in the presence of ligand: a stopped-flow NMR spectroscopy study. *Biochemistry.* 37:387–398.

33. McGarrity, J.F., Prodolliet, J.W., and Smyth, T. 1981. Rapid-injection NMR: a simple technique for the observation of reactive intermediates. *Org. Magn. Reson.* 17:59–65.

34. McGarrity, J.F., and Prodolliet, J.W. 1982. Hydrolysis of very reactive methylation agents: A rapid-injection NMR investigation. *Tetrahedron Lett.* 23(4):417–420.

35. McGarrity, J.F., and Prodolliet, J. 1984. High-field rapid-injection NMR: observation of unstable primary ozonide intermediates. *J. Org. Chem.* 49:4465–4470.

36. McGarrity, J.F., Ogle, C.A., Brich, Z., and Loosli, H.R. 1985. A rapid-injection (RI) NMR study of the reactivity of butyllithium aggregates intetrahydrofuran. *J. Am. Chem. Soc.* 107(7):1810–1815.

37. Bertz, S.H., Carlin, C.M., Deadwyler, D.A., Murphy, M.D., Ogle, C.A., and Seagle, P.H. 2002. Rapid-injection NMR study of iodo- and cyano-Gilman reagents with 2-cyclohexenone: observation of π-complexes and their rates of formation. *J. Am. Chem. Soc.* 124(46):13650–13651.

38. Bayer, E., and Albert, K. 1984. Continuous-flow carbon-13 nuclear magnetic resonance spectroscopy. *J. Chromatogr. A.* 312:91–97.

39. Albert, K., Kruppa, G., Zeller, K.P., Bayer, E., and Hartmann, F. 1984. In vivo metabolism of [4–^{13}C] phenacetin in an isolated perfused rat liver measured by continuous flow 13C NMR spectroscopy. *Z. Naturforsch.* 39:859–862.

40. Keifer, P.A. 2007. Flow techniques in NMR spectroscopy. *Ann. Rep. On NMR Spect.* 62:1–47.

41. Grimaldi, J.J., and Sykes, B.D. 1975. Concanavalin A: a stopped flow nuclear magnetic resonance study of conformational changes induced by Mn++, Ca++, and alpha-methyl-D-mannoside. *J. Biol. Chem.* 250:1618–1624.

42. Sykes, B.D., and Grimaldi, J.J. 1978. Stopped-flow nuclear magnetic resonance spectroscopy. *Methods Enzymol.* 49:295–321.

43. Brown, A.J., Couch, D.A., Howarth, O.W., and Moore, P. 1976. A direct study of the rate of solvent exchange between the Al(DMSO)$^{3+}$ ion and d6-DMSO by stopped-flow pulse Fourier-transform nuclear magnetic resonance. *J. Magn. Reson.* 21(3):503–505.

44. Brown, A.J., Howarth, O.W., and Moore, P. 1978. A stopped-flow pulse Fourier transform nuclear magnetic resonance investigation of the rates of chlorination of metal acetylacetonates by N-chlorosuccinimide. *J. Am. Chem. Soc.* 100(3):713–718.

45. Moore, P. 1985. Probing ligand substitution reactions in non-aqueous solvents by NMR methods: stopped-flow and high-pressure flow-Fourier-transform NMR and natural abundance ^{17}O NMR line broadening. *Pure Appl. Chem.* 57(2):347–354.

46. Merbach, A.E., Moore, P., Howarth, O.W., and McAteer, C.H. 1980. Stopped-flow Fourier-transform NMR and NMR line broadening studies of the rates of dimethyl sulphoxide exchange with the hexakis(dimethylsulphoxide) complexes of aluminium(III), gallium(III) ions in nitromethane solution. *Inorg. Chim. Acta.* 39:129–136.

47. Fyfe, C.A., Cocivera, M., and Damji, S.W.H. 1973. High resolution nuclear magnetic resonance study of chemical reactions using flowing liquids: kinetic and thermodynamic intermediates formed by the attack of methoxide ion on 3,5-dinitrocyanobenzene. *J. Chem. Soc. Chem. Commun.* (19):743–744.

48. Fyfe, C.A., Cocivera, M., Damji, S.W.H., Hostetter, T.A., Sproat, D., and O'Brien, J. 1976. Apparatus for the measurement of transient species and effects in flowing systems by high-resolution nuclear magnetic resonance spectroscopy. *J. Magn. Reson.* 23(3):377–384.

49. Fyfe, C.A., Koll, A., Damji, S.W.H., Malkiewich, C.D., and Forte, P.A. 1977. Low temperature flow nuclear magnetic resonance detection of transient intermediates on the actual reaction pathway in nucleophilic aromatic substitution. *J. Chem. Soc.* (10):335–337.

50. Fyfe, C.A., and Van Veen, Jr., L. 1977. Flow nuclear magnetic resonance investigation of the intermediates formed during the bromination of phenols in acetic acid. *J. Am. Chem. Soc.* 99(10):3366–3371.

51. Fyfe, C.A., Cocivera, M., and Damji, S.W.H. 1978. Flow and stopped-flow nuclear magnetic resonance investigations of intermediates in chemical reactions. *Acc. Chem. Res.* 11:277–282.

52. Cocivera, M., Woo, K.W., and Livant, P. 1978. Flow nuclear magnetic resonance study of the addition, dehydration, and cyclization steps for reaction between hydrazine and ethyl acetoacetate. *Can. J. Chem.* 56:473–480.

53. Tan, L.K., and Cocivera, M. 1982. Flow ^{1}H NMR study of the rapid nucleophilic addition of amino acids to 4-formylpyridine. *Can. J. Chem.* 60:772–777.

54. Cocivera, M., and Effio, A. 1976. Flow nuclear magnetic resonance study of the rapid addition of hydroxylamine to acetone and the rate-determining dehydration of the carbinolamine. *J. Am. Chem. Soc.* 98(23):7371–7374.

55. Cocivera, M., and Effio, A. 1976. Methyl–proton exchange involving transient species: flow nuclear magnetic resonance study of the addition of hydroxylamine to acetone. *J. Chem. Soc., Chem. Commun.* 11:393–394.

56. Spraul, M. 2006. Developments in NMR hyphenation for pharmaceutical industry. In *Modern Magnetic Resonance*, ed. G.A. Webb, 1221–1228, The Netherlands: Springer.

57. Eldridge, S.L., Korir, A.K., Merrywell, C.E., and Larive, C.K. 2000. Hyphenated chromatographic techniques in nuclear magnetic resonance spectroscopy. *Advances in Chromatography.* 46:351–390.

58. Dorn, H.C. 1996. Flow NMR. In *Encyclopedia of Nuclear Magnetic Resonance*, eds. D.M. Grant and R.K. Harris, 2026–2036, West Sussex, England: John Wiley & Sons.

59. Keifer, P.A. 2000. NMR spectroscopy in drug discovery: tools for combinatorial chemistry, natural products and metabolism research. In *Progress in Drug Research*, ed. E. Jucker, 137–211, Basel, Switzerland: Verlag.

60. Keifer, P.A. 2008. Flow NMR techniques in the pharmaceutical sciences. In *Modern Magnetic Resonance*, ed. G.A. Webb, 1213–1219, The Netherlands: Springer.

61. Patt, S.L. 1992. Single- and multiple-frequency-shifted laminar pulses. *J. Magn. Reson.* 96:94–102.

62. Keifer, P.A., Smallcombe, S.H., Williams, E.H., Salomon, K.E., Mendez, G., Belletire, J.L., and Moore, C.D. 2000. Direct-injection NMR (DI-NMR): a flow NMR technique for the analysis of combinatorial chemistry libraries. *J. Comb. Chem.* 2:151–171.

63. Kupce, E., and Freeman, R. 1993. Techniques for multisite excitation. *J. Magn. Reson. A.* 105:234–238.

64. Kupce, E., and Freeman, R. 1994. "Template excitation" in high-resolution NMR. *J. Magn. Reson. A.* 106:135–139.

65. Keifer, P.A. 2009. Chemical-shift referencing and resolution stability in gradient LC-NMR (acetonitrile: water). *J. Magn. Reson.* 199:75–87.

66. Laude, Jr., D.A., and Wilkins, C.L. 1987. Reverse-phase high-performance liquid chromatography/nuclear magnetic resonance spectrometry in protonated solvents. *Anal. Chem.* 59:546–551.

67. Lacey, M.E., Tan, Z.J., Webb, A.G., and Sweedler, J.V. 2001. Union of capillary high performance liquid chromatography and microcoil nuclear magnetic resonance spectroscopy applied to the separation and identification of terpenoids. *J. Chromatogr. A.* 922:139–149.

68. Griffiths, L., and Horton, R. 1998. Optimization of LC-NMR, III: increased signal-to-noise ratio through column trapping. *Magn. Reson. Chem.* 36:104–109.

69. de Koning, J.A., Hogenboom, A.C., Lacker, T., Strohschein, S., Albert, K., and Brinkman, U.A.T. 1998. Online trace enrichment in hyphenated liquid chromatography–nuclear magnetic resonance spectroscopy. *J. Chromatogr. A.* 813:55–61.

70. Lindon, J.C., Nicholson, J.K., and Wilson, I.D. 2000. Directly coupled HPLC–NMR and HPLC–NMR–MS in pharmaceutical research and development. *J. Chromatogr. B Biomed. Sci. Appl.* 748:233–258.

71. Jayawickrama, D.A., Wolters, A.M., and Sweedler, J.V. 2003. Mobile phase compensation to improve NMR spectral properties during solvent gradients. *Analyst.* 128:421–426.

72. Cloarec, O., Campbell, A., Tseng, L., Braumann, U., Spraul, M., Scarfe, G., Weaver, R., and Nicholson, J.K. 2007. Virtual chromatographic resolution enhancement in cryoflow LC-NMR experiments via statistical total correlation spectroscopy. *Anal. Chem.* 79:3304–3311.

73. Spraul, M., Freund, A.S., Nast, R.E., Withers, R.S., Maas, W.E., and Corcoran, O. 2003. Advancing NMR sensitivity for LC-NMR-MS using a cryoflow probe: application to the analysis of acetaminophen metabolites in urine. *Anal. Chem.* 75:1536–1541.

74. Lindon, J.C., Farrant, R.D., Sanderson, P.N., Doyle, P.M., Gough, S.L., Spraul, M., Hofmann, M., and Nicholson, J.K. 1995. Separation and characterization of components of peptide libraries using on-flow coupled HPLC–NMR spectroscopy. *Magn. Reson. Chem.* 33:857–863.

75. Keifer, P.A. 2010. Chemical-shift referencing and resolution stability in methanol: watergradient LC–NMR. *J. Magn. Reson.* 205:130–140.

76. Blechta, V., Kurfurst, M., Sykora, J., and Schraml, J. 2007. High-performance liquid chromatography with nuclear magnetic resonance detection applied to organosilicon polymers, Part 2: comparison with other methods. *J. Chromatogr. A*. 1145:175–182.

77. Bloom, A.L., and Shoolery, J.N. 1953. High resolution of nuclear resonance signals in flowing samples. *Phys. Rev.* 90:358.

78. Edzes, H.T. 1990. The nuclear magnetization as the origin of transient changes in the magnetic field in pulsed NMR experiments. *J. Magn. Reson.* 86:293–303.

79. Levitt, M.H. 1996. Demagnetization field effects in two-dimensional solution NMR. *Concepts Magn. Reson.* 8:77–103.

80. Huang, S.Y., Anklin, C., Walls, J.D., and Lin, Y.Y. 2004. Sizable concentration-dependent frequency shifts in solution NMR using sensitive probes. *J. Am. Chem. Soc.* 126:15936–15937.

81. Keifer, P.A. 2007. Flow techniques in NMR spectroscopy. *Annu. Rep. NMR Spectrosc.* 62:1–47.

82. Keifer, P.A. 2003. Flow injection analysis NMR (FIA-NMR): a novel flow NMR technique that complements LC-NMR and direct injection NMR (DI-NMR). *Magn. Reson. Chem.* 41:509–516.

2 NMR Theory

2.1 MAGNETIC PROPERTIES OF NUCLEI

To better understand and interpret the results from a nuclear magnetic resonance (NMR) experiment, it is important to acquire a fundamental knowledge of the theory and principles governing the NMR phenomenon. The following provides a basic review of NMR theory with emphasis on practical factors and mechanisms influencing spectroscopic data and is intended as a guide for the applications chemist engaged in liquid chromatography-nuclear magnetic resonance (LC-NMR) studies.

NMR is a phenomenon that exists when certain nuclei are placed in a magnetic field and are subsequently perturbed by an orthogonal oscillating magnetic field. This phenomenon occurs for nuclei that possess a property called spin. This property induces a spinning charge that creates a magnetic moment, μ. The magnetic moment μ of a nucleus is proportional to its spin, I, its *magnetogyric ratio*, γ (a property intrinsic to each nucleus), and Planck's constant, h.

$$\mu = \frac{\gamma I h}{2\pi} \tag{2.1}$$

Because of their magnetic properties, certain nuclei may be likened to tiny bar magnets. When such nuclei are placed in a magnetic field, some will align with the field and others against the field. Those nuclei aligned with the field are at a lower energy level than those aligned against the field. The separation in energy levels is proportional to the external magnetic field strength.

The energy of a nucleus at a particular energy level is given by

$$E = -\frac{\gamma h}{2\pi} m B \tag{2.2}$$

where B is the strength of the magnetic field at the nucleus and m is the magnetic quantum number.

The difference in energy between the transition energy levels is

$$\Delta E = \frac{\gamma h B}{2\pi} \tag{2.3}$$

Hence, as the magnetic field, B, is increased, so is ΔE (Figure 2.1). Also, if a nucleus has a relatively large magnetogyric ratio, then ΔE will be correspondingly large.

Because the separation in energy levels (ΔE) affects the Boltzman distribution of the populated energy states, higher magnetic fields and larger magnetogyric properties of nuclei will increase the population of nuclei in the lower energy state; hence, the number of possible NMR transitions will increase. These properties will correspondingly enhance the sensitivity of the NMR experiment [1].

As noted previously, only certain atoms possess a magnetic moment. This is because, for many atoms, the nuclear spins are paired against each other such that the nucleus of the atom has no overall spin. However, for atoms where the nucleus does possess unpaired spins, such atoms can experience NMR transitions. The rules for determining the net spin of a nucleus are as follows:

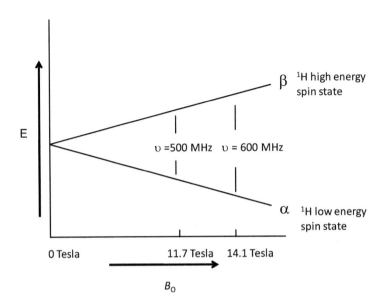

FIGURE 2.1 Graphical relationship between magnetic field B_o and frequency v for ^{1}H-NMR energy absorptions.

1. If the number of neutrons and the number of protons are both even, then the nucleus has no spin.
2. If the number of neutrons plus the number of protons is odd, then the nucleus has a half-integer spin (i.e., 1/2, 3/2, 5/2).
3. If the number of neutrons and the number of protons are both odd, then the nucleus has an integer spin (i.e., 1, 2, 3).

Nuclei most often observed in pharmaceutical applications are shown in Table 2.1. Because of sensitivity issues related to natural abundance, sample size, and concentration, LC-NMR studies primarily focus on the observation of proton nuclei. For highly concentrated samples, however, improvements in technology have enabled indirect detection of carbon.

TABLE 2.1
Common Nuclei Used in NMR Experiments

	I	γ 10^7 $T^{-1}s^{-1}$	v/MHz (at 14.1 T)	Natural Abundance
^{1}H	1/2	26.752	600.00	99.99%
^{13}C	1/2	6.728	150.85	1.1%
^{15}N	1/2	−2.713	60.81	0.37%
^{17}O	5/2	−3.628	81.33	0.04%
^{19}F	1/2	25.182	564.51	100%
^{27}Al	5/2	6.976	156.32	100%
^{29}Si	1/2	−5.319	119.19	4.7%
^{31}P	1/2	10.839	242.86	100%
^{11}B	3/2	8.584	192.48	80.1%

Nuclei can be imagined as spheres spinning on their axes. The phenomenon of nuclear resonance can be illustrated using a quantum mechanical description. Quantum mechanics shows that a nucleus of spin I will have $2I + 1$ possible orientations. For a nucleus with spin 1/2, two possible orientations exist, occupying discrete energy levels in the presence of an external magnetic field (Figure 2.2). Each level is given a *magnetic quantum number, m.*

The initial populations of the energy levels are determined by thermodynamics, as described by the Boltzmann distribution. The lower energy level will contain slightly more nuclei than the higher level. It is possible to excite these excess nuclei into the higher level with a band of radio frequency (RF) waves (Figure 2.3). The frequency of radiation needed for an energy transition to occur is determined by the difference in energy between the energy levels (ΔE).

A classical mechanical description has also been used to describe the resonance phenomenon. This description invokes the use of vectors to illustrate the magnitude and direction of the magnetic properties of the nucleus (Figure 2.4). A nucleus with spin when placed in a magnetic field will tend to precess about the direction of the field at a discrete frequency. This frequency is known as the Larmor precessional frequency. For a magnet of 11.75 tesla, a proton will precess at 500 MHz; and for 14.1 tesla, a proton precesses at 600 MHz. NMR spectrometers are commonly named based upon the Larmor precessional frequency of protons at a particular field strength.

If a sample is placed in an 11.75-tesla magnet and a RF field is applied at 500 MHz, the protons will absorb that energy; and the phenomenon known as resonance occurs. Hence for nuclei of spin 1/2 in a magnetic field, the nuclei in the lower energy level will be aligned in the direction of the external field and the nuclei occupying the higher energy level will align against the field. Because the nucleus is spinning on its axis in the presence of a magnetic field, this axis of rotation will precess around the direction of the external magnetic field.

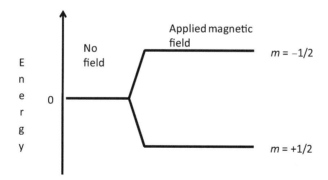

FIGURE 2.2 Zeeman interaction: In the absence of a magnetic field, magnetic dipoles are randomly oriented and there is no net magnetization. Upon application of a magnetic field, nuclei occupy discrete energy levels. Two energy levels exist for nuclei with spin quantum number of 1/2.

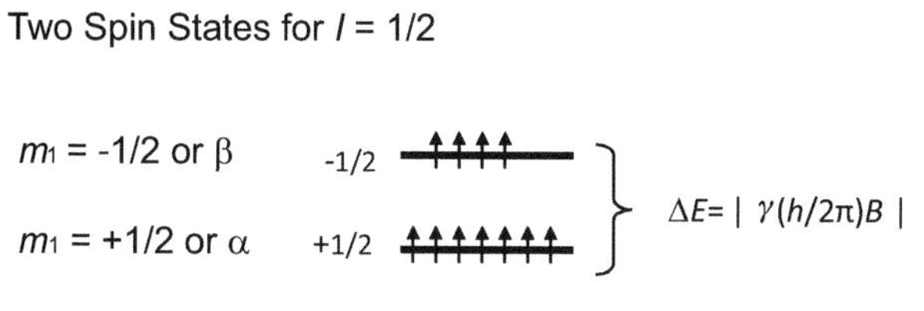

Two Spin States for I = 1/2

m_1 = -1/2 or β   -1/2

m_1 = +1/2 or α   +1/2 $\Delta E = | \gamma(h/2\pi)B |$

FIGURE 2.3 Illustration of spin states for nuclei aligned with α or against β the external magnetic field (B_o). Transitions between these spin states can occur by applying a RF field B_1.

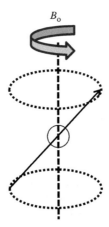

B_o

FIGURE 2.4 Classical mechanical illustration of a nucleus (of spin 1/2) in an applied magnetic field (B_o). When the nucleus is in a lower energy state, it is aligned with the field. The nucleus may be viewed as a sphere spinning on its axis. In the presence of a magnetic field, the axis of rotation will precess around the direction of B_o.

The frequency of precession, the Larmor frequency, is identical to the transition frequency. The potential energy of the precessing nucleus is given by

$$E = -\mu B \cos \theta \qquad (2.4)$$

where θ is the angle between the direction of the applied field and the axis of nuclear rotation.

When energy is absorbed by the nucleus, the angle of precession, θ, will change. For a nucleus of spin 1/2, absorption of radiation "flips" the magnetic moment so that it opposes the applied field (the higher energy state). Since only a small excess of nuclei exists in the lower energy state to absorb radiation, only a small number of transitions can occur. The low number of energy transitions is what gives rise to the observed NMR signal; hence, this condition negatively affects the overall sensitivity of the technology. By exciting the low energy nuclei, the populations of the higher and lower energy levels will become equal, resulting in no further absorption of radiation. The nuclear spins are then considered *saturated*. When the RF pulse is turned off, the nuclei undergo a relaxation process, returning the nuclei to thermodynamic equilibrium [2].

2.2   DATA ACQUISITION

The classical description of NMR spectroscopy invokes a vector representation of individual nuclei and it is this description that is commonly used by spectroscopists to illustrate what happens in the NMR experiment. When nuclei are placed in a magnetic field, they will tend to align with and against the field. The Boltzmann distribution gives rise to a small excess of nuclei aligned with the field. If one considers the sum of all the vectors aligned with and against the field, the result is a small magnetization vector (M_0) aligned in the direction of the external magnetic field. This small resultant magnetization is responsible for the observed NMR signal. To understand the sensitivity issue, consider that, in a field of 100 MHz for every 32 million nuclei, there are two excess nuclei aligned with the field. It is thissmall excess that produces the NMR signal (Figure 2.5a).

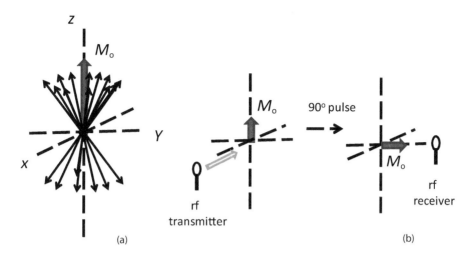

FIGURE 2.5 (a) Precessing nuclear dipoles. The sum of the dipole vectors is represented by the resultant magnetization M_o that is aligned along the $+Z$-axis. (b) Direction of magnetization vector M_o following a 90° pulse along the X-axis. The transverse magnetization induces a signal in the receiver coil, which is oriented along the Y-axis.

The magnetization vector (M_o) is used to visualize what happens to nuclei when a sequence of RF pulses is applied. If one considers an XYZ coordinate system with the direction of the external magnetic field aligned with the $+Z$-axis, upon the application of a RF pulse along the X-axis, the resultant magnetization (M_o) begins to precess about the applied RF field. In the NMR experiment, the RF pulse is applied for a duration long enough to tip the magnetization 90 degrees into the XY plane and then is turned off. Following the RF transmitter pulse, the RF receiver is turned on (where the phase of the transmitter is orthogonal to the receiver) along the Y-axis (Figure 2.5b).

A signal is induced from the relaxing nuclei that may be likened to a decaying sine wave. This signal arises as the result of the decay of transverse magnetization and is detected by the spectrometer's RF receiver (Figure 2.6a). The term describing this signal is the free induction decay or FID. Because NMR samples usually contain nuclei with different resonance frequencies, the decay curves arising from the different magnetization frequencies are superimposed, causing the FIDs to interfere with each other. This collection of signals is also described as an interferogram. The FID represents a time-domain spectrum; however, because it is not possible to directly interpret the FID signals, a mathematical function, known as Fourier transformation (FT), is applied [3]. When FT is applied to the FID, a frequency-domain spectrum is produced that, upon phase and baseline correction, may be interpreted by the analyst (Figure 2.6b). Because NMR is an inherently insensitive technique, the FID signals are small compared to the noise. This is especially problematic for nuclei with low natural abundance, such as ^{13}C. To compensate for these weak signals, the FIDs of many pulses are acquired and added together. However, although the signal is additive, the noise is random and both adds and subtracts. Therefore, the signal-to-noise (S/N) does not increase linearly with the number of scans (NS) but instead increases in proportion to the square root of NS.

$$S:N \sim \sqrt{NS} \qquad (2.5)$$

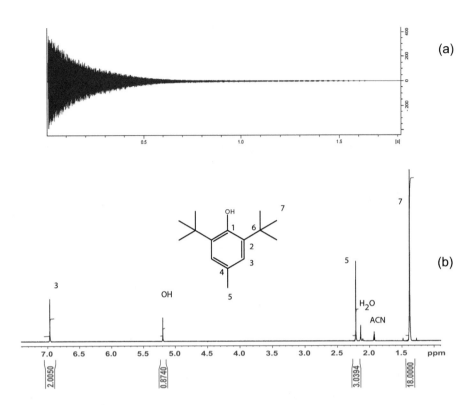

FIGURE 2.6 500-MHz ^{1}H-NMR spectrum of 2,6-di-tert-butyl-4-methylphenol. Spectral width 9014 Hz, 32K data points, acquisition time 1.8 s in solvent acetonitrile-d_3. (a) A time-domain spectrum (FID) and (b) frequency-domain spectrum of 2,6-di-tert-butyl-4-methylphenol obtained after Fourier transformation of the FID.

In cases where concentrations are low or nuclei have low natural abundance, thousands of scans are required. The nature of the decaying acquisition produces a higher degree of noise at the end of the FID than at the beginning. The rate at which the FID decays is a function of spin-spin relaxation or T_2 and the change in field homogeneity (see Section 2.3 on Relaxation of Nuclei). Rapid decay of the FID leads to line broadening and loss of sensitivity (loss of S/N).

The signals that are produced from FT may be integrated to give the area under each peak or set of peaks. For proton spectra, the area under each pattern that is obtained from integration of the signal is proportional to the number of hydrogen atoms whose resonance is giving rise to the pattern. The integration may be displayed as a step function on top of the peak, with the relative integrated area values displayed below each peak. In Figure 2.6, the integration of the patterns at δ 6.97, 2.21, and 1.38 ppm for H-3, H-5, and H-7 in 2,6-di-tert-butyl-4-methylphenol is approximately 2:3:18, respectively. It is important to exercise care with integrated values in proton spectra as significant errors in quantitation can occur depending on factors such as instrument optimization and discrepancies in relaxation of the nuclei. For typical ^{1}H-NMR spectra, the integration gives a measure of the number of protons per assigned peak. The compound 2,6-di-tert-butyl-4-methylphenol has molecular symmetry; hence, the integration gives the combined proton count per assignment. For example, the plane of symmetry bisecting the molecule makes the two t-butyl groups (H-7) equivalent and the two aromatic protons (H-3) equivalent, so only one singlet is observed for each set with an integration ratio 18:2. The integration displays are generated by the instrument using an arbitrary scale but can then be normalized by the operator to correlate with the expected number of protons per peak. It should be noted that these values are not exact integers and need to be rounded to the nearest integer to obtain the proper value.

2.3   RELAXATION OF NUCLEI

In NMR spectroscopy, two primary relaxation processes occur. They are termed spin-lattice (longitudinal) relaxation (T_1) and spin-spin (transverse) relaxation (T_2). Spin-lattice relaxation, T_1, is the time it takes an excited nucleus to release its energy to its lattice or surrounding environment. It is dependent on the magnetogyric ratio of the nucleus and the mobility of the lattice. Spin-spin relaxation, T_2, describes the interaction between neighboring nuclei with identical precessional frequencies resulting in a decrease of the average lifetime of a nucleus in the excited state. This can result in line broadening. Both these relaxation processes need to be taken into account in the design and execution of NMR experiments [2]. A more detailed description of these processes is discussed next.

2.3.1   SPIN-LATTICE RELAXATION TIME T_1 (LONGITUDINAL)

Longitudinal relaxation (T_1) is due to energy exchange between the spins and surrounding lattice (spin-lattice relaxation), reestablishing thermal equilibrium. As spins go from a high-energy state back to a low-energy state, RF energy is released back into the surrounding environment. The recovery of longitudinal magnetization follows an exponential curve.

The relaxation time T_1 represents the "lifetime" of the first-order rate process that returns the magnetization to the Boltzman equilibrium. From a classical mechanical standpoint, this may be viewed as the time it takes for the bulk magnetization vector (M_o) to align itself along the direction of the external magnetic field (+Z-axis).

T_1 relaxation time can be measured by various techniques. The inversion recovery technique [4] is one method commonly used for small molecules. This technique involves applying a 180° pulse followed by a variable delay (τ) and then a 90° read pulse. The time delay between each transient (D) must be sufficiently long to enable full relaxation of all the nuclei. This is a slow sequence and preevaluation of T_1 is necessary. The inversion recovery sequence may be described as follows:

$$D - 180° - \tau - 90°_{\text{FID}} \tag{2.6}$$

The T_1 is calculated from a semilogarithmic plot of the signal intensity relative to the variable delay, τ, and modern spectrometers have programs incorporated to provide automatic calculations. A rapid approximate calculation of individual T_1s may be obtained by taking the delay where the signal is nulled (τ_{null}) and dividing by ln2 (0.693).

An alternate pulse sequence is saturation recovery, which enables faster repetition ≥1 s and, hence, a major reduction in overall experiment time. This sequence involves the incorporation of a homospoil pulse Z^* [5].

$$Z^* - 90° - Z^* - \tau - 90°_{\text{FID}} \tag{2.7}$$

In routine data acquisition, following a 90° pulse and a delay of $1*T_1$, approximately 63% of the magnetization is recovered. To recover about 99% of the magnetization, a delay of $5*T_1$ is needed. Because the longitudinal relaxation process (T_1) will affect signal intensities, it is important to adjust the delay between acquisition transients. Following a 90° pulse, if the interval between transients is shorter than $5*T_1$, the accuracy of the integration may be compromised. To enable more accurate integrations, the pulse width may be shortened to adjust for the length of the relaxation process. For proton NMR, a pulse width of 10° and a delay of about 2 s will enable accurate integrations for a majority of small molecules where the T_1s are less than 20 s. Such considerations are critical when relative quantities of sample components need to be measured. In LC-NMR applications, chromatographically isolated samples often contain impurities or multiple isomers. The need to accurately calculate relative quantities can be critical for separating and identifying multiple species that may be present in an isolated mixture.

It should be noted that the magnitude of the relaxation time depends on the type of nucleus being studied. For example, nuclei with spin 1/2 and low magnetogyric ratio have long relaxation times, whereas nuclei with spin greater than 1/2 have very short relaxation times. In addition to the intrinsic properties of the nucleus, there are other factors that contribute to relaxation. These are associated with the motion of the molecule.

2.3.1.1 Relaxation and Molecular Motion

Spontaneous T_1 relaxation of spin 1/2 (that is relaxation in the absence of external influences) is essentially non-existent. For T_1 relaxation to occur, there must be magnetic field fluctuations. Such fluctuations are most effective when they occur at the Larmor precessional frequency; hence, T_1 relaxation can be field dependent, since the Larmor precessional frequency varies with the field.

The principal source of fluctuating magnetic fields in most molecules is due to Brownian motion. One source of motion is correlation time, τ_c, which can be defined for a molecule (assuming the molecule behaves like a sphere) as the average time it takes the molecule to rotate through one radian (360°). The correlation time for small molecules is on the order of 10^{-12} s in non-viscous solvents.

In general, there are several mechanisms by which molecular motions can influence nuclear relaxation. These include direct dipole-dipole (DD) interactions with nearby magnetic nuclei, paramagnetic interactions, chemical shift anisotropy effects (CSA), scalar coupling (SC) that includes rapid modulation of J-coupling, and quadrupole-electric field gradient interactions. In addition, spin rotational (SR) transitions can also be the source of fluctuating magnetic fields.

Below are descriptions of the relaxation mechanisms that influence NMR observable nuclei. Table 2.2 outlines the mechanism, modulation source, and associated parameters.

2.3.1.2 Dipole-Dipole Interaction "Through Space"

A direct dipole-dipole coupling interaction is very large (often kilohertz) and depends upon the distance between nuclei and the angular relationship between the magnetic field and the internuclear vectors. This coupling is not seen for mobile molecules in solution because it is averaged to zero by tumbling of the molecule. However, as the molecule tumbles in solution, the dipole-dipole coupling is constantly changing as the vector relationships change. This creates a fluctuating magnetic field at each nucleus. When such fluctuations are at the Larmor precessional frequency, they can cause nuclear relaxation. This is a primary relaxation mechanism for nuclei such as proton (i.e., nuclei with high natural abundance and a large magnetogyric ratio).

For small molecules, the dipole-dipole relaxation rate depends on the strength of the dipolar coupling (γ, magnetogyric ratio of the nucleus), on the orientation/distance between the interacting nuclei (r_{IS}), and on the tumbling motion (correlation time, τ_c).

$$R_{1(DD)} = 1/T_{1(DD)} = k * \gamma_I^2 * \gamma_s^2 * (r_{IS})^{-6} * \tau_c \qquad (2.8)$$

TABLE 2.2

Spin Relaxation Mechanisms

Mechanism	Modulation Source	Determining Parameters
1. Dipolar coupling	Reorientational motion	• Abundance of magnetically active nuclei • Size of the magnetogyric ratio
2. Paramagnetic	Paramagnetic species: translational and rotational motion	• Concentration of paramagnetic impurities
3. Chemical shift anisotropy (CSA)	Reorientational motion	• Size of the chemical shift anisotropy • Symmetry at the nuclear site
4. Scalar coupling	Modulation of spin-spin coupling	• Size of the scalar coupling constants
5. Electric qadrupolar relaxation	Reorientational motion	• Size of quadrupolar coupling constant • Electric field gradient at the nucleus
6. Spin rotation	Angular momentum	• Spin rotation (T_1) shorter at high temperature

As shown in Equation (2.8), the distance dependence for DD relaxation is significant. For example, when spin "I" is ^1H and spin "S" is ^{13}C, a carbon-13 NMR spectrum would show that protonated carbons relax more rapidly than quaternary carbons.

The dependence on the gyromagnetic ratio (γ) of the nucleus is also important. If a proton is replaced by deuterium, the ^{13}C nuclei relax much slower for C-D than the corresponding C-H, since the gyromagnetic ratio for proton γ-H is 6.5 times larger than the gyromagnetic ratio for deuterium γ-D.

At typical spectrometer frequencies, small molecules (MW < 1000) are tumbling too fast for the most effective relaxation (τ_c is short). Thus, the more rapidly a molecule or part of a molecule tumbles, the less effective DD relaxation becomes, resulting in longer T_1. Large molecules (e.g., proteins), on the other hand, are usually moving very slowly (τ_c is long), and they have the opposite association between molecular motion and T_1. With macromolecules, relaxation is more effective when the molecule moves faster and the dipole-dipole (T_1) relationship no longer applies (Figure 2.7).

The effect of molecular tumbling on relaxation depends upon spectral density. Spectral density is defined as the concentration of fields at a given frequency of motion. For small molecules in solution, the tumbling motion is rapid; hence, there is a larger population of different frequencies existing for a shorter period of time, so the population of specific frequencies required to induce relaxation at a specific field strength is small. Hence, for small molecules, relaxation would not be significantly affected by field strength. The opposite is true for large slow-moving molecules where the concentration of fields induced by molecular tumbling is field dependent. In such cases, as the external field increases, the population of nuclei tumbling at the accelerated rate needed to induce relaxation decreases. This results in a lower concentration of fields available to induce relaxation and T_1 becomes longer.

For certain nuclei, such as carbon, anisotropic motion can have an impact on DD relaxation. Long, thin molecules, for example a biphenyl, that do not move isotropically in solution will have correlation times (τ_c) that would be different for rotation around different axes of the molecule. Hence, significantly different DD relaxation would be expected for nuclei that are on the axis of rapid rotation. In the case of biphenyl, the para carbons would be expected to relax faster than the ortho- or meta-carbons due to rapid "on axis" rotation.

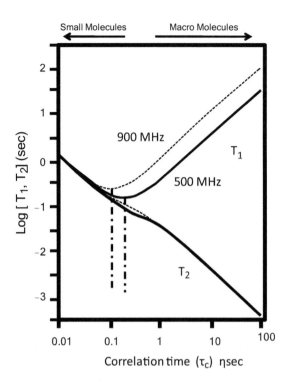

FIGURE 2.7 The effect of molecular weight on proton relaxation. T_1 and T_2 values are plotted as a function of the rotational correlation time, τ, using a log scale. The curves for T_1 and T_2 are labeled. The solid line indicates relaxation at a proton frequency of 500 MHz, while the dotted line represents relaxation at 900 MHz. T_1 minima relative to correlation times are indicated with the dot-dash line.

2.3.1.3 Paramagnetic Relaxation

When paramagnetic nuclei are present in solution, they produce a fluctuating field that induces rapid T_1 relaxation. This relaxation is proportional to the square of the magnetic moment. Because the magnetic moment of an unpaired electron is larger than a nuclear magnetic moment by a factor of 10^3, paramagnetic substances significantly accelerate the spin-lattice relaxation process. The paramagnetic relaxation mechanism may be denoted as a spin-dipolar interaction modulated by translational diffusion. Spin-dipolar modulation by rotational reorientation is another relaxation mechanism that applies in cases where a substrate is bound to a paramagnetic relaxation reagent. In practical terms, paramagnetic relaxation contributions are strongly dependent on the concentration of paramagnetic ions, bonded distances, and solution viscosity.

Atmospheric oxygen is a common source of paramagnetic relaxation. Relaxation of proton and carbon nuclei are induced by dipolar interactions with the unpaired spins on molecular oxygen. Therefore, when initiating T_1 studies, it is important to ensure that dissolved oxygen is removed from the sample. This may be accomplished by bubbling nitrogen or argon gas through the sample and/or pulling a vacuum and sealing the tube. In LC-NMR studies, mobile phase solvents are typically degassed and stored over a blanket of dry nitrogen gas, prior to and during use. The removal of dissolved oxygen from the LC-NMR sample helps to reduce the associated paramagnetic relaxation effects that lead to line broadening.

2.3.1.4   Chemical Shift Anisotropy Relaxation

The chemical shift is dependent on the orientation of a molecule in the magnetic field. This effect, called chemical shift anisotropy (CSA), is partially responsible for the very wide lines observed in solid-state NMR. In solution, CSA is averaged out by molecular tumbling and sharp lines are observed; however, the modulation of the shielding can sometimes introduce an observable relaxation mechanism.

As the molecule tumbles in solution, the chemical shift (and hence the magnetic field at the nucleus) is constantly changing and this can cause relaxation of the nucleus. The relaxation rate is proportional to the square of the magnetogyric ratio and the magnetic field strength, molecular motion, and chemical shift anisotropy. The equation below applies to cylindrically symmetric molecules.

$$R_{1(CSA)} = 1/T_{1(CSA)} = 2/15 \; \gamma^2 B_0{}^2 (\sigma_{\parallel} - \sigma_{\perp})^2 \tau_c \qquad (2.9)$$

where

γ = magnetogyric ratio
B_0 = magnetic field
σ = nuclear shielding tensor
τ_c = correlation time

Because the rate of relaxation is proportional to the square of the field strength, it becomes more important at high field strengths. For some spin 1/2 nuclei with large chemical shift ranges (phosphorus-31 or cadmium-113), lines become sufficiently broadened by CSA relaxation at high field, causing loss of coupling information. Molecules in which the nucleus is at the center of tetrahedral, octahedral, or higher symmetry (^{29}Si in Me$_4$Si) have no chemical shift anisotropy and so do not relax by this mechanism. The mechanism does not occur for protons and therefore would not be a consideration in the vast majority of LC-NMR studies. It can, however, be observed in ^{13}C NMR for certain carbonyl carbon atoms.

2.3.1.5   Scalar Coupling Relaxation

The effect of scalar coupling (SC) relaxation on T_1 is significant only when the two interacting nuclei have very close frequencies. This condition is rare. Scalar (J) coupling of a nucleus A to a quadrupolar nucleus B can provide a relaxation mechanism for A if B is undergoing very rapid T_1 relaxation. Under these conditions, A is subject to a fluctuating magnetic field because of the rapid spin reorientation of B. For this mechanism to be effective, the Larmor precessional frequencies of nuclei B and A must be nearly the same. For example, short T_1 values are observed for bromine-bearing carbons. On a 500-MHz spectrometer, the Larmor precessional frequency of ^{13}C is 125.68 MHz, while that of ^{79}Br (I = 3/2) is 125.23 MHz; therefore, a short T_1 for ^{13}C bonded to ^{79}Br may be attributed to SC relaxation.

2.3.1.6   Electric Quadrupolar Relaxation

Electric quadrupolar relaxation operates for spins >1/2 only, and only for nuclei which are not at the center of tetrahedral or octahedral symmetry. Spin 1/2 nuclei can be considered to have spherical charge distribution, but for spins greater than 1/2, the charge distribution has the shape of an ellipsoid. The electric field gradients in molecules exert a torque on their quadrupolar nuclei. Tumbling of the molecule can initiate transitions among the spin states. The effectiveness of this relaxation mechanism is critically dependent on quadrupolar coupling. If the electric quadrapole moment (Q) is small (e.g., ^{2}H and ^{6}Li), the nucleus behaves like a spin 1/2 nucleus; however, if Q is large (e.g., ^{35}Cl or ^{79}Br), the nucleus can have a very short T_1 and observation can be very difficult.

Quadrupolar relaxation of a nucleus can also have effects on nearby magnetic nuclei (X), since rapid relaxation can broaden or entirely remove J-coupling between the two nuclei. This effect is especially common for X nuclei in ^{14}N-X and ^{11}B-X groups (see section 2.3.2 below on spin–spin relaxation, T_2).

2.3.1.7   Spin Rotation

Intramolecular dynamic processes (like the rotation of a methyl group) can also contribute to longitudinal relaxation. A local magnetic field is generated by the circular motion of electrons in a rapidly rotating molecule or part of a molecule, such as a methyl group. The magnitude of this field changes when the rotational energy levels change as a result of molecular collisions. When these changes occur at the Larmor precessional frequency, they can cause relaxation of nearby nuclei. The spin rotation mechanism is effective for small molecules, or freely spinning portions of larger molecules. It is more effective at higher temperatures.

2.3.2   Spin-Spin Relaxation Time T_2 (Transverse)

Spin-spin relaxation (T_2) is a complex phenomenon corresponding to a decoherence of the transverse nuclear spin magnetization. The T_2 relaxation is observed when nuclei in an excited state begin to precess about the external magnetic field at slightly different frequencies. This happens as spins move together and their magnetic fields interact (spin-spin interaction), thereby slightly modifying their precession rate. These interactions are temporary and random. Thus, spin-spin relaxation causes a cumulative loss in phase resulting in transverse magnetization decay and a broadened signal.

Transverse relaxation (T_2) is always at least slightly faster than longitudinal relaxation (T_1). For small molecules in non-viscous liquids, usually T_2 approximately equals T_1. However, processes like scalar coupling with quadrupolar nuclei, chemical exchange, and interaction with paramagnetic centers can accelerate the T_2 relaxation such that T_2 becomes shorter than T_1. Transverse magnetization decay is described by an exponential curve, characterized by the time constant T_2. After time T_2, transverse magnetization has lost 63% of its original value.

Similar mechanisms that induce spin-lattice relaxation also induce spin-spin relaxation. However, there are other mechanisms contributing to spin-spin relaxation. As noted above, these include scalar relaxation and relaxation induced by quadrupolar nuclei. Scalar relaxation occurs when two spins interact through bond (electron mediated) J-coupling. For example, in the case of chemical exchange where a proton may be in dynamic equilibrium between two chemically distinct sites, the fluctuating fields from neighboring spins will induce transverse relaxation. Resonance lines will broaden due to partial scalar coupling. The resultant line-broadening effect is typically observed for exchangeable protons like OH or NH.

Spin-spin relaxation may be induced by quadrupolar nuclei when the relaxation rate ($1/T_1$) of the quadrupolar nucleus is rapid. In such cases, resonance lines may be broadened by coupling to the quadupolar nucleus. This line broadening is typically observed for ^{1}H attached to nuclei such as ^{14}N or ^{11}B, where T_1s are approximately tens of milliseconds in length. However, for ^{1}H next to ^{35}Cl, where the T_1 is approximately 1 μs, the line broadening is insignificant.

A major contributing factor to transverse or spin-spin relaxation (T_2) is magnetic field inhomogeneity. Therefore, it is important to ensure that the magnetic field is as homogeneous as possible throughout the NMR sample. This involves a process called shimming the magnet, which requires adjusting the current in coils within the magnet to compensate for field discrepancies across a sample. In modern NMR spectrometers, this is an automated process; however, in many LC-NMR applications, some manual shimming is necessary. For LC-NMR studies, shimming may be a challenge due to solvent gradients across a sample often seen with an HPLC gradient solvent system of 2%–3%/min. Technology advancements reducing flow-cell volume and enabling elution into a flow cell with a single solvent have drastically improved the quality of spectra that may be obtained. More detailed discussion is provided in Chapters 1 and 5.

As noted previously, longer T_1 and hence T_2 relaxation times are observed for small molecules in the absence of oxygen; however, this is also found for quaternary carbons, heavier spin 1/2 nuclei, and compounds in the gas phase. Shorter relaxation times are observed when there is medium-to-fast chemical exchange, paramagnetic nuclei are present, and for interactions with quadrupolar nuclei. Since the nuclear T_1 relaxation time is magnetic-field dependent for compounds with longer correlation times (e.g., molecular weight >1000 Da), the T_1 relaxation time measured on a 300-MHz NMR instrument will be different than that measured on a 600-MHz instrument in such cases. The T_2 relaxation, however, rapidly decreases for molecular weights above 1000 Da and remains field independent (see Figure 2.7).

The relaxation time is an intrinsic property of the nucleus that reflects its nature and environment; however, in practice, it is routinely used to decide the length of the relaxation delay between acquisitions. As noted previously, for quantitative work, this needs to be at least five times T_1 to achieve a 1% integration accuracy.

Because numerous NMR applications focus on the identification of a molecular structure, the rigor in setting the relaxation delay for quantitative analysis is not needed. For maximum sensitivity within a limited time frame for a qualitative 1D NMR experiment, the repetition rate should be set to the longest T_1 in the sample with a pulse angle of 68°. Most relaxation times set in routine proton NMR are between 1.0 and 10 s. When ample compound is available for good S:N, the delay between proton acquisitions is usually set to about 1.5–2 s. In practice, this will produce good quality spectra with integrations accurate enough for compound identification.

2.4 THE CHEMICAL SHIFT

All nuclei in a molecule do not experience the same local magnetic field. This is because the magnetic field at the nucleus is not equal to the applied magnetic field. Electrons around the nucleus create an opposing magnetic field that shields a nucleus from the applied field (Figure 2.8). The difference between the applied magnetic field and the field at the nucleus is called *nuclear shielding*.

Nuclear shielding leads to the property known as the chemical shift. Chemical shift is a function of the nucleus and its environment. It is measured relative to a reference compound causing the calculated value to be independent of the external magnetic field

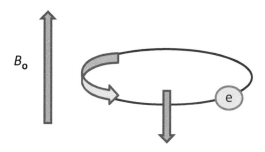

FIGURE 2.8 The electrons around a nucleus (e.g., a proton) create a magnetic field that opposes the applied field. This reduces the field experienced at the nucleus; hence, the electrons provide local shielding effect.

strength. For ^{1}H and ^{13}C NMR, the reference is usually tetramethylsilane, Si $(CH_3)_4$. This reference was selected because it is inert (i.e., non-reactive and stable). In addition, its resonance falls outside the region of the resonances for most molecules of interest and its volatility enables it to be easily removed from the sample. Equation (2.10) is used to calculate the chemical shift.

$$\text{Chemical shift,} \, \delta = \frac{\text{Frequency of signal} \, - \text{frequency of reference}}{\text{spectrometer frequency}} \times 10^6 \qquad (2.10)$$

If one considers a field of 11.75 T (500 MHz), the protons of benzene are found to resonate 3670 Hz downfield from TMS set at 0.0 Hz; hence, the chemical shift for benzene protons would be calculated to be 7.34 ppm. Although the resonance frequency (Hz) will vary with magnetic field strength, the calculated chemical shift relative to TMS will always be the same regardless of which field strength is used. Because the chemical shift is dependent upon the electronic environment surrounding each nucleus, it provides a wealth of information that may be used in defining and characterizing a chemical structure [1,2].

Small molecule structure identification and characterization depend heavily on proton and carbon assignments. Although protons have high natural abundance and excellent magnetic properties, they possess a rather narrow chemical shift range. Most protons resonate in a 0–20 ppm range relative to a TMS reference. For ^{13}C, the chemical shift range is about 20 times larger with carbons resonating in a 0–210 ppm range relative to TMS.

The chemical shift range for protons is smaller because proton resonances are primarily shielded by electronic effects from σ-bonding orbitals, whereas carbon nuclei experience effects from both σ- and p-orbitals. This condition may be attributed to the fact that electrons in p-orbitals have no spherical symmetry. They produce comparatively large magnetic fields at the nucleus, which give a low field shift also termed "deshielding." Protons become deshielded when electrons are pulled away from the protons by neighboring electromagnetic nuclei. Carbon chemical shifts are influenced by electronegative substituents as well as π-bonds, and experience strong deshielding from delocalization of electrons in non-spherical p-orbitals. Hence, the effect from electrons in p-orbitals produces large downfield shifts of carbon resonances where the range of chemical shifts spans 0–220 ppm. In ^{1}H NMR, p-orbitals are not involved in bonding hydrogen atoms to other nuclei, which is why only a small chemical shift range (0–20 ppm) is observed.

Some typical chemical shift ranges for ^{1}H and ^{13}C are given in Figure 2.9(a) and (b), respectively.

2.5 SPIN COUPLING

Spin-spin coupling is another property that plays a critical role in structure elucidation. Consider proton-proton interactions. When signals for single protons appear as multiple lines, this is due to ^{1}H-^{1}H coupling, also known as spin-spin splitting or J-coupling. The spin-spin splitting arises as a result of inter-nuclear magnetic influences.

As mentioned previously, protons may be viewed as tiny magnets that can be oriented with or against the external magnetic field. For a molecule which contains a proton (H_A) attached to a carbon that is attached to another carbon containing a proton (H_B), H_A will feel the presence of the magnetic field of H_B. When the field created by H_B reinforces the magnetic field of the NMR instrument (B_0), H_A experiences a slightly stronger field, but when the field created by H_B opposes B_0, H_A experiences a slightly weaker field. The same situation occurs for H_B relative to magnetic influences from H_A. The result is two resonance lines for H_A and two resonance lines for H_B, commonly termed a doublet.

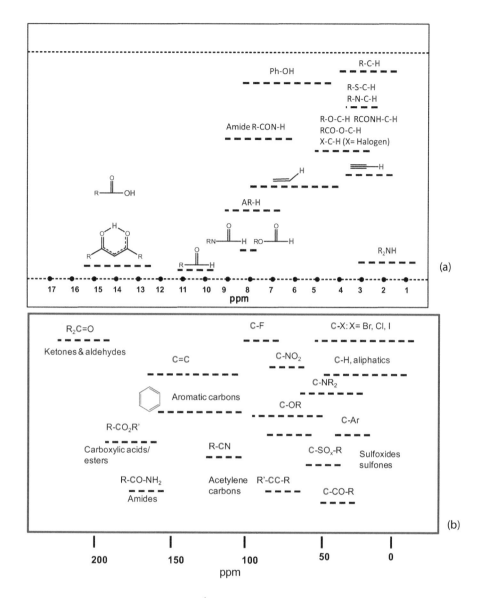

FIGURE 2.9 (a) Chemical shift ranges for ^{1}H nuclei in organic molecules. (b) Chemical shift ranges for ^{13}C nuclei in organic molecules.

The magnitude of the observed spin-splitting depends on many factors and is given by the coupling constant J (units of Hz). J is the same for both partners in a spin-splitting interaction and is independent of the external magnetic field strength. Equivalent nuclei or those nuclei having the same chemical shift do not interact with each other. For example, the three methyl protons in ethanol cause splitting of the neighboring methylene protons; they do not cause splitting among themselves.

For nuclei with the quantum spin number of 1/2, the multiplicity of a resonance is given by the number of equivalent protons in neighboring atoms plus one, i.e., "the $n + 1$ rule." The coupling pattern follows Pascal's Triangle shown in Figure 2.10.

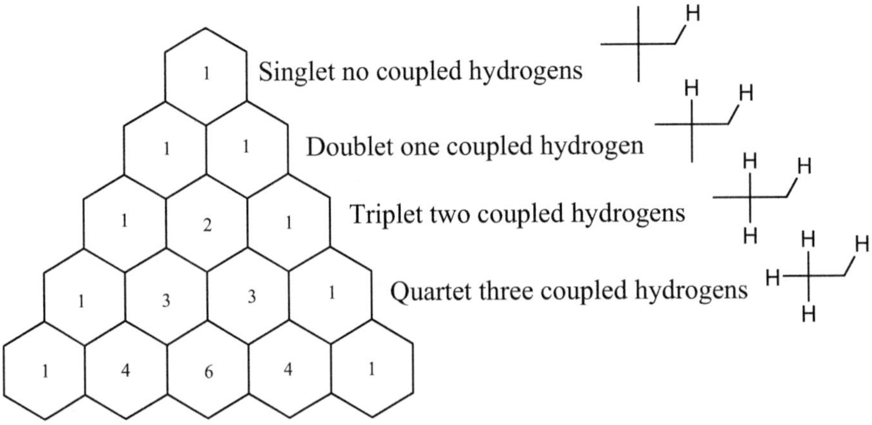

FIGURE 2.10 The splitting pattern of a given nucleus (or set of equivalent nuclei) with spin 1/2 can be predicted by the $n + 1$ rule, where n is the number of neighboring spin-coupled nuclei with the same or similar coupling constants. The intensity ratio of these resonance lines follow Pascal's triangle. A doublet has equal intensities; a triplet has an intensity ratio of 1:2:1; a quartet has a ratio of 1:3:3:1.

The power of spin-spin coupling in structure elucidation arises from the correlation between coupling constants and molecular structure. Coupling constants for protons fall into three categories. These include germinal or two-bond coupling (J^2), vicinal or three-bond coupling (J^3), and long-range or four- or five-bond coupling (J^4, J^5).

The magnitude of the coupling constant can provide information that may be used to unambiguously define structural elements of a compound. For example, trans-alkenes show larger vicinal coupling than cis-alkenes allowing one to identify the regio-chemistry around a double bond. In fact, spin-spin coupling offers one of the best ways of determining the stereochemistry around a double bond.

Sometimes, coupling does not occur between protons on heteroatoms, such as the OH proton of an alcohol and the adjacent protons on carbon. In such cases, the absence of coupling is caused by the rapid exchange of the OH protons, usually with residual water in the organic solvent. Because of this rapid exchange, the influence of the α or β spin states that induces coupling is lost. Some exchangeable protons, however, do not exchange rapidly, for example, the NH of amide protons in a peptide. This may be due to intramolecular hydrogen bonding interactions. In such cases, coupling interactions between the exchangeable protons and neighboring protons may be observed.

The choice of organic solvent can also affect the nature and rate of proton exchange. Solvents like methanol that possess an exchangeable proton will tend to induce rapid exchange between the solvent and the compound of interest. For those chemists engaged in structural studies, observation of exchangeable protons can be critical to structure elucidation. To enable such observations, a solvent commonly used is deuterated dimethyl sulfoxide (DMSOd$_6$). With DMSOd$_6$, the higher viscosity and ability to hydrogen bond with the solute will tend to slow down the exchange rate. This condition allows exchangeable protons to be observed as well as preserving the coupling information.

Geminal or two-bond coupling of protons will exist when the protons are diasteretopic. This condition will occur when the structure of the molecule yields a different environment for each proton and, hence, different chemical shifts. Geminal coupling may be observed from protons on a saturated or sp^3 carbon when the environment that the methylene group experiences cannot be averaged by rapid rotation.

Vicinal coupling is also commonly observed in NMR spectroscopy and has played a major role in defining the three-dimensional (3D) structures of molecules. A critical contribution to the understanding of the relationship between vicinal couplings and structural conformation or configuration was first defined by Martin Karplus, and is known as the Karplus relationship [6,7]. This relationship establishes a correlation between the dihedral angle Φ between two protons separated by three bonds and the magnitude of the coupling constant. Dihedral angles between vicinal protons that are either near 0 or 180° have large coupling between 9 and 13 Hz. Dihedral angles for protons that are near 90°, however, give small couplings 0–2 Hz. The Karplus relationship is highly valued and widely used for structure elucidation studies. This relationship is often applied in determining the conformations of ethane derivatives and saturated six-membered rings; however, it also works for establishing the configuration of alkene moieties (Figure 2.11).

Four- or five-bond coupling (J^4, J^5) is very small and depends upon spin-spin interactions that occur through both σ and π orbitals. The presence of π electrons often has a greater contribution to the observation of long-range coupling; hence, long-range coupling is usually more pronounced in olefinic molecules and highly conjugated aromatic ring systems.

The letters used in describing the types of observed spin systems are A, B, M, and X. The nomenclature for the proton coupling depends on the similarity of the chemical shifts of the interacting protons. For example, protons that are labeled H_A and H_B instead of H_A and H_X are protons with similar chemical shifts because A and B are close in the alphabet. More complex splitting patterns can be observed for ^{1}H-NMR spectra than are predicted by the "$n + 1$" coupling rule. This occurs when coupling of one proton or set of

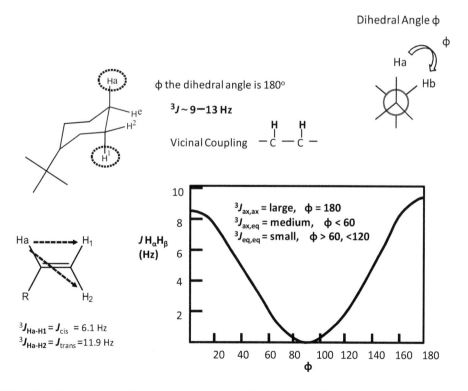

FIGURE 2.11 Karplus curve showing a range of coupling constants for different values of the dihedral angle Φ.

equivalent protons occurs to two different protons or sets of protons with different size coupling constants or when coupling occurs between protons with similar but not identical chemical shifts. The former situation can be analyzed in terms of overlapping "$n + 1$" pattern spectral diagrams (Figure 2.12). This is shown for the spectrum of a hypothetical oxirane moiety with three oxirane protons having different chemical shifts. All three protons are coupled to each other. The protons are labeled H_A, H_M, and H_X, reflecting the fact that their chemical shifts are not close to each other. Each resonance would appear as a doublet of doublets, with the overall pattern called an "AMX pattern."

The situation of protons with close chemical shifts coupled to each other, such as an AB spin system, will give a pattern of two doublets, but the intensities are not 1:1, with the inner signals having larger intensity than the outer signals. However, the separation between the lines of each doublet is still the coupling constant J. If more than two protons of close chemical shift are coupled to each other, more complex patterns – often described as complex multiplets – are observed. Multiplets still provide useful structural information because they indicate the presence of coupled protons of similar chemical shift. The AB pattern and complex multiplet patterns result from what is called "second-order effects." Second-order effects occur when the ratio of the chemical shift separation in Hz to the measured coupling constant is less than approximately 10. A more in-depth discussion of the types of coupled spin systems is beyond the scope of this book and may be found in other manuscripts [8].

Examples of some typical coupling types and their associated geometries are shown below (Figure 2.13).

Although this section has focused on ^{1}H-^{1}H coupling, coupling interactions can occur with other spin 1/2 nuclei (e.g., ^{13}C, ^{19}F, and ^{31}P). Because of their high natural abundance, compounds containing ^{19}F or ^{31}P may show coupling interactions with neighboring protons. For example, if ^{19}F is spin-coupled to ^{1}H, the ^{19}F resonance signal will appear as a doublet and the same J constant will be observed in both the ^{1}H-NMR spectrum and the ^{19}F-NMR spectrum. Spin coupling with nuclei having spin other than 1/2 is more complex and the reader is advised to consult other references on quadrupolar nuclei [9].

2.6   NUCLEAR OVERHAUSER EFFECT (NOE)

The nuclear Overhauser effect (NOE) is a phenomenon that occurs from the transfer of nuclear spin polarization from one spin population to another via cross-relaxation. This

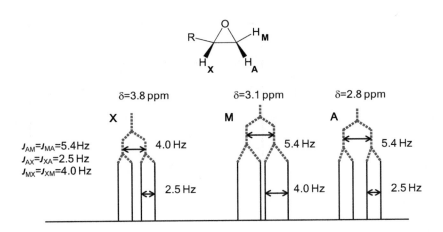

FIGURE 2.12   Stick diagram of AMX splitting patterns.

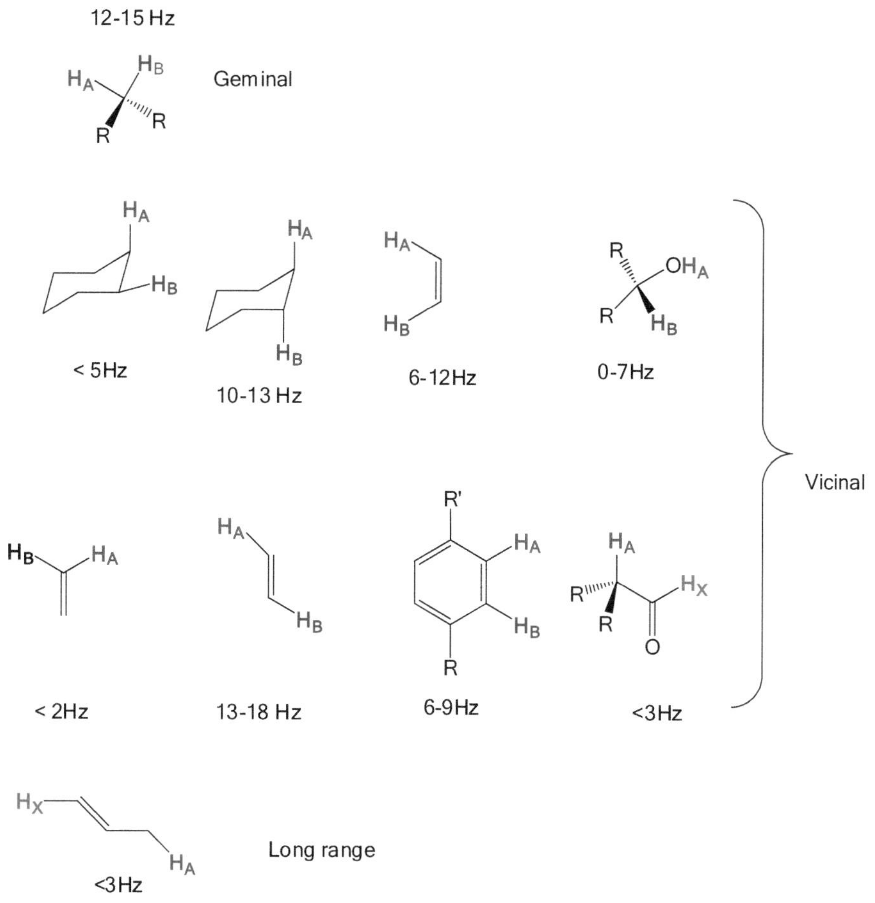

FIGURE 2.13 Typical ^{1}H coupling constant ranges for geminal, vicinal, and long-range interactions.

magnetization-transfer phenomenon was named after American physicist, Albert Overhauser, who first proposed it in the early 1950s [10]. The phenomenon was demonstrated experimentally by T. R. Carver and C. P. Slichter in 1953 [11]. Although the Overhauser effect was first described in terms of polarization transfer between an electron and nuclear spins, the recognition that magnetization could also be transferred between nuclear spins caused the technique to become a dominant force in its application to 3D structure elucidation [12].

The NOE differs from spin-spin coupling in that the NOE is observed through space phenomenon and not through bond. An NOE transfer occurs between atoms that are spatially separated by distances up to 5 Å, whereas spin coupling is observed only when the atoms are directly or indirectly bonded to neighboring atoms. Hence, if one considers a polypeptide chain with protons at opposite ends of the chain – if the chain is folded such that those two protons come within 5 Å of each other – then an NOE interaction will be observed between the two protons but a coupling interaction will not be observed.

Magnetization transfer will occur when a nucleus, such as a proton, is excited by an RF pulse. When this happens, spatially close nuclei, for example neighboring protons, may experience an intensity enhancement, or NOE. Because the NOE does not depend upon through-bond or J-couplings but only on the spatial proximity between nuclei, this interaction not only assists in confirming the structure but provides a powerful mechanism for obtaining detailed information on conformation and absolute configuration. In addition, the strength of the NOE may be proportional to the distance between two protons. Relative distance ranges may be entered into molecular dynamics programs to generate ensembles of structures that satisfy the NOE distance constraints. For small molecules, NOEs may be observed between protons that are up to 4 Å apart, while the upper limit for large molecules is about 5Å. The correlation between NOE and distance has played a critical role in the 3D structure elucidation of peptides, proteins, and protein ligand complexes.

There are numerous NMR experiments that give rise to through-space interactions. A common application is nuclear Overhauser effect spectroscopy (NOESY), a two-dimensional (2D) NMR technique that is often used for the 3D structure determination of macromolecules. The NOESY experiment is also very important in structure elucidation studies of small organic molecules providing data for the unambiguous assignment of nuclei. Other experimental techniques exploiting the NOE include but are not limited to rotating frame nuclear Overhauser effect spectroscopy (ROESY), transferred nuclear Overhauser effect (TRNOE), NOE difference, and heteronuclear Overhauser effect spectroscopy (HOESY) [6]. Because of their power in structure elucidation, the NOESY and ROESY experiments require closer examination and will be discussed in more detail in Chapter 6.

Although there are many different possible NOE experiments, in order to ensure that the correct experiment is selected and the resulting data is interpreted properly, some understanding of the theory of NOE is necessary. Some of the primary mechanisms affecting NOE are described next [13].

2.6.1 MOLECULAR WEIGHT AND MAXIMUM NOE

The maximum possible NOE depends on the molecular correlation time (or the inverse of the rate of molecular tumbling), which is primarily determined by the molecular weight (MW) and solvent viscosity. Larger molecular weights and higher viscosities lead to larger correlation times. The NOE is positive for small molecules (MW < 600), goes through zero for medium-sized molecules (MW range 700–1200), and becomes negative for large molecules (MW > 1200). These molecular weight ranges are approximate. For medium-sized molecules, the NOE may be theoretically zero (see the Figure 2.14) [14]. The ROESY experiment (rotating frame NOE) is preferred for medium-sized molecules, since the ROE is always positive.

The change in phase of NMR signals in the NOESY spectrum is associated with the induction of certain allowed quantum energy transitions. These energy transitions may be described by the consideration of a two-spin system, I^1 and I^2. Since NOE involves polarization and does not involve coherences, population differences between energy states α and β states in Figure 2.15 can be used to describe the phenomenon.

The possible transitions can be classified as follows:

- The W_1 transitions corresponding to T_1 relaxation of the spin involve a spin flip of only one of the two spins (either I^1 or I^2).
- W_0 transitions involve a zero-quantum transition where the α spin simultaneously flips to the β spin and the β spin simultaneously flips to the α spin.
- W_2 corresponds to a net double-quantum transition involving a simultaneous spin flip of both spins.

When spins absorb RF radiation, populations of the α and β spin states change. Saturating the resonances creates equal populations of both spin states. Relaxation will reestablish the equilibrium Boltzmann distribution. For the W_1 mechanism, the relaxation of spin I^1 will occur without affecting spin I^2.

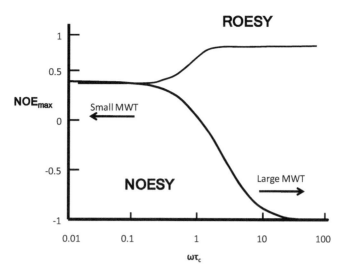

FIGURE 2.14 Maximum NOE versus molecular correlation time, τ (or the inverse of the rate of molecular tumbling). Larger molecular weights and higher viscosities lead to larger correlation times. The NOE is positive for small molecules (MW < 600), is nearly nulled for medium-sized molecules (MW range 700–1200), and becomes negative for large molecules (MW > 1200). (MW ranges are approximate.) The ROESY experiment (rotating frame NOE) is preferred for medium-sized molecules, since the ROE is always positive.

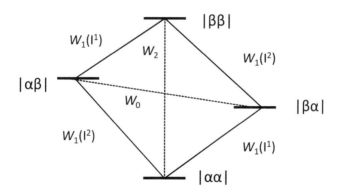

Possible transitions for a two-spin system

FIGURE 2.15 The energy transitions in a two-spin system.

However, for the double-quantum and zero-quantum transitions, the relaxation of spin I^1 will affect I^2. When reestablishing the Boltzmann equilibrium for a spin (e.g., I^1), the W_0 mechanism causes the neighboring (unperturbed) spin I^2 to deviate from its Boltzmann equilibrium, pushing nuclei from the α to the β state toward a decrease in $\alpha\beta$ population difference. Following a 90° RF pulse, a *decrease* in signal intensity for I^2 will occur resulting in a "negative NOE effect." For the W_2 mechanism, the population difference of the unperturbed spin I^2 will increase with a transfer of nuclei from the β state to the α state. More nuclei in the lower energy state result in more nuclei able to transition to the higher energy state following an RF pulse. This leads to an increase in signal intensity and a positive NOE effect (see Figure 2.16).

On a 500-MHz NMR spectrometer, for a small molecule (molecular weight below 1000 Da) in a non-viscous solvent, the molecule would tumble rapidly so the double-quantum transition dominates; hence, the relaxation ratios for $W_1:W_0:W_2$ would be 3:2:12. For large molecules such as proteins (molecular weight > 10 kDa), the zero-quantum transitions dominate and the transition rate ratio for $W_1:W_0:W_2$ would be 1:28:1. Hence, because small molecules tumble rapidly in solution, the population of tumbling frequencies that induce double-quantum transitions (W_2) predominate and give rise to positive NOEs. In a 2D NOESY spectrum, this means the NOE cross peaks would be opposite to the diagonal peaks. Macromolecules tumble more slowly; hence, the population of tumbling frequencies that induce zero-quantum transitions (W_0) predominate. This produces negative NOEs in a NOESY spectrum and all peaks have the same phase [14].

2.6.2 TIME DEPENDENCE OF NOE – MIXING TIMES

In transient experiments, such as NOESY and ROESY, the NOE dynamically builds and then decays due to relaxation during the mixing time. This is illustrated below in the plot of NOE versus mixing time (Figure 2.17). The NOE goes through a maximum as a function of mixing time. The maximum NOE and rate of buildup depend on the correlation time, the molecular weight, and the distance between protons for a particular NOE. In general, large molecules build up NOE quickly, while small molecules build up NOE more slowly. For large molecules, the maximum NOE typically occurs at shorter mixing times. A shorter distance between protons will also lead to faster build-up of NOE and shorter mixing times are recommended.

There is only one mixing time specified per NOE experiment, and it is one of the most important parameters in the NOE experiment. For small molecules, a mixing time that

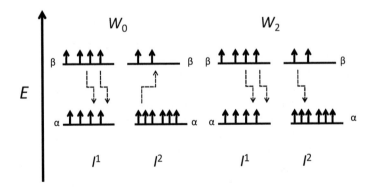

FIGURE 2.16 Diagram showing the effects of perturbation of spin I^1 and subsequent relaxation back to the equilibrium Boltzmann distribution. For the W_0 and W_2 mechanisms, relaxation of spin I^1 will affect I^2. The W_0 mechanism will cause the neighboring spin I^2 to deviate from its Boltzmann equilibrium toward a decrease in αβ population difference. After a 90° pulse, this will result in a decrease in signal intensity for I^2 and produce a "negative NOE effect." The W_2 mechanism will cause the population difference of the undisturbed spin I^2 to increase, corresponding to an increase in signal intensity resulting in a "positive NOE effect."

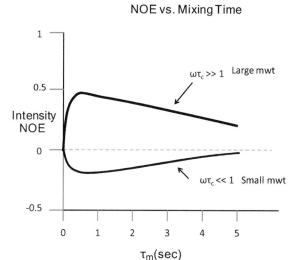

NOE vs. Mixing Time

FIGURE 2.17 Graph illustrating correlation of NOE and mixing time. For small molecules, a mixing time that maximizes the NOE is desirable. For large molecules, the mixing time must be kept small so that the buildup obeys the linear approximation and spin diffusion is avoided.

maximizes the NOE is desirable, provided calculation of an actual distance is not needed. Generally, one is interested in a range of distances so the choice depends on molecular weight. For large molecules, the mixing time must be kept small so that the build-up reflects the linear approximation and spin diffusion is avoided.

The following guidelines may be used as a starting point for setting mixing times:

1. Small molecules, 0.5–1 s; start with 0.5 s
2. Medium-size molecules, 0.1–0.5 s; start with 0.25 s
3. Large molecules, 0.05–0.2 s; start with 0.1 s

Artifacts may appear in 2D NOE experiments that often provide additional information. Zero-quantum peaks are a common artifact in NOESY spectra. They occur between nuclei that are *J*-coupled, such as ortho-protons on an aromatic ring, and can be identified by their up-down type of pattern.

In ROESY spectra, a common occurrence is long-range through-bond magnetization transfer between protons that are *J*-coupled or symmetric with respect to the center of the spectrum. This latter artifact can be removed by proper positioning of the spectral center frequency. Also, the cross-peak intensities have an offset dependence.

If protons are undergoing chemical or conformational exchange, corresponding cross peaks will occur in 2D NOE and ROE experiments. For small molecules, the phase of the exchange cross peaks will be the same as the diagonal peaks and opposite to the NOE cross peaks. This is often an effective means of identifying chemically exchangeable protons in small molecules, since they will show pronounced correlation with the residual water peak in the sample.

The choice of 2D NOESY versus 2D ROESY requires some knowledge of the sample to be studied. Although the NOESY experiment is powerful, as noted previously, the phase of the NOE cross peaks changes with molecular weight. Hence, macromolecules give rise to negative NOE cross peaks, while small molecules give positive NOEs. There is a molecular size (~1000 Da) where W_0 equals W_2 and NOE is negligible in the NOESY spectrum. In such cases, the ROESY experiment is required. ROESY cross peaks are always positive and the experiment suffers less from spin diffusion and the resulting interpretation errors; however, ROESY is less sensitive for large molecules.

Overall when selecting a 2D NOE experiment, the following guidelines will aid in maximizing the likelihood of a successful experiment.

1. Small molecules (MW < 600) use NOESY or ROESY.
2. Medium-sized molecules (700 < MW < 1200) use ROESY.
3. Large molecules (MW > 1200) use NOESY.

2.7 COUPLING OF HPLC WITH NMR

Unlike conventional NMR spectroscopy that requires dissolving a compound in a deuterated solvent, placing it in an NMR tube and then inserting the tube into an NMR probe in the spectrometer, the LC-NMR system uses a modified probe that allows either continuous flow of the solution or a stopped flow condition. As described in Chapter 1, when employing a flow probe, however, complications with respect to flow rate, solvent gradient, and magnetic field homogeneity, as well as S/N need to be considered. Each of these issues is linked to the inherent insensitivity of the NMR experiment and the properties influencing relaxation.

The challenges in LC-NMR can be summarized using four main issues. The first is related to the limits on the dynamic range of the receiver. The second involves selecting the appropriate volume of the LC-NMR flow probe to enhance spectral quality. The third issue consists of accurate synchronization of HPLC and NMR systems. And the fourth concerns the low sensitivity of the NMR technology relative to other hyphenated technologies such as LC-UV or LC-MS.

In LC-NMR studies, there is often a need to detect signals from low-level molecules in the presence of large ^{1}H-NMR signals from the HPLC solvents. Because of this, experimental techniques were developed to circumvent dynamic range issues that affect observation of signals of interest. Dynamic range problems occur following detection by the NMR coil when the signal must be digitized for computer processing to occur. The analog-digital-converters or ADCs in modern NMR spectrometers typically have 16 to 18 bits. The receiver that records the signal, prior to digitization, must be adjusted to allow the strongest signal to fit to the largest bit. In an LC-NMR experiment, the largest signal will be that of the solvent, since solvent has the highest concentration relative to the analyte of interest. For example, if the concentration of a compound is 1 mM in aqueous solution (H_2O), there is approximately 55,000 times more solvent. Efficient digitization of the solvent will cause the compound of interest to be digitized in the lowest bit, similar to the noise. As a result the analyte signals will be very weak and not easily distinguishable from the noise. To address this problem, the relatively large solvent signal needs to be reduced. One way to reduce the solvent signal is to deuterate the solvent. In LC-NMR (reverse phase), the solvents typically used are acetonitrile, methanol, and water. The problem with deuterating such mobile phase solvents is that deuteration may be prohibitively expensive and or could result in loss of important exchangeable protons. A second approach involves the development of NMR pulse sequences that can selectively reduce or eliminate large resonances from the solvent. Methods to address this issue are discussed in later chapters.

When selecting a flow cell, the volume of the chromatographic peak versus the volume of the NMR flow cell must be taken into account [15]. If a relatively high volume flow probe is used, samples on the order of several micrograms need to be captured in the NMR flow cell. This would mean that analytical HPLC columns would have to be saturated to inject samples in that range. Saturating the column would cause loss of chromatographic resolution and poor separation. Reducing the volume can prevent the need for sample over-load and improve chromatographic peak separation; however, the number of nuclei in the NMR flow cell is reduced and hence NMR sensitivity suffers. An additional factor that can affect chromatographic peak performance is the use of deuterated mobile phase solvents. In numerous cases, compounds have shown changes in retention time and peak broadening. Hence, direct transitioning from non-deuterated solvents to deuterated solvents is not recommended. When employing smaller flow-cell volumes, a primary way to increase the S/N of a spectrum is to increase concentration. Different flow-cell sizes are commercially available with volume sizes as low as 5 µL. For very small limited amounts of isolates, the microcoil probes are best, provided one can concentrate the material in the volume of the flow cell (i.e., 5–30 µL). For continuous flow or stopped flow LC-NMR separations with abundant analyte, a larger 120 µL flow cell would enable capture by the probe of the max-imum amount of chromatographically isolated material.

The integration of LC and NMR requires careful consideration of flow rates, timing, and tubing length. Sophisticated commercial software has been developed that effectively automates the integration of NMR and HPLC systems. For continuous flow LC-NMR, selecting the optimum flow rate is critical. The selection must be a compromise between the rate required for the best chromatographic resolution and the best NMR sensitivity [16]. The resolution in a continuous flow ^{1}H-NMR spectrum is strongly dependent upon the flow rate/detection volume ratio. Under stationary conditions, the line width at half height (LW) is proportional to the transverse relaxation (T_2). The relationship is as follows:

$$LW = (1/\pi)T_2 \tag{2.11}$$

In a continuous flow experiment, the NMR signal has a linewidth (LW_f) that is propor-tional to the stationary linewidth (LW) plus the residence time (t) of the sample within the NMR probe.

$$LW_f = (1/\pi)T_2 + 1/t \tag{2.12}$$

Hence, because the residence time of the sample in the flow cell is proportional to line-width, the cell volume and flow rate will affect the line broadening of the sample. It has been shown that larger flow-cell volumes can tolerate faster flow rates without compromis-ing the NMR spectral resolution. For example, a flow-cell volume of 120 µL at a flow rate of 1 mL/min would produce spectral line widths of 0.14 Hz, whereas a 60 µL flow cell at 1 mL/min would produce a linewidth of about 0.28 Hz [17]. If the flow rate drops to 0.5 mL/min, the linewidth for a 60 µL flow cell is about 0.14 Hz. The net result is that for continuous flow studies, the larger flow-cell volumes are preferred.

In a stopped-flow experiment, the LC pump is stopped so that the selected chromato-graphic peak sits motionless in the probe allowing extended NMR acquisition. In most cases, the peak is identified with a UV detector that resides between the column and the NMR flow probe. The time to allow transfer from the UV detector to the flow probe must be carefully calculated for maximum capture of the chromatographic peak in the flow cell. In the stopped-flow experiment, it is important to account for the chromatographic peak width. If peaks are stopped on the column for excessively long periods, the peak width could become severely broadened. If the chromatographic peak width exceeds the cell width (both represented in seconds), the NMR sensitivity is inversely proportionally to the chro-matographic peak width and sensitivity is compromised. Optimization of chromatography and gradient elution are possible options to address the peak-width problem [18].

Improving sensitivity is a major challenge in NMR spectroscopy. One means of increasing signal is to increase the number of scans. With continuous-flow NMR, the measurement time for each analyte depends upon the residence time in the flow probe. If the flow rates are too fast, this will result in poor S/N and line broadening from field inhomogeneity for resonances in the NMR spectrum. Reduction of the flow rate by a factor of 3–10 or temporarily stopping the flow at short intervals (time slicing) as a chromatographic peak passes through the probe enables acquisition of sequential spectra through the chromatographic peak. This technique can be useful in cases of poor separation of analytes and will increase the residence time as well as the measurement time and S/N for each component. The problem with reducing the flow is the effect of diffusion at slow flow rates that can reduce the chromatographic separation of the individual peaks eluting from the LC column into the NMR flow cell.

While very slow flow rates are not usually recommended, continuous flow measurements with flow rates of 0.05 mL/min have been reported. The reports showed that spectra could be acquired with 128 scans and good S/N per spectrum was achieved [19]. The continuous flow mode may employ rapid screening (less scans) with ^{1}H-NMR when examining intense signals from major constituents; however, the short residence time in the flow probe will not yield quality ^{1}H-NMR data with good S/N for minor components.

There are two methods by which NMR measurements can be carried out under static conditions. Besides stopped flow, which involves use of a valve to stop the elution when the analyte reaches the flow cell, one may also divert the peaks to sample loops to store the individual analyte fractions obtained from the chromatographic separation before sending them to the flow cell. In both cases, the analytes can be examined with more time-intensive one-dimensional (1D) and 2D NMR experiments.

For the stopped-flow mode, the delay time or time required for transport of the analyte from the UV (or comparable) detector of the flow probe must be determined. The delay time depends on the flow rate of the chromatographic separation and the length and diameter of the tubing that is used to connect the UV detector with the NMR flow cell. Once the tubing size, length, and time have been calibrated for the hardware at hand, the system software can be programed to automatically stop the chromatographic run elution at the time after the analyte has passed the LC detector. After NMR data acquisition, the chromatographic run is restarted and the procedure is repeated for the next analyte. A number of chromatographic peaks can be studied with a sequence of stopped-flow data acquisition modes during the chromatographic run. However, frequent stops may disturb the quality of chromatographic separation, which will, in turn, compromise the quality of the spectra.

The advantage of the loop-storage mode is that the chromatographic run is not interrupted. Each analyte peak is stored in an individual capillary loop for transfer to a flow probe and NMR data acquisition at a later stage. With loop collection, the delay time between the LC detector and loop-storage device and the delay time for transport from the loop to the NMR flow cell must be calibrated. An important prerequisite for the loop-storage mode is that the stored analytes must be stable during the extended period of residence in the loop prior to NMR analysis [20–28].

An additional complication occurs when solvent gradients are used for LC separation because the NMR chemical shifts of the solvent and analyte resonances depend on the solvent composition and vary continuously as spectra are acquired during the chromatographic run [18]. This can be addressed with the use of solid phase extraction (SPE). With SPE, a single deuterated solvent is used to transfer the compound from a storage cartridge to the flow probe.

SPE allows the use of normal protonated solvents for the LC run. This method enables multiple trapping of the same analyte from repeated LC injections to the same cartridge to enhance the low concentration. A subsequent drying step with nitrogen enables all solvents that were used in the chromatographic separation to be removed. Analytes may then be transferred with the deuterated solvent of choice (e.g., methanol, acetonitrile, or chloroform) to the NMR flow-cell probe for spectral acquisition. The use of a single deuterated solvent reduces the need for solvent suppression with highly concentrated samples.

Although considerable advances have been made in flow-probe technology, direct observation of ^{13}C (a critical nucleus for structure elucidation) is still a challenge. Because of its low natural abundance, the direct acquisition of natural abundance ^{13}C-NMR spectra, even for the main constituents, is not feasible with continuous-flow NMR. Since quantity and concentration of material is often at the lower detection limits for LC samples, the natural abundance and magnetic properties of the observed nucleus becomes a critical factor for NMR detection. Because of this, most LC-NMR studies primarily observe either ^{1}H or ^{19}F nuclei.

In summary, since LC-NMR typically deals with small quantities of compounds and low concentrations. The observed nucleus is usually proton. Because of its high sensitivity and natural abundance, sometimes ^{19}F may be used, but this is primarily in applications where known metabolites are being formed and relative concentrations are being analyzed.

For structural characterization of organic molecules, however, ^{13}C is an essential nucleus. Since ^{13}C has natural abundance of 1.1%, direct observation by continuous-flow LC-NMR is not feasible for dilute samples. The development of sophisticated experiments that include indirect detection of ^{13}C has played a major role in circumventing this problem. When coupled with solvent suppression techniques, organic structure elucidation using hyphenated technologies such as LC-NMR becomes possible because the indirect methods address the challenges of low natural abundance and the suppression of mobile phase solvents allow analyte signal detection. Such experiments are covered in Chapter 6.

Because the primary nuclei in structure elucidation are ^{1}H and ^{13}C, it is useful to compare and contrast the properties of these nuclei as there are certain differences and similarities with respect to NMR detection.

- ^{13}C has only about 1.1% natural abundance (of carbon atoms), whereas ^{1}H has 99% natural abundance. ^{12}C has high natural abundance but cannot be observed by NMR spectroscopy (nuclear spin, $I = 0$).
- The ^{13}C nucleus is about 400 times less sensitive than ^{1}H nucleus in NMR spectroscopy. This is due to the intrinsic magnetic properties of the nucleus as well as the low natural abundance for ^{13}C. Practical implications are that significantly more sample and longer acquisition times are required for direct observation of carbon.
- Due to low natural abundance, ^{13}C-^{1}H coupling is usually not observed, which greatly simplifies the ^{1}H spectrum.
- ^{13}C has a chemical shift range that spans 0–220 ppm, whereas ^{1}H has a chemical shift range of 0–20ppm. The larger spectral width for ^{13}C results in greater peak dispersion, which tends to simplify the spectrum. This has favorable implications for structure elucidation and is notably important for organic compound characterization using indirect detection experiments where proton is the observed nucleus.
- The number of peaks in a ^{13}C spectrum usually correlates with the number of types of carbons in the molecule.
- Chemical shifts for both ^{1}H and ^{13}C are measured with respect to tetramethylsilane, $(CH_3)_4Si$ (i.e., TMS).
- ^{13}C spectra are acquired in a "broadband, proton decoupled" mode, so the peaks appear as single lines. This simplifies the spectrum, but ^{1}H-NOE enhancements will affect the ability to quantitate the integrated intensities.

- Shorter relaxation times are typically experienced for ^1H resonances; hence, integration of ^1H spectra is quantitative. ^{13}C, on the other hand, requires long relaxation times so obtaining a quantitative ^{13}C spectrum would result in prohibitively long acquisition times. In addition, to eliminate the discrepancy in integrated intensity caused by ^1H-NOE enhancement, undecoupled spectra would be required, which would negatively impact S/N.

The general implications in these comparisons are that although ^{13}C spectra may appear simpler, they take longer to acquire at the natural abundance level and are not quantitative when employing ^1H decoupling. Overlap of peaks is much less common than for ^1H-NMR, which makes it easier to determine how many types of carbons are present. With LC-NMR, the concentrations of analytes are often at the microgram and sub-microgram levels. This would make direct observation of ^{13}C a prohibitively long experiment and/or requiring incorporation of isotopic label.

Overall, NMR is an inherently insensitive experiment. Issues with respect to natural abundance, magnetic properties, energy level populations, and relaxation all impact the ability to observe an NMR signal. When interfaced with HPLC, these limitations are exacerbated by properties such as flow, concentration, and dynamic range. To enable and enhance the capabilities of LC-NMR and related studies, significant advances in NMR instrumentation, experiments, method development, and probe technology were necessary. Many improvements in instrument sensitivity, range of capability, component integration, and automation design have resulted in significant breakthroughs in the limits of detection for structural characterization. These advancements are described in the following chapters.

REFERENCES

1. Martin, M.L., Martin, G., and Delpuech, J.-J. 1980. *Practical NMR Spectroscopy*. London: Heyden & Son Ltd.
2. Freibolin, H. 2005. *Basic One and Two-Dimensional NMR Spectroscopy*. Wiley-VCH Verlag GmbH & Co.
3. Levitt, M.H. 2001. *Spin Dynamics – Basics of Nuclear Magnetic Resonance*. Chichester: John Wiley and Sons.
4. Vold, R.L., Waugh, I.S., Klein, M.P., and Phelps, D.E. 1968. Measurement of spin relaxation in complex systems. *J. Chem. Phys.* 48:3831–3832.
5. Markley, J.L., Horsley, W.H., and Klein, M.P. 1971. Spin-lattice relaxation measurements in slowly relaxing complex spectra. *J. Chem. Phys.* 55:3604–3605.
6. Karplus, M. 1959. Contact electron-spin coupling of nuclear magnetic moments. *J. Chem. Phys.* 30:11–15.
7. Karplus, M. 1963. Vicinal proton coupling in nuclear magnetic resonance. *J. Am. Chem. Soc.* 85:2870–2871.
8. Gunther, H. 1998. *NMR Spectroscopy: Basic Principles, Concepts and Applications in Chemistry*, 2nd edition. New York: John Wiley and Sons.
9. Eliav, U., and Navon, G. 1996. Measurement of dipolar interaction of quadrupolar nuclei in solution using multiple-quantum NMR spectroscopy. *Journal of Magn. Reson., Series A.* 123 (1):32–48.
10. Overhauser, A.W. 1953. Polarization of nuclei in metals. *Phys. Rev.* 92 (2):411–415.
11. Carver, T.R., and Slichter, C.P. 1953. Polarization of nuclear spins in metals. *Phys. Rev.* 92 (1):212–213.
12. Noggle, J.H., and Schirmer, R.E. 1971. *The Nuclear Overhauser Effect: Chemical Applications*. New York: Academic Press, Inc.
13. Jacobsen, N.E. 2007. *NMR Spectroscopy Explained: Simplified Theory Applications and Examples for Organic Chemistry and Structural Biology*. New York: John D. Wiley and Sons, 410–414.
14. Neuhaus, D., and Williamson, M.P. 2000. *The Nuclear Overhauser Effect in Structural and Conformational Analysis*, 2nd edition. New York: Wiley-VCH.
15. Elipe, M.V.S. 2003. Advantages and disadvantages of nuclear magnetic resonance spectroscopy as a hyphenated technique. *Anal. Chim. Acta.* 497:1–25.

16. Korhammer, S.A., and Bernreuther, A. 1996. Hyphenation of high-performance liquid chromatography (HPLC) and other chromatographic techniques (SFC, GPC, GC, CE) with nuclear magnetic resonance (NMR). *J. Anal. Chem.* 354:131–135.

17. Albert, K. 2002. LC-NMR: theory and Experiment. In *On-line LC–NMR and Related Techniques*, ed. K. Albert, 1–4, Chichester: John Wiley and Sons, Ltd.

18. Griffiths, L. 1995. Optimization of NMR and HPLC conditions for LC-NMR. *Anal. Chem.* 67:4091–4095.

19. Sandvoss, M. 2002. Application of LC-NMR and LC-NMR-MS hyphenation to natural products analysis. In *On-line LC–NMR and Related Techniques*, ed. K. Albert, 111, Chichester: John Wiley and Sons, Ltd.

20. Bringmann, G., Gunther, C., Schlauer, J., and Rucket, M. 1998. HPLC-NMR online coupling including the ROESY technique: direct characterization of naphthylisoquinoline alkaloids in crude plant extracts. *Anal. Chem.* 70:2805–2811.

21. Strohschein, S., Rentel, C., Lacker, T., Bayer, E., and Albert, K. 1999. Separation and identification of tocotrienol isomers by HPLC-MS and HPLC-NMR coupling. *Anal. Chem.* 71:1780–1785.

22. Bringmann, G., Messer, K., Wohlfarth, M., Kraus, J., Dumbuya, K., and Rucket, M. 1999. HPLC-CD on-line coupling in combination with HPLC-NMR and HPLC-MS/MS for the determination of the full absolute stereostructure of new metabolites in plant extracts. *Anal. Chem.* 71:2678–2686.

23. Bailey, N.J.C., Cooper, P., Hadfield, S.T., Lenz, E.M., Lindon, J.C., Nicholson, J.K., Stanley, P.D., Wilson, I.D., Wright, B., and Taylor, S.D. 2000. Application of directly coupled HPLC-NMR-MS/MS to the identification of metabolites of 5-trifluoromethylpyridone (2-hydroxy-5-trifluoromethylpyridine) in hydroponically grown plants. *J. Agric. Food Chem.* 8:42–46.

24. Bringmann, G., Wohlfarth, M., Rischer, H., Heubes, M., Saeb, W., Diem, S., Herderich, M., and Schlauer, J.A. 2001. Photometric screening method for dimeric naphthylisoquinoline alkaloids and complete on-line structural elucidation of a dimer in crude plant extracts, by the LC-MS/LC-NMR/LC-CD triad. *Anal. Chem.* 73:2571–2577.

25. Tseng, L.H., Braumann, U., Godejohann, M., Lee, S.S., and Albert, K. 2000. Structure identification of aporphine alkaloids by on-line coupling of HPLC-NMR with loop storage. *J. Chin. Chem. Soc.* 47:1231–1236.

26. Dachtler, M., Glaser, T., Kohler, K., and Albert, K. 2001. Combined HPLC-MS and HPLC-NMR on-line coupling for the separation and determination of lutein and zeaxanthin stereoisomers in spinach and in retina. *Anal. Chem.* 73:667–674.

27. Andrade, F.D.P., Santos, L.C., Datchler, M., Albert, K., and Vilegas, W. 2002. Use of on-line liquid chromatography-nuclear magnetic resonance spectroscopy for the rapid investigation of flavonoids from sorocea bomplandii. *J. Chromatogr. A.* 953:287–291.

28. Louden, D., Handley, A., Lafont, R., Taylor, S., Sinclair, I., Lenz, E., Orton, T., and Wilson, I.D. 2002. HPLC analysis of ecdysteroids in plant extracts using superheated deuterium oxide with multiple on-line spectroscopic analysis (UV, IR, 1H NMR, and MS). *Anal. Chem.* 74:288–294.

3 Separation Methods

3.1 MODES OF SEPARATION

In high-performance liquid chromatography (HPLC) there are four primary modes of separation: normal-phase chromatography (NPC), reverse-phase chromatography (RPC), ion-exchange chromatography (IEC), and size-exclusion chromatography (SEC) [1].

NPC involves the use of a polar stationary phase consisting of porous particles (silica or alumina) with hydroxyl groups on the surface or inside the pores and a nonaqueous mobile phase [1–3]. Separation occurs when polar analytes slowly migrate through the solid support carried by nonpolar mobile phases such as hexane modified with small amounts of alcohol. NPC works best in the separation of nonpolar compounds (e.g., analytes readily soluble in nonpolar solvents) and may be used for the separation of complex samples by functional groups. NPC attains separation when the analyte associates with and is retained by the polar stationary phase. Adsorption strengths increase with increased analyte polarity, and the interaction between the polar analyte and the polar stationary phase (relative to the mobile phase) increases the elution time. The interaction strength does not rely entirely on the analyte's functional groups. Steric factors also contribute to interaction strength allowing NPC to resolve and separate structural isomers.

A major disadvantage of NPC is that the polar surfaces are easily contaminated by bound analyte components. This problem may be addressed in part by the bonding of polar functional groups, such as amino or cyano, to the silanol moieties. Also with NPC, increasing solvent polarity in the mobile phase will decrease the retention time of the analytes, but more hydrophobic solvents tend to increase retention times. Furthermore, very polar solvents will tend to deactivate the stationary phase by creating a bound water layer on the stationary phase surface. Use of NPC became less attractive because of a lack of reproducibility of retention times as water or protic organic solvents changed the hydration state of the silica or alumina chromatographic media. The normal-phase separation mode subsequently regained popularity with the development of hydrophilic interaction chromatography (HILIC) bonded phases that improve reproducibility [4].

With respect to the interface of NPC with NMR spectroscopy, this combination has drawbacks of interference from the eluting solvent. Typically, the solvents employed are hexane or heptanes mixed with isopropanol, ethylacetate, or chloroform. These solvents produce too many ^1H resonances that are mainly multiplets, and such signals obliterate a major portion of the proton spectrum. For this reason, NPC has extremely limited utility in LC-NMR applications.

RPC is the separation of analytes based on a hydrophobic nonpolar stationary phase and polar mobile phase. The separation mode is called RPC because the introduction of alkyl chains bonded covalently to the stationary support surface reversed the elution order relative to NPC (Figure 3.1) [5]. Hydrophobic groups such as octadecyl (C18) bonded groups on silica support are common RPC systems where polar analytes elute first while nonpolar analytes interact tightly with the solid support. Elution order of the polar first and the nonpolar last is the reverse of NPC, giving rise to the term "reverse-phase chromatography." Mobile phases in RPC typically consist of methanol or acetonitrile and water. The separation mechanism is attributed to hydrophobic or solvophobic interactions [6,7]. RPC is the most common HPLC mode that is interfaced with NMR spectroscopy, and such modes have been readily integrated into the NMR laboratory environment [1]. The

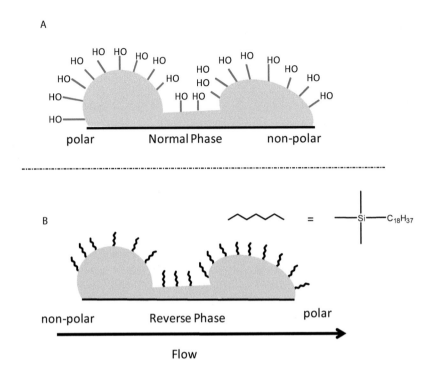

FIGURE 3.1 Illustration of separation modes for normal-phase chromatography (NPC) and reversed-phase chromatography (RPC). (A) Diagram of NPC showing silanol groups (Si-OH) in the pores and on the surface. Polar compounds will migrate slowly due to strong interactions with silanol groups. (B) Diagram of RPC where polar analytes elute first and nonpolar analytes elute later due to strong interactions with the hydrophobic C18 moieties. This interaction (polar first and nonpolar last) is the reverse of NPC.

attraction of RPC use in LC-NMR spectroscopy is the availability of deuterated mobile phase components such as acetonitrile, methanol, and water. These deuterated mobile phase solvents have an added advantage of giving rise to proteo resonances that are in NMR spectral regions outside the area of interest of most organic molecules; hence, complications from overlay with NMR peaks of interest are minimized. RPC is suitable for the analysis of water soluble polar compounds, compounds with medium polarity, some nonpolar compounds, and accounts for the vast majority (>70%) of analysis performed in liquid chromatography.

Hydrophilic interaction chromatography (HILIC) was invented and so named in 1990 to distinguish it from normal-phase chromatography. HILIC is a *variation* of NPC without the disadvantage of requiring solvents that are not miscible with water. This separation mode has also been referred to as "reversed-phase" or "aqueous normal-phase" chromatography. With HILIC, the stationary phase is a polar material such as silica, cyano, amino, and so on, while the mobile phase is highly organic (>80%) with a small amount of aqueous/polar solvent. Water or similar polar solvents are the strong, eluting solvent [4].

The HILIC mechanism on silica is the result of the polar analyte partitioning into and out of an adsorbed water layer. A charged polar analyte can also undergo cation exchange with charged silanol groups. The combination of these mechanisms results in enhanced retention of polar analytes. Lack of either of these mechanisms results in no polar retention. HILIC offers greater retention than RPC for very polar bases.

There are several benefits to using HILIC. This method retains highly polar analytes that would be un-retained by RPC. Because of this, it has complementary selectivity to RPC. There are also distinct advantages with respect to the enhanced sensitivity in mass spectrometry. Utilizing high organic mobile phases (>80%) promotes enhanced electrospray ionization-mass spectrometry (ESI-MS) response and shortens sample preparation procedures. Finally, samples collected from SPE, protein precipitation, or liquid-liquid extraction can be directly introduced onto a HILIC column since these are already in organic solvent (e.g., acetonitrile and isopropanol). With RPC, the organic solvent would need to be evaporated to dryness and reconstituted in mobile phase before injecting onto reversed-phase HPLC column. Elimination of the evaporation/reconstitution step by directly injecting the organic eluent provides a major advantage in reducing sample loss.

When using HILIC, the pH can be adjusted to reduce the selectivity toward functional groups with the same charge as the column, or enhanced for oppositely charged functional groups. However, changing the pH also affects the polarity of the solutes and hence retention on the column. An alternative would be employing column surface chemistries that are strongly ionic, and thus resistant to pH values in the mid-range of the pH scale (pH 3.5–8.5). In such cases, separations will be reflective of the polarity of the analytes alone, thereby simplifying method development. Because HILIC chromatography can be developed using a water miscible mobile phase (typically acetonitrile/water or a volatile buffer replaces the NP hydrocarbon mobile phase), this separation mode may be well suited for online coupling with NMR spectroscopy.

For ionic analytes, however, ion-suppression or ion-pairing techniques are needed. IEC is the exchange of ionic analytes with the counter ion of the ionic groups on the solid support. IEC invokes a process that allows the separation of ions and polar molecules based on the molecular charge. It can be used for almost any kind of charged molecule. This mode of separation operates on the principle that analyte molecules are retained on the column based on coulombic (ionic) interactions. The stationary phase surface displays ionic functional groups that interact with analyte ions of the opposite charge. This mode of separation may be further divided into cation exchange chromatography and anion exchange chromatography. Cation exchange chromatography retains positively charged cations because the stationary phase displays a negatively charged functional group, whereas anion exchange chromatography retains anions using a positively charged functional group in the stationary phase. The ion strength of either the cationic or anionic analyte in the mobile phase can be adjusted to shift the equilibrium position and thus retention time.

In IEC, the mobile phase (a buffered aqueous solution) carries the sample from the injection port onto a column that contains the stationary phase material. The stationary phase is typically a resin or gel matrix consisting of agarose or cellulose beads with covalently bonded charged functional groups. Stationary phases for solid support include sulfonates for cationic exchange or quaternary ammonium groups for anionic exchange that may be bonded to polymeric- or silica-based materials. The mobile phases used in these systems are buffers with a gradient of increasing ionic strength. The target analytes (anions or cations) are retained on the stationary phase but can be eluted by increasing the concentration of a similarly charged species that will displace the analyte ions from the stationary phase. For example, in cation exchange chromatography, the positively charged analyte could be displaced by the addition of positively charged sodium ions. This separation mode is often used in the isolation of compounds such as amino acids, proteins/peptides, and nucleotides. One should note, however, that because many amino acids do not absorb strongly in the UV or visible region, other types of detection (e.g., conductivity) would need to be implemented.

When IEC is interfaced with NMR, two conditions need to be addressed. The first includes suppression of the water peak from the mobile phase. This may be alleviated with the use of D_2O, thereby adding increased expense from the deuterated solvent. The second condition is the effect of ions in salts and buffers leading to the formation of a "lossy" sample. In "lossy" solutions, rf power is absorbed by the electric fields created by the ions in the sample which results in longer pulses and less efficient excitation of nuclei. Sensitivity is also compromised since the electric fields in the sample induce noise in the receiver coil. The electric field effects are more strongly coupled with coils in cryogenically cooled probes than with room temperature probes; hence, the deleterious effects are even more pronounced with cryo NMR technology. Overall, this condition compromises NMR performance.

SEC is the separation of molecules based entirely on the molecular size of the compound, not by molecular weight. In such systems, large molecules are excluded from the pores of the solid support and therefore migrate more quickly. Small molecules, however, can penetrate the pores and migrate more slowly. SEC is usually applied to large molecules or macromolecular complexes such as proteins and industrial polymers. In fact, SEC is widely used for polymer characterization because of its ability to correlate with the molecular size of polymers. For polymers, the molecular size correlates with the molecular weight; hence, this mode of separation has also been termed gel permeation chromatography (GPC) when used to determine molecular weights of organic polymers. GPC employs an organic solvent as a mobile phase. When an aqueous solution is used to transport the sample through the column, the technique is known as gel filtration chromatography (GFC). GFC is used to separate biological compounds.

One of the advantages of SEC is the good separation of large molecules from the small molecules with a minimal volume of eluate [8]. Since this is a filtration process, solutions can be applied that preserve the biological activity of the particles to be separated. The SEC mode is usually combined with other chromatography to further separate molecules by other characteristics, such as acidity, basicity, and charge. SEC also has short and well-defined separation times and narrow bands that produce good sensitivity. There is no sample loss with this method because solutes do not interact with the stationary phase. Disadvantage includes the fact that only a limited number of component bands can be accommodated because the time scale of the chromatogram is short. In addition, a 10% difference in molecular mass is required to have good resolution [8].

There are other types of separation modes that may be amenable as an interface with NMR spectroscopy. These include supercritical fluid chromatography (SFC) and capillary electrophoresis (CE). Unlike HPLC, CE [9] is a separation mode in which the mobile phase is driven by electromotive force from a high-voltage source instead of a mechanical pump. The separations are based upon the different electrophoretic mobilities of the molecules to be separated in an applied electric field. Because the coupling of CE with NMR is an emerging integration of technologies, this system is described subsequently in a separate section.

SFC is a form of NPC that is used for the analysis and purification of low-to-moderate molecular weight, thermally labile molecules. It can be used for the separation of chiral compounds. The principles are similar to those of HPLC; however, SFC typically utilizes CO_2 as the mobile phase. Hence the entire chromatographic flow path must be pressurized. The SFC-pressurized mobile phase (CO_2) may be modified with polar organic solvents such as methanol. However, the SFC mobile phase has limited polarity even in the presence of polar modifiers. For example, at low pressures, CO_2 has the solvating power equivalent to aliphatic hydrocarbons, whereas at higher pressures its solvating power is similar to methylene chloride.

Because SFC generally uses CO_2, it contributes no new chemicals to the environment; therefore, the use of this separation mode supports "Green Chemistry" initiatives. SFC separations can be done faster than HPLC separations because the diffusion of solutes in supercritical fluids is about ten times greater than that in liquids. This results in a decrease in resistance to mass transfer in the column and allows for fast high-resolution separations.

Compared with GC (gas chromatography), capillary SFC can provide high-resolution chromatography at lower temperatures, thus allowing a fast analysis of thermo-labile compounds. This method is useful for nonpolar analytes and preparative applications requiring full recovery of analytes.

SFC can be interfaced with an NMR spectrometer but requires a special high-pressure system configuration and high-pressure flow probes [10]. A major advantage of SFC is the use of CO_2 as the mobile phase solvent, which is ideal for ^1H NMR because it is free of protons that may require solvent suppression. A drawback of SFC with NMR is that T_1 of ^1H resonances becomes long and this can affect signal intensity. In addition, it is known that the dissolution power of a solvent varies as a function of pressure in the supercritical state. Because chemical shift depends on the solvating state, this condition can severely affect the chemical shifts of proton resonances in the NMR spectrum. It has been shown that increasing the density of CO_2 with increasing pressure will shift the ^1H resonances to a higher field with the largest effects observed for exchangeable protons. This condition necessitates the need to avoid steep pressure gradients; otherwise, severe line broadening can occur in the NMR spectrum [11].

3.2 GENERAL METHOD DEVELOPMENT STRATEGIES

In performing HPLC, some basic corollaries need to be taken into account. The major goal of HPLC is separation of one or more components in a sample mixture; hence, certain properties need to be present. One major property is solubility of the analyte of interest. Samples with low solubility are difficult to extract, and this will lead to low recoveries. The extent of sample recovery becomes a critical factor when attempting to observe very dilute systems via NMR spectroscopy. One must also be mindful of the fact that in order for separations to occur, compounds must have different column retention times. To enable the separation process, an analyte must exihibt sufficient interaction with the stationary phase. The mobile phase is equally important in analyte separation. While the stationary phase controls analyte-media interaction, the mobile phase affects the overall separation. For primary HPLC methods such as RPC, finding appropriate mobile phase conditions is critical for analyte separation. When introducing the separation mixture, it is important to dissolve the analyte in either mobile phase or a solvent weaker than the starting mobile phase. Problems with peak fronting are caused by injecting samples with stronger solvents than mobile phase. When stronger solvents must be used to dissolve the analytes, smaller injection volumes are necessary. Typically, <5 μL of injection volume is required when analytes are dissolved in strong solvents. In addition to peak fronting, exponential tailing can occur from secondary interactions with acidic silanols. Mass overload is also problematic for good separation and results in the creation of a shark fin peak shape. The optimal chromatography peak shape is Gaussian, which results from a single interaction with the stationary surface. Examples of common peak shapes are illustrated in Figure 3.2. Controlling the chromatographic peak shape is not only important in effecting good separation but can also have a profound effect on the concentration of an isolated sample for LC-NMR detection.

One must also give significant attention to the type of stationary phase that is used. For example, there are a variety of C18 reverse-phase columns commercially available; however, these columns vary significantly in their retention and silanol properties. Finding the best column for a specific separation can depend to a large extent on method development experimentation.

In developing a new separation method, it is important to first identify the goal of the separation. A key question is whether it is necessary to separate and isolate multiple components or just one component. If there are multiple components, it is important to determine the number of components and their respective solubilities. Next, one might need to

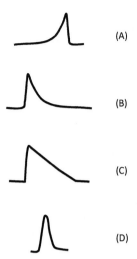

FIGURE 3.2 (A) Peak fronting caused by injections of compounds in solvents stronger than the mobile phase. (B) Peak tailing can be caused by activity with acidic silanols. (C) Peaks in the shape of a shark fin result from mass overload. (D) Good gaussian peak shape.

determine whether a gradient or isocratic method is needed. For many HPLC separations, a gradient method provides the best separation and is usually employed. Resolution is important in effecting good separation. This property is dependent on both physical and chemical parameters that are described in Equation 3.1.

$$R_s = \frac{\sqrt{N}}{4} \times \frac{\alpha - 1}{\alpha} \times \frac{k}{k+1} \quad (3.1)$$

where
  R_s = resolution
  N = plate number (column length and particle size)
  α = selectivity (stationary and mobile phase, temperature)
  k = retention (stationary and mobile phase)

Based on Equation 3.1, it is readily apparent that no separation is possible without retention. If k is zero, all analytes will co-elute. Next, selectivity (α) is shown to be maximized by optimizing column and mobile phase conditions. Small changes in selectivity (e.g., mobile phase and column stationary phase) can have a major effect on resolution. Finally, the plate number (N) appears as a secondary means of maximizing efficiency. This may be achieved by increasing the length of the column or using a more efficient column. Since R_s is proportional to the $\sqrt{N}$, doubling N will only increase resolution by a fraction of the increased value. For example, if the column length is doubled, the analysis time will increase by a factor of 2 but the resolution increases by $\sqrt{2}$ or 41%. When compared, with an increase in α from 1.1 to 1.2, one sees almost a doubling in resolution, showing that optimization of the mobile phase will provide higher resolution efficiency. Because of the various factors that influence resolution, it is important to determine at the outset how much resolution is needed. Properties of common chromatographic solvents are given in Table 3.1 [8].

The polarity of the analyte relative to the mobile phase and the stationary phase is also critical. The pH of the mobile phase should be selected such that pKa of the analyte (i.e.,

TABLE 3.1

Properties of Chromatographic Mobile Phases[a]

Solvent	Refractive Index[b]	Viscosity cP[c]	Boiling Point, °C	Polarity Index, P'	Eluent Strength[d], ϵ^0
Fluoroalkanes[e]	1.27–1.29	0.4–2.6	50–174	<−2	−0.2
Cyclohexane	1.423	0.90	81	0.04	−0.16
n-Hexane	1.372	0.30	69	0.1	0.008
1-Chlorobutane	1.400	0.42	78	1.0	0.208
Carbon Tetrachloride	1.457	0.90	77	1.6	0.144
i-Propyl ether	1.365	0.38	68	2.4	0.224
Toluene	1.494	0.55	110	2.4	0.232
Diethyl ether	1.350	0.24	35	2.8	0.304
Tetrahydrofuran	1.405	0.46	66	4.0	0.456
Chloroform	1.443	0.53	61	4.1	0.320
Ethanol	1.359	1.08	78	4.3	0.704
Isopropanol	1.383	2.04	82	3.9	0.82[g]
Ethyl acetate	1.370	0.43	77	4.4	0.464
Dioxane	1.420	1.2	101	4.8	0.448
Methanol	1.326	0.54	65	5.1	0.76
Acetonitrile	1.341	0.34	82	5.8	0.52
Nitromethane	1.380	0.61	101	6.0	0.512
Ethylene glycol	1.431	16.5	182	6.9	0.888
Acetic acid	1.372	1.1–1.26[f]	117.9	6.2	High
Water	1.333	0.89	100	10.2	High

[a] Adapted from reference [8] and other sources.

[b] At 25°C.

[c] centiPoise (cP) is a unit of viscosity (1 cP = 1 mN – s – m^{-2}).

[d] ϵ^0 on SiO_2.

[e] Range given/specific values dependent on molecular weight.

[f] At 20°C.

[g] On Al_2O_3/multiplication by 0.8 gives ϵ^0 on SiO_2.

the condition where 50% of the analyte is ionized) is avoided in order to obtain good separation.

When creating a new method, numerous factors need to be considered and tested. A general method development protocol can be summarized in four basic steps. The first step is column selection. Selection of the appropriate stationary phase may be guided by vendor-supplied selectivity charts and the chemical properties of the analytes. Next, a pH range (acidic, neutral, and basic) should be determined. For RPC (especially when interfaced with NMR), the mobile phase will be either methanol or acetonitrile; however, modifiers such as formic acid may be added to effect separation. In general, acetonitrile is preferred because it is less viscous than methanol. Methanol is weaker in solubilizing analytes, which translates into longer retention times and higher pressures. Finally, the temperature of the column can be adjusted to improve selectivity. However, when using elevated temperatures, one must be careful to avoid compromising the performance and lifetime of the column.

A basic protocol for performing HPLC method development may be summarized as follows:

1. Column selection (perform method development trials with three different columns). LC-NMR typically employs RPC.
2. Identify the optimal pH (choose conditions at pH 3, 6, and 10, adjust accordingly).
3. Select mobile phase and modifiers (LC-NMR mobile phases are usually MeOH/ACN and water; common modifiers are formic acid and TFA). To ensure optimal performance and reproducibility, it is recommended that additives or buffers be added to both aqueous and organic mobile phases although this can sometimes be problematic with respect to precipitation of salts from organic solvents or baseline distortion at low UV values. For LC-NMR, modifiers should be chosen to minimize introduction of ^{1}H NMR signals that can overlap with NMR peaks of interest.
4. Set column temperature (temperatures from 30°C to 50°C are a good starting point). Increasing the temperature will reduce mobile phase viscosity. This will lower backpressure when the flow rate is held constant. Temperature can be used to optimize linear velocity and improve analyte diffusivity. This can improve analyte retention and selectivity.

Once the pH, mobile phase, column, and temperature are selected, the method may be fine-tuned by optimizing the gradient slope. A shallow gradient slope may improve resolution at the expense of a decrease in sensitivity. Increasing the gradient slope will increase sensitivity but compress the peaks and hence reduce resolution. Adjusting the gradient slope requires a balance between peak height and resolution.

The time needed for method development trials may be significantly reduced by employing ultra-performance liquid chromatography (UPLC) technology. Run times may be reduced by a factor of 3. For example, a method development protocol that would take 21 hours to complete with an HPLC system may be completed on a UPLC system in 7 hours. If the UPLC method subsequently needs to be transferred to an HPLC (as may be required for an LC-NMR application), there are software packages available from vendors that will automatically convert the system (column, mobile phase, modifier) to enable the same separation and resolution.

3.3 COLUMN PACKING TYPES

A primary component of the HPLC system is the column. Key parameters defining the column are size and packing materials. Column sizes typically range from 50 to 250 mm in length and 2 to 4.6 mm inner diameter. Column packing materials are critical to separation success. Packing for small molecule separations consists of support types such as silica, polymer, or hybrid. Bonding groups that are typically used for RPC are alkyl, amino, cyano, and phenyl. For NPC, silica (OH) and amino-capped solid support are used, while for IEC diethylaminoethyl, sulfonate, quarternary ammonium, and carboxylate groups are used. Particle sizes range from 2 to 5 μm in analytical columns but increase up to 10 μm for preparative columns.

The major support material in column packing is silica (SiO_2). Unbonded silica has intensely strong absorptive characteristics and therefore is modified through bonding with alkyl silanes. Such attachments create a hydrophobic binding surface that is used in RPC. Typical bonded phases used in RPC, NPC, and IEC are shown in Table 3.2.

RPC is the most common chromatography used in the isolation of small organic molecules. As noted previously, it is the primary mode of separation interfaced with NMR in pharmaceutical applications. Hence, some general guidelines for selecting bonded phases for small molecule separation are summarized below.

Because a fundamental consideration in choosing a column is selection of suitable packing, careful attention must be made in the selection of the stationary phase. It is known that the

TABLE 3.2
Bonded Phases in HPLC Solid Support*

Reversed-Phase (RPC)		
C18	Octadecyl	$Si–(CH_3)_2–(CH_2)_{17}–CH_3$
C8	Octyl	$Si–(CH_3)_2–(CH_2)_7–CH_3$
CN	Cyano	$Si–(CH_3)_2–(CH_2)_3–CN$
Phenyl		$Si–(CH_3)_2–(C_6H_6)$
Low bleed alkyl amide		$Si–(CH_3)_2–(CH_2)_3–NHCO–(CH_2)_{14}–CH_3$
Normal-Phase (NPC)		
Si	Silica	$Si–OH$
NO2	Nitro	$Si–NO_2$
NH_2	Amino	$Si–(CH_3)_2–(CH_2)_3–NH_2$
Ion-Exchange (IEC)		
SP	Sulfo propyl	$Si–(CH_3)_2–(CH_2)_3–SO_2^-$
CM	Carboxymethyl	$Si–(CH_3)_2–CO_2^-$
DEAE	Diethylaminoethyl	$Si–(CH_3)_2–(CH_2)_3–NH^+–(C_2H_5)_2$
SAX	Triethylaminopropyl	$Si–(CH_3)_2–(CH_2)_3–N^+–(C_2H_5)_3$

* Adapted from Reference [1].

C18 column packing is a very hydrophobic, retentive, and stable stationary phase. Because of this, it is the first choice for most separations. C8 has similar selectivity as C18 but is much less retentive. The C8 sorbent is used in the separation of more polar compounds and with lower organic mobile phase. C3 and C4 may also be used; however, they are less retentive than C8 or C18 and are usually used for protein separations on wide pore supports. CN (cyano) phases have different selectivity than C8 and C18, but they are less retentive than C8 and less stable than C18. Phenyl sorbents are used with medium polarity components with preferred selectivity for aromatics.

Column selection guidelines for RPC are often based upon personal experience; however, some general guidelines for separation of small organic compound mixtures may be considered as a starting point. Recommendations include selecting columns with 3 to 5 µm packing of high purity silica bonded phases. Columns ranging from 50 to 150 mm with 3.0 to 4.5 mm diameter are a good starting point. Initial attempts at separation can begin with C18, C8, cyano- or phenyl-bonded phases. Column selectivity charts provided by HPLC vendors can be used to assist in column selection. If pH is a consideration, low pH (<2.0) requires columns resistant to acid hydrolysis. For high pH studies (pH >8) columns should be selected that are stable at high pH.

3.4 DETECTOR SELECTION

Detectors equipped with a flow-through cell were a major breakthrough in the development of liquid chromatography. Modern LC detectors possess a wide dynamic range, allowing both analytical and preparative scale runs on the same instrument. They have high sensitivities, often allowing the detection of nanograms of material. More advanced models are very flexible, allowing rapid conversion from one mobile phase to another and from one mode to another.

Almost all LC detectors are on-stream monitors. While on-stream detectors are generally the system of choice, it is possible to interface an HPLC system with an off-line detector. An example of an off-line detector is the FTIR spiral disk monitor, which requires sample transfer followed by scanning with the FTIR instrument. Because HPLC detectors are used under continuous flow conditions, the sample is dissolved in the eluent during detection. The actual sample may be present in only nano gram quantitities in the detector, but this quantity could even be femto grams. The high sensitivity of HPLC detectors (such as UV/Vis or diode array detector [DAD]) makes them ideal for trace analysis studies where low levels of sample impurities or degradants are present.

Given that UV/Vis and DAD are the most widely used detectors in HPLC studies, it needs to be recognized that there is no universal detector, so the liquid chromatographer must expect to eventually use more than one type of detector. A number of detector variations have been created and used with hyphenated technologies such as "LC-NMR." Some of the most common HPLC detectors are listed as follows:

Detectors Interfaced with HPLC

1. UV/Vis
   a. Variable wavelength
   b. Diode array
2. Refractive index
3. Fluorescence
4. Conductivity
5. Mass-spectrometric (LC/MS)
6. Evaporative light scattering detector (ELSD)

While each of these detectors employs a different principle of operation, an ideal LC detector should have the following common properties:

- Low drift and noise level (particularly crucial in trace analysis)
- High sensitivity
- Fast response
- Wide linear dynamic range (this simplifies quantitation)
- Low dead volume (minimal peak broadening)
- Cell design which eliminates remixing of the separated bands
- Insensitivity to changes in type of solvent, flow rate, and temperature
- Operational simplicity and reliability
- Tuneable so that detection can be optimized for different compounds
- Nondestructive.

There are, however, differences with respect to the capabilities of the individual detectors. UV/Vis detectors are often used, which allow the selection of the operating wavelength. Such systems are called variable wavelength detectors and are particularly useful in three cases. First, such systems offer improved sensitivity for any absorptive component by enabling selection of an appropriate wavelength. Second, because individual sample components have high absorptivity at different wavelengths, operation at multiple wavelengths enhances the system's sensitivity. Third, depending on the sophistication of the detector, wavelength change may be programmed on a "time basis" into the memory of the system.

A diode array detector is a UV/Vis variant that consists of a number of photosensitive diodes used to monitor light that has passed through a liquid sensor cell as in a multi-wavelength liquid chromatography detector. The light source is usually polychromatic (e.g., light from a deuterium lamp) and after passing through the cell, the light is dispersed by a quartz prism or a diffraction grating onto the surface of the diode array. The wavelengths most useful in liquid chromatography range from about 210 to 330 nm (i.e., UV light); hence, a sufficient number of diodes are needed to cover this range. Many organic compounds have characteristic spectra in the UV; therefore, when a given substance is passed through a sensor cell, all the outputs from the array can be acquired and used to construct an absorption spectra. Wavelengths and the associated functional groups are given in Table 3.3.

TABLE 3.3
UV Chromophores and Related Functional Groups*

Chromophore	Chemical Configuration	λ_{max} (nm)	ϵ_{max} (L/m/cm)	λ_{max} (nm)	ϵ_{max} (L/m/cm)
Ether	–O–	185	1000		
Thioether	–S–	194	4600	215	1600
Amine	–NH$_2$	195	2800		
Thiol	–SH	195	1400		
Disulfide	–S–S–	194	5500	255	400
Bromide	–Br	208	300		
Iodide	–I	260	400		
Nitrile	CN	160	–		
Acetylide	–CC–	175–180	6000		
Sulfone	–SO2	180	–		
Oxime	–NOH	190	5000		
Azido	>C=N–	190	5000		
Ethylene	-C=C-	190	8000		
Ketone	>C=O	195	1000	270–285	18–30
Thioketone	>C=S	205	Strong		
Esters	–COOR	205	50		
Aldehyde	–CHO	210	Strong	280–300	11–18
Carbonyl	–COOH	200–210	50–70		
Sulfoxide	>S–O	210	1500		
Nitro	NO$_2$	210	Strong		
Nitrite	–ONO	220–230	1000–2000	300–400	10
Azo	–N=N–	285–400	3–25		
Nitroso	–N=O	302	100		
Nitrate	–ONO$_2$	270 shoulder	12		
Allene	–(C=C)$_2$– acyclic	210–230	21,000		
Allene	–(C=C)$_3$–	260	35,000		
Allene	–(C=C)$_4$–	300	52,000		
Allene	–(C=C)$_5$–	330	118,000		
Allene	–(C=C)$_2$– alicyclic	230–260	3000–8000		
Ethylenic/ Acethylenic	C=C–CC	219	6500		
Ethylenic/Amido	C=C–C=N	220	23,000		
Ethylenic/Carbonyl	C=C–C=O	210–250	10,000–20,000		
Ethylenic/Nitro	C=C–NO$_2$	229	9500		

* Adapted from reference [12] and other sources.

For biological studies, chromophores and their associated nucleotide bases are given in Table 3.4 [12].

Because solvents chosen can affect detection, one also needs to exercise caution in selecting the appropriate mobile phase solvents. The UV cutoffs for some common chromatographic solvents are given in Table 3.5.

By selecting the appropriate diode, the wavelength of the light at which there is maximum absorption by the molecule can be monitored to provide maximum detector sensitivity. DADs help to identify analytes beyond simple identification by retention time. DADs have an advantage related to the problem of peak purity. Since peak shape often does not reveal that it may actually correspond to two or more components, absorbance rationing at several wavelengths can help determine if a composite peak is present.

The ELSD is a valuable complement to spectroscopic detectors for HPLC. This type of detector works by measuring the light scattered from the solid solute particles remaining after nebulization and evaporation of the mobile phase. Because the detector's response is independent of the light absorbing properties of molecules, it can reveal weakly chromophoric

TABLE 3.4
Chromaphores and Nucleotide Bases*

UV/Vis chromophore	λ_{max}	Em × 10^{-3} @ λ_{max}
Adenine	260.5	E = 13.4
Guanine	275	E = 8.1
Cytosine	267	E = 6.1
Thymine	264.5	E = 7.9
Uracil	259.5	E = 8.2
NADH	340	E = 6.23
NAD	260	E = 18

* Adapted from reference [12] and other sources.

TABLE 3.5
UV Cutoffs for Some Common Solvents*

Solvent	UV Cutoff (nm)	Solvent	UV Cutoff (nm)
Water	180	N-Heptane	197
Methanol	205	Cyclohexane	200
N-Propanol	205	Carbon tetrachloride	265
Acetonitrile	190	Chloroform	245
THF	225	Benzene	280
Acetone	330	Toluene	285
Methyl acetate	260	Methylene chloride	232
Ethyl Acetate	260	Tetrachloroethylene	280
Nitromethane	380	1,2-Dichloroethane	225

* Adapted from reference [12] and other sources.

sample components that UV detectors miss and provide a more accurate profile of relative component abundance than is possible with a spectroscopic detector.

The fluorescence detector used in liquid chromatography is a detector that senses only those substances that fluoresce. A flow cell is used as the sensor through which the excitation light passes and a photocell receives the radially emitted light. The cell wall is often made of Pyrex glass to prevent the excitation light (usually UV light) from reaching the photocell. The excitation light may be UV at 254 nm produced by a mercury lamp, or it may be light of any wavelength selected from the light produced by a deuterium lamp using a monochrometer. The fluorescence detector is one of the most sensitive LC detectors and is often used for trace analysis. Unfortunately, although the detector is sensitive, its response is linear over a relatively limited concentration range. Also, because the majority of substances do not naturally fluoresce, this type of detector has only very specific use.

The electrical conductivity detector also has limited use as an HPLC detector since it measures the conductivity of the mobile phase. There is usually background conductivity, which must be attenuated by suitable electronic adjustments. If the mobile phase contains buffers or ionic additives, the detector gives a base signal that completely overwhelms that from any solute usually making detection impossible. Hence, the electrical conductivity detector senses all ions whether they are from a solute or from the mobile phase, making it of limited use in LC-NMR applications.

In recent years, significant progress has been made in the development of LC/MS interfacing systems. MS as an online HPLC detector is said to be the most sensitive, selective, and at the same time the most universal detector, but it is still the most expensive one. Interfacing an LC-MS system with NMR may have some challenges with respect to buffer/modifier compatibilities, but these issues are not insurmountable and such configurations have been successfully implemented [13].

3.5   RPC METHOD DEVELOPMENT AND COMPATIBILITY WITH NMR

LC-NMR is by convention assumed to be run in the reversed-phase HPLC mode. There are also primarily two mobile phases employed in LC-NMR, which include acetonitrile and water or methanol and water. For structure elucidation and chemical characterization by LC-NMR, the observed signal is a proton. From the standpoint of NMR analysis, the number of proton signals from sources other than the compound of interest need to be kept to a minimum. This is because proton signals from other components in the sample will overlap with signals of interest and create ambiguity. In addition, large signals from solvents or additives create dynamic range issues leading to a reduction in receiver gain and compromising the observation of the more dilute sample signals. It is for this reason that the mobile phase in an LC-NMR study needs to be as simple as possible. When selecting a solvent system, acetonitrile is often preferred since it produces one ^{1}H singlet in proton NMR spectrum, which would affect only one region of the spectrum. Since residual water is also present, the mobile phase gives rise to two large NMR signals that need to be suppressed. Methanol, on the other hand, produces two ^{1}H resonances – one broad and one sharp. When combined with residual water, three regions of the NMR spectrum are affected. Another disadvantage for methanol and water are the exchangeable protons. Deuteration or suppression of exchangeable protons can prevent such critical protons in the molecule of interest from being observed. This condition can adversely affect chemical shift assignment and structure elucidation.

Because chromatographic peak widths need to be adjusted in continuous-flow LC-NMR experiments to match the active volume of the flow cell, careful attention needs to be given to chromatographic conditions. Approaches used for the reduction of HPLC peak width include the addition of an acid modifier. Typical acid modifiers in HPLC studies include acetic acid, formic acid, trifluoroacetic acid, and phosphoric acid. Although modifiers such as acetic acid and formic acid provide good chromatographic peak widths, they are less desirable in ^{1}H LC-NMR flow studies due to the introduction of unwanted resonances that may require suppression and can overlap with critical analyte peaks required for structure elucidation [14].

As noted previously, normal phase HPLC is not amenable to LC-NMR studies since the use of organic mobile phases such as hexane or heptane produces too many ^{1}H resonances. Solvent suppression of the multiple resonance signals, while possible, becomes counterproductive when key analyte signals needed for structure elucidation are eliminated.

3.6   INTEGRATION OF CE AND NMR

The coupling of capillary electrophoresis (CE) with NMR was first reported by Sweedler et al. in 1994 [15,16]. Although capillary electrophoresis NMR can sometimes be a useful alternative to HPLC, few papers have been published combining CE with NMR [17–19]. This is partially due to the lack of commercially available instrumentation coupled with the fact that smaller volumes and shorter retention times of analytes make NMR detection challenging.

Despite its drawbacks, microcoil NMR spectroscopy provides a viable way to combine the powerful separation capabilities of capillary electrophoresis with the detailed structural and molecular dynamic information afforded by NMR spectroscopy. A capillary CE system can be coupled with a CapNMR microcoil probe with some nonroutine instrument modifications required. The reported CE system was fabricated with nonmagnetic plastic buffer vials and Pt electrodes, which do not present a problem and could be positioned within the magnet bore to perform CE-NMR [20].

In CE-NMR studies, either solenoid microcoils or saddle coils may be used. To prevent magnetic susceptibility effects from degrading line shape and signal to noise, thicker-walled capillaries are employed to bring the coil closer to the sample. Larger observe volumes can be realized by expanding the length of the capillary flow cell; however, an expanded detection flow cell will improve the signal at the expense of the separation efficiency.

Problems associated with coupling electrophoresis and NMR result from electrophoretic currents that can induce a second local magnetic field gradient and perturb the NMR magnetic field homogeneity. This can occur when using solenoid coils positioned parallel with the external magnetic field. The resultant magnetic field inhomogeneity may not be easily restored by shimming procedures employed in NMR. This issue can be addressed using a saddle coil parallel to the external magnetic field. In this case, the current-induced magnetic field does not contribute to magnetic field inhomogeniety along the z-axis. An additional consideration with CE-NMR is that one must be careful to avoid sample heating. As current is increased, Joule heating can occur, which might affect chemical shifts and compromise spectral integrity.

One of the most promising variants of microcoil CE-NMR is capillary isotachophoresis-NMR. Capillary isotachophoresis (cITP) separates and concentrates charged species based on their electrophoretic mobilities through the application of a high voltage across a capillary and using a two-buffer system composed of a leading electrolyte (LE) and a trailing electrolyte (TE). When hyphenated to capillary NMR spectroscopy, cITP allows focusing of sample components precisely in the active volume of the capillary probe, affording active volumes on the order of tens of nanoliters. This provides great gains in mass-sensitivity with the potential for separation and analysis of nanomolar quantities of biologically relevant small molecules [21].

Capillary electrophoresis-NMR (CE-NMR) is still in the early stages of development. Most results reported to date are based on custom-built probes, which often use internally modified capillaries and very small active volumes. Nonetheless, the preliminary results seem promising and suggest that CE-NMR may eventually become the technique of choice for specific applications where only submicrogram quantities of a compound are available.

3.7 TRANSITIONING FROM ANALYTICAL TO PREPARATIVE CHROMATOGRAPHY

In the pharmaceutical industry, it is often necessary to isolate and structurally characterize impurities in drug substances or drug products in sufficient quantities to enable full structural characterization of such impurities. These impurities often exist at 0.1% level or lower. To obtain sufficient material for NMR investigations, this requires isolating tens of micrograms to milligrams of material for structure elucidation. One approach to address this challenge is to take several grams of material (if available) and scale up the isolation of the impurity from an analytical platform to a preparative platform. Because scale-up from analytical liquid chromatography (LC) to preparative-scale chromatography can be time consuming and wasteful of materials, it requires an optimized scale-up strategy. A recommended method-development strategy is to develop and optimize the initial separation on an analytical column, overload the column while maintaining adequate separation of components of interest, and then scale up accordingly to a preparative column of appropriate dimensions based upon the amount of purified compound needed. With this approach, the choice of the analytical column is determined by the availability of preparative columns containing the same column-packing material. Using Equation 3.2, a linear scale-up to the preparative method may be achieved. Hence, both analytical and preparative columns from the same line of packing material must be available before beginning the preparative method development and optimization process.

$$V_{ac}/V_{pc} = r_{ac}/r_{pc}$$
$$X_{ac}/\left(\pi + r_{ac}^2\right) = X_{pc}/\left(\pi + r_{pc}^2\right) * 1/C_L \tag{3.2}$$

where
V_{ac} = flow analytical column
x_{ac} = maximum amount of analytical column
r_{ac} = radius analytical column
L_{ac} = length analytical column
V_{pc} = flow preparative column
x_{pc} = maximum amount preparative column
r_{pc} = radius preparative column
L_{pc} = length preparative column
C_L = ratio of lengths of columns L_{pc}/L_{ac}

In preparative scale-up, the same separation modes that were used in analytical-scale chromatography can be employed for many different chromatographic systems. However, the cost and availability of high-performance preparative columns, the cost of solvents and additives, and the need to recover isolated fractions in a high purity state often limit the available choices; hence, reversed-phase chromatography is most frequently used. SEC and IEC are sometimes used for preparative scale-up but are often applied to protein purifications.

Another important consideration in making the transition from analytical chromatography to preparative chromatography is particle size. Smaller particle sizes used in analytical columns generally allow greater efficiency and permit the use of shorter columns to increase separation speed. In preparative chromatography, the particle size is important, but the column is often used in an overloaded state; hence, the smaller and more expensive particles of <3.5 μm that are used in analytical columns are generally not used in larger-scale preparative columns. For complex mixtures with poor resolution (and selectivity), overloading may be difficult and particle size would need to be kept small. For well-resolved samples, larger particles of 7 and 10 μm can be used. Often when using larger less expensive particles, the yield and purity can be compromised. Because pressure drop is inversely proportional to the particle diameter squared, larger particles produce lower pressure drop, allowing higher flow rates, which in turn enhances the throughput of preparative columns. Although the cost of packing material is inversely proportional to particle size, the need for high purity of analytes in complex mixtures with poor resolution between the important components often necessitates small-particle packings.

Column dimensions are also important in scaling up to a preparative column. The amount of sample that can be injected increases with column internal diameter (and length), so using Equation 3.2, one can calculate the column diameter that fits the sample size required. Diameters commonly selected are 4.6 mm i.d. for small-scale preparative work, 7.8 mm i.d. columns for semipreparative work, and 21.2 mm i.d. columns for larger-scale preparative applications.

Many of the success factors in analytical HPLC are also prevalent in preparative HPLC, but some issues present even greater challenges. This is because samples in preparative applications, which are often crude mixtures, contain impurities that can accumulate at the head of the column. If these impurities are not removed, they can cause changes in peak shape and retention time. Accumulated impurities also present problems since they can lead to clogs, and while they may not affect retention, they can increase the column pressure causing leaks in the system if not monitored carefully. The recommended practice is to flush the column occasionally with increasingly stronger solvents to remove bound impurities. The build-up of material in a packed column occurs most frequently when the injection solvent is weaker than the mobile phase and is especially noticeable when using isocratic elution. Gradient elution helps with removal of strongly held impurities due to the stronger solvents used in such cases. Reversed-phase packings tend to hold on to hydrophobic impurities and hence should be flushed with strong solvents. Therefore, the history of the preparative column is of vital importance when selecting a column. When attempting to use preparative chromatography for separating samples for NMR studies, especially when scaling-up low levels of impurities, residual compounds from previous samples can show up unexpectedly, resulting in deleterious effects on the spectral quality. Therefore, in cases where preparative separation conditions are being developed, it may be advisable to start with a fresh column, or if this is not feasible, a thorough solvent-washing procedure should be employed. Cost considerations of the solvents required to regenerate the column versus column replacement would need to be considered.

3.8 GENERAL CONSIDERATIONS

The vast majority of LC-NMR experiments (~97%) "directly detect" ^{1}H nuclei. Proton spectra provide important information based on indirect interactions with neighboring nuclei. This information may be captured with one-dimensional or multidimensional experiments. For small molecules containing magnetic nuclei such as ^{19}F or ^{31}P in high natural abundance, coupling information in 1D spectra can provide valuable information for structure elucidation studies. With nuclei having low natural abundance such as ^{13}C and ^{15}N, indirect detection experiments can provide the information necessary for a full structural assignment. It should also be noted that nuclei such as ^{19}F, ^{31}P, and ^{29}Si can be rapidly observed due to their high magnetic abundance. Because such nuclei are often chemically dilute, their spectra are less complicated, selective, and highly sensitive. No solvent suppression is needed nor are extra resonances present from solvent or even mobile phase additives. A disadvantage, however, is that such spectra are lacking much of the detail needed for full structure elucidation. While studies have been reported where there has been direct observation of magnetically abundant nuclei such as ^{19}F or magnetically dilute nuclei such as ^{13}C, these studies address a specific purpose and are typically not widely applicable.

When considering the integration of LC with NMR, one must also recognize the challenges. This combined instrumentation brings with it the high costs of capital equipment and the complexity of achieving the appropriate compromise to obtain efficient LC separation while collecting a concentrated sample in sufficient quantity for NMR detection. Often significant method development and prework involving preparative isolation are needed to build sample concentration. The NMR experiment then may still require long experiment times. With continuous-flow or stop-flow NMR, the extensive use of deuterated solvents is essential and can be expensive.

Other considerations in implementing LC-NMR technology include operator training requirements. Doing LC-NMR requires a unique set of skills that often may require collaboration between LC and NMR specialists. Even with collaborations, each specialist must gain familiarity with the requirements and limitations of both technologies. The difficulty in finding an HPLC method that makes LC and NMR experts all uniformly happy is generally acknowledged, so it is known that compromises must happen. Modifiers such as triethylamine may work well in an HPLC method but would introduce signals in the NMR that would mask critical peaks from the analyte of interest. In addition, stopping the pump for NMR signal averaging is important to obtain sufficient signal needed for an interpretable NMR spectrum; however, stopping the flow introduces peak dispersion for compound residing on the column. This becomes an issue when multiple peaks need to be isolated and structurally characterized. Introduction of in-line solid phase extraction or loop collection can be used to address this issue (for more on SPE and loop collection, see Chapter 5).

Any flow system can clog up, get dirty, and be hard to clean. For cases where an analyte needs to be isolated in pure form, it becomes critical to ensure the system is free of any materials that could contaminate the sample. NMR flow cells are not disposable elements; hence, these systems need to be washed before and after each use to preserve the integrity of the system. To assist in removing particulate matter that can lead to clogs, guard columns (or pre-columns) can be used. Such systems are commercially available as disposable units with direct connections to the column.

It is also good practice to appropriately prepare the sample prior to injection onto the HPLC unit. Typically, sample filtration is applied. Other modes of sample preparation may include centrifugation, pre-column SPE, lyophylization, liquid–liquid extraction, and sublimation. Removing lipophilic impurities is also recommended as this will enhance the chromatographic separation and lead to a more concentrated sample of the analyte of interest.

When injecting a sample onto an HPLC column, it is preferable to use the starting mobile phase system. However, analytes of interest may have limited solubility in the starting eluent. In such cases, fraction systems such as online precolumn SPE or multidimensional chromatography may be employed [22,23].

Overall, NMR coupling to many different kinds of chromatographic separation systems is possible. Off-the-shelf or specialty designed NMR flow probes are commercially available. With the standard LC-NMR probes, it is possible to use standard analytical scale chromatography (2 × 5 mm i.d. columns).

All separation techniques such as RP (reversed phase), NP (normal phase)-HPLC (high-performance liquid chromatography), GPC (gel permeation chromatography), and SEC (size-exclusion chromatography) are applicable. With special equipment, SFC (supercritical fluid chromatography) and CE (capillary electrophoresis) are also possible.

REFERENCES

1. Dong, M.W. 2006. *Modern HPLC for Practicing Scientists*, New York: John Wiley and Sons.
2. Snyder, L.R., and Kirkland, J.J. 1979. *Introduction to Modern Liquid Chromatography*, 2nd edition. New York: John Wiley and Sons.
3. Meyer, V.R. 2004. *Practical HPLC*, 4th edition. New York: Wiley-Interscience.
4. Alpert, A.J. 1990. Hydrophilic-interaction chromatography for the separation of peptides, nucleic acids and other polar compounds. *J. Chromatogr.* 499:177–196.

 5. Molnar, I., and Horvath, C. 1976. 42. Reverse-phase chromatography of polar biological substances: separation of catechol compounds by high-performance liquid chromatography. *Clin. Chem.* 22:1497–1502.
 6. Melander, W.R., and Horvath, C. 1980. Reversed-phase chromatography. In *High Performance Liquid Chromatography: Advances and Perspectives*, ed. C. Horvath, Vol. 2, 113–319, New York: Academic Press.
 7. Carr, P.W., Li, J., Dallas, A.J., Eikens, D.I., and Tan, L.C. 1993. Revisionist look at solvophobic driving forces in reversed-phase liquid chromatography. *J. Chromatogr. A.* 656(1–2):113–133.
 8. Skoog, D.A., Holler, F.J., and Crouch, S.R. 2007. *Principles of Instrumental Analysis*, 6th edition. Belmont, CA: Thompson Brooks/Cole.
 9. Weinberger, R. 2000. *Practical Capillary Electrophoresis*, 2nd edition. New York: Academic Press.
10. Allen, L.A., Glass, T.E., and Dorn, G.C. 1988. Direct monitoring of supercritical fluids and supercritical chromatographic separations by proton nuclear magnetic resonance. *Anal. Chem.* 60:390–394.
11. Bai, S., and Yonker, C.R. 1998. Pressure and temperature effects on the hydrogen-bond structures of liquid and supercritical fluid methanol. *J. Phys. Chem. A.* 102:8641–8647.
12. Kaye and Laby. 1995. *Tables of Physical and Chemical Constants*, 16th edition. Vol 2.1.4. Hygrometry. online version 1.0,2005/. www.kayeLaby.npt.co.uk.
13. Exarchou, V., Krucker, M., van Beek, T.A., Vervoort, J., Gerothanassis, I.P., and Klaus Albert, K. 2005. LC–NMR coupling technology: recent advancements and applications in natural products analysis. *Magn Reson. Chem.* 43:681–687.
14. Taylor, S.D., Wright, B., Clayton, E., and Wilson, I.D. 1998. Practical aspects of the use of high performance liquid chromatography combined with simultaneous nuclear magnetic resonance and mass spectrometry. *Rapid Commun. Mass Spectrom.* 12:1732–1736.
15. Wu, N., Peck, T.L., Webb, A.G., Magin, R.L., and Sweedler, J.V. 1994. Nanoliter volume sample cells for ^{1}H NMR: application to online detection in capillary electrophoresis. *J. Am. Chem. Soc.* 116:7929–7930.
16. Wu, N.A., Peck, T.L., Webb, A.G., Magin, R.L., and Sweedler, J.V. 1994. ^{1}H NMR: spectroscopy on the nanoliter scale for static and online measurements. *Anal. Chem.* 66:3849–3857.
17. Pusecker, K., Schewitz, J., Gfrorer, P., Tseng, L.H., Albert, K., Bayer, E., Wilson, I.D., Bailey, N.J., Scarfe, G.B., Nicholson, J.K., and Lindon, J.C. 1998. On-flow identification of metabolites of paracetamol from human urine by using directly coupled CZE-NMR and CEC-NMR spectroscopy. *Anal. Commun.* 35:213–215.
18. Schewitz, J., Gfrorer, P., Pusecker, K., Tseng, L.H., Albert, K., Bayer, E., Wilson, I.D., Bailey, N. J., Scarfe, G.B., Nicholson, J.K., and Lindon, J.C. 1998. Directly coupled CZE-NMR and CEC-NMR spectroscopy for metabolite analysis: paracetamol metabolites in human urine. *Analyst.* 123:2835–2837.
19. Schewitz, J., Pusecker, K., Gfrorer, P., Gotz, U., Tseng, L.H., Albert, K., and Bayer, E. 1999. Direct coupling of capillary electrophoresis and nuclear magnetic resonance spectroscopy for the identification of a dinucleotide. *Chromatographia.* 50:333–337.
20. Olson, D.L., Lacey, M.E., Webb, A.G., and Sweedler, J.V. 1999. Nanoliter-volume 1H NMR detection using periodic stopped-flow capillary electrophoresis. *Anal. Chem.* 71:3070–3076.
21. Jayawickrama, D.A., and Sweedler, J.V. 2003. Hyphenation of capillary separations with nuclear magnetic resonance spectroscopy. *J. Chromatogr. A.* 1000:819–840.
22. Mars, C., and Smit, H.C. 1990. Sample introduction in correlation liquid chromatography application, properties and working conditions for a novel injection system. *Anal. Chim. Acta.* 228:193–208.
23. de Koning, J.A., Hogenboom, A.C., Lacker, T., Strohschein, S., Albert, K., and Brinkman, U.A. T. 1998. On-line trace enrichment in hyphenated liquid chromatography-nuclear magnetic resonance spectroscopy. *J. Chromatogr. A.* 813:55–61.

4 NMR Instrumentation and Probe Technologies

4.1 INSTRUMENTATION CONFIGURATION

LC-NMR integrates the separation capabilities of high-performance liquid chromatography (HPLC) with the structure elucidation capabilities of NMR to produce a hyphenated analytical instrument. With this system, a sample mixture may be separated into individual components by injection into the HPLC unit. The separated fractions are then either directly sent to a flow probe or may be transferred to a variety of different collection devices for eventual transport to a flow probe in the NMR spectrometer. Because compound identification and structural elucidation using NMR are nondestructive, the compounds can be easily recovered. Both the HPLC and NMR systems can be operated as an integrated unit or independently. This versatility enables the system to adapt to the needs of the experiment.

The LC-NMR system is comprised of an NMR spectrometer console, a superconducting magnet, a workstation, and a flow probe, all under the operation of specialized LC-NMR software. The HPLC portion consists of an HPLC pump, HPLC column(s), variable-wavelength UV detector and/or photo diode array detector, and an LC workstation that may be set up for stop-flow or continuous flow. Timing for movement of a peak between the different positions in the hyphenated system must be carefully calibrated. The time required for a peak to reach the NMR probe or a designated collection unit depends upon the void volume between the LC unit and the collection unit or probe flow cell. This will depend upon the flow rate. In order to allow selection of desired peaks, the separation is monitored by an LC detector, usually a UV detector, which displays a chromatogram of the separation. The chromatography software allows certain positions in the chromatogram to be manually or automatically selected for further measurement. The NMR probe or storage compartment is located downstream of the UV detector and is estimated to be reached about 10 to 40 s after the first peak appears in the LC detector [1]. The software calculates the appropriate delays to capture the peak at its desired position where the necessary actions for storage or measurement are initiated. Software is commercially available from instrument vendors that allow automated or interactive selection of the peaks from the chromatogram and automatic calculation of the time delays. A typical LC-NMR system configuration is shown schematically in Figure 4.1.

The illustration shows that the HPLC system and the NMR spectrometer can exist as two separate units that can be operated independently. The interface allows transfer of a peak eluting from a column directly to a flow probe. Other collection modes that are frequently used include loop collection and solid-phase extraction (SPE). With both loop collection and SPE, eluting compounds are parked in a loop or on a cartridge where the compound may reside until it is sent to the flow cell chamber for NMR detection.

When conducting an LC-NMR study, analytical chromatographic separation of small organic molecules is usually carried out using reversed phase chromatography. In considering column selection, silica-based C18 or C8 columns are commonly chosen because of their high efficiency and stability. Factors such as flow rate, temperature, and gradient composition will affect separation of the analyte mixture and need to be adjusted to obtain optimal conditions. The appropriate method for good analyte separation may be developed on an independent HPLC method development system and transferred to the LC-NMR at a later time.

LC-NMR Configuration

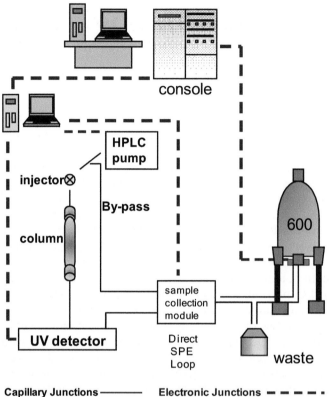

FIGURE 4.1 Illustration of a typical LC-NMR configuration. The NMR spectrometer is integrated with an LC system. Sample may be transported directly from the column to the probe. Other common collection modules include a solid-phase extraction (SPE) system or loop collector. Capillary junctions are designated with solid lines and electronic junctions with dashed lines.

The basic conditions of a chromatographic separation conducted for an LC-NMR experiment are the same as those required for ordinary analytical chromatography. However, the transfer of chromatographic methods from a method-development system to the LC-NMR system requires attention to the type of chromatography used, the buffer, and solvent considerations depending on the mode of collection and the column. In fact, in many cases, it may be advantageous to transfer the same column used in method development to the LC-NMR system to obtain the same chromatographic separation resolution.

Buffer selection can be an important consideration since highly protonated buffers such as ammonium acetate will produce large proton signals that can affect the receiver gain settings and create dynamic range issues that will interfere with observation of the low-level analytes of interest. This is further compounded by the large NMR signals from the mobile phase solvents (typically H_2O and acetonitrile or methanol). Although the proteo solvent and buffer signals can be suppressed with sophisticated pulse sequences (see section on NMR experiments, Chapter 6), problems arise due to excessive overlay of the suppressed peaks with analyte peaks of interest. Phosphate buffers or TFA additives are well suited for NMR evaluation since they do not contain non-exchangeable protons to complicate the spectrum;

however, such buffers may compromise other modes of detection. With respect to solvent suppression and the dynamic range issues, the large solvent signal problem may be partially alleviated with the use of D_2O since this deuterated solvent is not cost-prohibitive. However, D_2O will cause deuterium exchange with the analyte eliminating observation of exchangeable protons that may be critical for structure elucidation. Hence, when developing a method for transfer to an LC-NMR system, the composition of the mobile phase must be carefully determined.

As noted previously, additives such as phosphate and TFA may be ideal for NMR detection; however, if the method needs to be compatible with mass spectrometry (MS), such buffers can be problematic. This is because phosphate buffers tend to crystallize out and contaminate the MS ion source and can be very difficult to remove. In addition, TFA is not usually compatible with ESI-MS since it causes ion suppression as a result of strong ion pairing between the TFA anion and the protonated analyte. In the negative ion mode, TFA suppresses the analyte by competing for charge. Hence, use of such buffers may prevent the ability to correlate the LC-MS and LC-NMR chromatograms.

While careful control of buffer and mobile phase composition is important for the successful execution of an LC-NMR experiment, this is not a factor when the system employs SPE as a means of trapping the analyte. With SPE, LC methods can be directly transferred to the LC-NMR system without concern for such compatibility issues (see section on solid-phase extraction).

It is well known that NMR spectroscopy is inherently an insensitive technique. From an instrumentation standpoint, the sensitivity of NMR spectroscopy depends on the following three parameters:

1. The magnetic field strength (sensitivity increases with magnetic field strength)
2. The size and filling factor of the receiver coil (mass sensitivity increases with increasing fill factor and decreasing coil diameter)
3. The noise introduced during detection

Factors governing these parameters are addressed subsequently.

4.2 THE MAGNET

The first commercial magnets that gained popularity in NMR in the early 1960s were electromagnets. These magnets were constructed using a stable yoke terminated by two pole pieces. The magnetic field in electromagnets is generated by high-intensity electrical current running through two main field coils wound around the magnetic circuit. This type of magnet dissipates a significant amount of heat; consequently, saturation of the ferromagnetic material limits the field strength to about 2.3 T (100 MHz). The Varian A-60, introduced at the Pittsburgh Conference in 1961, was an electromagnet designed for continuous wave (CW) NMR at 60 MHz. The introduction of this system helped establish NMR spectroscopy as a standard tool for organic chemists [2].

The next major step in the early 1960s was the development of a superconducting magnet. A superconducting magnet is an electromagnet made from coils of superconducting wire. For superconducting magnets, the wire can conduct much larger electric currents than ordinary wire; hence, very stong magnetic fields can be created. Superconducting magnets have a number of advantages over resistive electromagnets. They can achieve an order of magnitude stronger field than ordinary ferromagnetic-core electromagnets, which are limited to fields of around 2 T. The field is generally more stable, resulting in less noise in experimental measurement, and its operation does not require expensive consumption of electrical power and cooling water as required for the electromagnets.

The basic construction of such superconducting magnets consists of a solenoid wound with superconducting wire, which comprises the fundamental design of every high-frequency NMR magnet [3]. The niobium-titanium alloy (NbTi) embedded in a copper matrix has been used to construct solenoids for field strengths up to 9.4 T (400 MHz). For magnetic fields 11.7 T (500 MHz) and above, niobium-tin (Nb_3Sn) alloy wires were developed. Although this alloy has the appropriate flexibility and tensile strength to allow precise winding into a solenoid form, the Nb_3Sn alloy is nonetheless more expensive, brittle, and difficult to wind relative to the NbTi wire, hence adding to the cost of higher field magnets. In 2007, a magnet with windings of yttrium barium copper oxide (YBCO) was reported, which achieved a world record field of 26.8 T [4]. As a result of this work, the US National Research Council set a goal of creating a 30 T superconducting magnet, and it is expected that fields exceeding 30 T may be achieved with this material.

With NMR superconducting magnets, in order to reach and maintain superconductivity, the coil must be immersed in a cryo coolant. This is required to keep the magnet operational over a long period. Nearly all superconducting magnets currently used for NMR spectroscopy are cooled by liquid helium (4.2 K) being at normal atmospheric pressure or at reduced pressure. Boil-off of the liquid helium cryogen is mediated by containment in a vacuum-jacketed canister with liquid nitrogen in a concentric vessel. Liquid nitrogen is replenished on a weekly basis, whereas liquid helium may last several months prior to replenishing supplies depleted by atmospheric boil-off.

The strength of the magnetic field is determined by the current density contained in the superconducting coil. The higher the current density, the higher is the magnetic field and, hence the corresponding Larmor frequency. Most commercially produced superconducting magnets have had ^1H frequencies ranging from 100 to 1000 MHz.

The room temperature bore size of modern superconducting magnets is usually 54 or 89 mm and those magnets are called SB (standard bore) or WB (wide bore) magnets, correspondingly. Superconducting magnets are not only used in NMR equipment but are also widely used in MRI machines, mass spectrometers, magnetic separation processes, and particle accelerators.

Although sensitivity can be increased by increasing the magnetic field, this is not a cost-effective solution. Currently, magnetic field strengths in NMR systems are commercially available at 900 MHz and beyond; however, these systems are quite expensive. Relative to a 900 MHz spectrometer, a 600 MHz spectrometer will possess half the sensitivity but can be purchased at a considerably lower cost [5]. Nonetheless, efforts continue to design and build increasingly stronger magnets for NMR investigations. Such systems are installed at regional centers to allow access by multiple investigators. One such center is the European Centre for High Field NMR (CRMN) in Lyon, France, which houses a 1 GHz NMR spectrometer. The intention is that a more powerful magnet will enable spectroscopists to observe tiny traces of chemicals in complex mixtures of body fluids, environmental samples, or even pharmaceutical samples. The magnet at the Lyon center possesses a massive 23.5 T field and represents a milestone in the expansion of NMR magnetic field strength. The cost of the system was estimated at US$16.3 million, purchased in 2010. However, it is expected that the cost should become more affordable with advances in magnet technology. Given the power of this system, this spectrometer should attract and benefit chemists and structural biologists on a global scale.

The types of studies that can benefit from a 1 GHz NMR instrument include studies of large proteins and membrane-bound proteins, which are extremely difficult to crystallize for X-ray analysis. In protein NMR-spectroscopy studies, researchers can use lipid constructs known as bicelles that can potentially mimic cell membranes; hence, bound proteins may assume a more physiologically relevant shape. For scientists using nanodiscs, which corral lipid and protein together in their natural conformation, the increased sensitivity and resolution of the higher field magnets can be critical in problem-solving.

Other applications include metabolic profiling work in chemistry applications. Analytical chemists will be able to perform chemical profiling for water and soil samples to measure industrial effects on these materials. High-field NMR spectroscopy could be used to study chemical reactions that may occur on large particle beads, for example, catalysts attached to grains of silica. In addition, trace quantities of isolated impurities may be characterized to provide more definitive structures for safety assessment in toxicology studies. In such studies, the concentration of the trace impurity is often only <0.1% of the total. Hence, higher fields such as 1 GHz will open up new horizons in the study of compounds and systems that were not amenable to NMR at lower field strengths [6]. It should be noted that in 1986, the discovery of high temperature superconductors by Georg Bednorz and Karl Müller created great interest and hope for future development of magnets that could be cooled by liquid nitrogen instead of liquid helium cryogen. This would be a highly welcome advancement since liquid helium is more difficult to work with and more expensive; however, this technology has not been realized in the evolution of NMR magnet technology [7].

Although there are significant benefits in building higher field magnets, more cost-effective efforts to further improve sensitivity have centered on improving probe performance [8,9]. Reduction of thermal noise was a major advancement that was addressed with the development of cryogenically cooled probes. In addition, development of probe geometries with smaller receiver coils and better fill factors have had a major impact on sensitivity. These advances in probe technology are particularly attractive since probes may be easily added or retrofitted to existing spectrometers at a fraction of the cost of replacing the magnet.

Various advancements in probe design used in supporting LC-NMR studies are discussed below.

4.3 ROOM TEMPERATURE FLOW PROBE

Flow probes for NMR spectroscopy have been designed for both manual and automated use. For a flow NMR probe to be optimally functional, it needs to incorporate a number of criteria. When a flow probe is interfaced with an HPLC/column unit, the system must be robust enough to sustain connection and flow transfer through capillary tubing. The flow probe must possess an enhanced "filling factor" to achieve maximum sensitivity on a chromatographically isolated peak. This becomes particularly important for low-level quantities of isolates where S/N is challenged by minimal sample concentration. In addition, care must be taken to eliminate leaks, clogs, and air bubbles in the NMR flow cell, leading to degraded NMR lineshape from magnetic-susceptibility inhomogeneities.

To enable maximum flow cell performance, NMR flow probes are designed with the inlet to the probe at the bottom of the flow cell (vertically aligned with the magnetic field) and the outlet from the top of the flow cell (Figure 4.2). This configuration is conducive to the elimination of air bubbles.

A variety of high-resolution NMR flow probes are commercially available and are considered to be standard items from NMR vendors. NMR flow probes are designed with optimized flow-cell size for chromatographic peaks and minimized band broadening during sample transfer. These probes possess the capability for high sample throughput. By linking the flow probe to a chromatography system, direct transfer of a sample into the NMR probe is possible. This direct transfer has the advantage of minimizing sample handling, and thus the danger of contamination and decomposition of samples. Because the probe is the only modification to the NMR spectrometer that is required for flow applications, a probe change can be done in minutes and preserve the versatility of the instrument.

Flow probes are available on spectrometers ranging in field strength of 300–900 MHz. Typical flow-cell sizes are 30, 60, and 120 μL although micro-capillary probes are also available

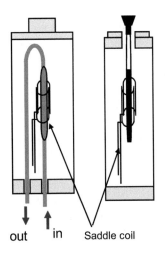

Flow probe vs. Capillary probe

FIGURE 4.2   Illustration of a flow probe and a capillary cryoprobe. Both probes may be configured with 30 μL flow-cell volume.

(see below). The NMR probe configurations are available with a variety of multinuclear capability including [19]F. Flow probes that have gradient shimming and automatic tune and match, which vastly increase the utility and ease of use of LC-NMR systems, are commercially available.

4.4   MICROCAPILLARY PROBES (ROOM TEMPERATURE)

To enable quality NMR data to be obtained from shrinking sample sizes, capillary room temperature NMR probes (CapNMR) were developed and are commercially available. These probes have a flow-cell capacity of either 5 or 10 μL. Popular probe configurations include ^{1}H with ^{13}C indirect detection or triple resonance inverse probes with ^{1}H, ^{13}C, and ^{15}N or [31]P. CapNMR probes enable hundreds of samples of isolated compounds to be run per day using 96 or 384 well microplates.

4.4.1   Microcoil Capillary Flow Probes (Room Temperature)

The CapNMR system (Protasis/MRM Corp., Savoy, IL) requires a different approach to sample preparation. Instead of preparing a sample in a conventional 5 mm-glass NMR tube containing 200–600 μL of deuterated solvent, the CapNMR sample is dissolved in only 5–10 μL of solvent. The solution is then transferred into the capillary probe either manually using a syringe or by using an automated liquid-handling system. It is important to note that this technology requires regularly scheduled cleaning protocols to ensure reproducible results.

The CapNMR probe's microflow probe characteristics include a swept flow path that is composed entirely of fused silica. This enables the probe to be compatible with the vast majority of NMR solvents. The flow-cell volume of this probe is 5 μL, V_{obs} = 1.5 μL; a feed capillary is included consisting of ~2 μL on both the inlet and outlet sections (i.d. of 50 μm), yielding a total probe volume of ~9 μL. The probe is reported to yield excellent line shape using a sample of 5% $CHCl_3$ in deuterated acetone, with values of <1 Hz at 50%, <10 Hz at 0.55%, and <20 Hz at 0.11% on a 500 MHz NMR spectrometer. Typically, shimming this probe requires using just four to eight first- and second-order shims. Mass sensitivity at 500 MHz was reported using a 3 µL of 0.33 mM sample of sucrose in D_2O (340 ng total mass) collecting 256 scans. S/N > 10 was determined for the anomenric proton at 5.4 ppm.

To prevent blockages, the capillary probe is equipped with an in-line microfilter consisting of a 2 µm screen and having a dead volume of just 0.085 µL. The narrow capillary inner diameter (i.d.) and small dead volume filter minimize sample dispersion and solvent requirements. The standard CapNMR probe is equipped with a deuterium lock, proton and carbon observe channels, and z-gradient. Unlike cryogenically cooled probes, this probe may be easily installed and removed and has a range of compatibility with instrument vendors.

Because of its small probe volume, the capillary NMR probe exhibits inherently less concentration sensitivity relative to a larger probe. However, this has a concomitant advantage with respect to solvent background. This is because a small-volume NMR also results in a smaller signal from protonated solvent, rendering any solvent interference comparatively decreased and less detrimental to the quality of the spectrum obtained. A comparison between the solvent background signal of residual HOD (partially deuterated water) in D_2O in a conventional 5 mm tube $^1H/^{13}C/^{15}N$ triple-resonance triax (inner coil ^1H, outer coil $^{13}C/^{15}N$) vs. the CapNMR flow probe (V_{obs} = 1.5 µL) was examined. Both the 5 mm tube and CapNMR probes were filled with identical solvent and optimized for line shape and 90° pulse width. Spectra were acquired using a single scan and allowing for T_1 recovery. Identical data processing parameters were used. The S/N of the HOD signal in the larger-volume probe was found to be 13-fold larger compared to the capillary probe.

Overall, the probe performance distinction indicates that the NMR analysis of a mass-limited sample is best accomplished using a probe with the smallest volume and highest mass sensitivity. Although a larger-volume probe having identical mass sensitivity as a smaller-volume probe (such as a cooled-coil probe) may yield a similar S/N for the analyte, spectral quality will be diminished due to interference from a larger solvent background signal.

Another advantage of the micro-capillary flow probes is that they possess high salt tolerance. Hence, the presence of dissolved salt in a sample has little to no effect on spectral quality. This is because in a microcoil probe, the major source of S/N is from the transceiver coil. As a result, even a large increase in the ionic character of a sample will have a minor effect compared to the S/N in the absence of salt. The presence of 500 mM KCl is reported to have reduced the S/N of a model compound <10%. Only a change in chemical shift was observed that might be attributed to different ionic environment of the molecule in the two solutions [10].

A comparison of CapNMR flow-cell probes to that of traditional 5 mm sample-tube-based probes revealed significantly higher mass sensitivity for the capillary configuration. CapNMR was claimed to improve the sensitivity of NMR spectroscopy beyond that achieved by using Shigemi-type tubes, where the sample volume of a 5 mm NMR tube is reduced by half and a two- to three-fold net increase in mass sensitivity is realized. Gronquist et al. directly compared spectra obtained with both probe designs using a 10 mM sucrose solution. In one series of experiments, one-dimensional ^1H NMR spectra as well as two-dimensional (^1H,^{13}C) HMQC and (^1H,^{13}C) HMBC spectra were acquired for a 5 µL sample of a solution of sucrose (10 mM) injected into the CapNMR probe by syringe. In a second series of experiments, the same amount of sucrose was dissolved in the volume of solvent (D_2O) required using a 5 mm Shigemi tube, followed by acquisition of an equivalent set of spectra using a conventional 5 mm H{C,N} probe and the same spectrometer.

A comparison of the ^{1}H spectra revealed that the signal-to-noise (S/N) ratio was about five times better for the CapNMR spectrum over that found using the Shigemi tube [11]. In addition, the residual water peak in the CapNMR spectrum was several orders of magnitude smaller than in the spectrum obtained using a 5 mm probe since the active volume is over 100 times greater in the 5 mm sample than that of the CapNMR 5 μL flow cell. The vastly reduced water peak enables higher receiver gain, which contributes to the observed sensitivity gain.

Indirect detection with carbon-proton correlation experiments, HMQC, HSQC, and HMBC, has also been reported using CapNMR. Because of the low natural abundance of ^{13}C, the sensitivity of these experiments in producing sufficient S/N for structure determination can often be challenging. However, because CapNMR probes are equipped with gradient coils, sensitivity-enhanced gradient versions of the HMQC and HMBC experiments are possible. This leads to improved signal intensity. Also, smaller solvent signals enable appropriate receiver gain adjustment and contribute to higher sensitivity for the indirect detection experiments [5].

The principles and parameters for choosing an NMR probe to be used in the analysis of volume- or mass-limited samples have been explored [10]. The focus of the comparison was on capillary-based micro-flow probes. The term micro-flow (as used in the study) refers to both the flow rate through the NMR probe, usually 1–50 μL/min, as well as the sample volume of 1–10 μL. In addition, the transceiver coil used in such probes commonly has a diameter of 1000 μm or less, yielding a flow cell volume of typically 5 μL. These practices have emerged to form a new spectroscopic size scale called micro-NMR.

Mass sensitivity of micro-flow probes has been described as comparable to cryoprobes (observe volume ~40 μL), but with several advantages. The micro-flow probe studied had a flow-cell volume of 5 μL, an observe volume of 1.5 μL, and was equipped with ^{1}H and ^{13}C observe channels, a deuterium lock, and z-gradient capability. The inlet and outlet capillary inner diameters were 50 μm to minimize sample dispersion, to make it amenable for volume-limited samples.

An example of probe performance was demonstrated using an injected sample of 1 nmol of sucrose (0.34 μg in 3 μL, 0.33 mM; MW = 342 g/mol), which yielded a 1D proton spectrum in 10 min on a 500 MHz spectrometer. A second case involved 15 μg of sucrose (in 3 μL; 15 mM, 45 nmol) injected and parked in the probe to yield a heteronuclear multiple-quantum coherence (HMQC) spectrum in less than 15 h. The natural product muristerone A (75 μg in 3 μL, 50 mM, 150 nmol; MW = 497 g/mol) was injected into a flow cell, and a gradient COSY spectrum was acquired in 7 min, a gradient HMQC in 4 h, and a gradient HMBC in 11 h, showing that with sufficiently concentrated samples a full range of NMR experiments required for unambiguous structure elucidation may be obtained.

There are four basic modes of sample injection into the probe that vary in the degree of user intervention, speed, solvent consumption, and sample delivery efficiency. The four basic modes include manual, manual-assisted (employing a micropump), automated (using an autosampler), and capillary HPLC. Manual injection involves injecting a sample via a small-volume inlet by hand using a syringe. A variety of solvents may be used, including dimethyl sulfoxide (DMSO), which has a fairly high viscosity. A few microliters of sample can be injected into the flow cell in a single step. Alternatively, a few microliters of injected sample can be followed by an injection of push solvent to position the analyte in the 5 μL flow cell. A manual injection module can be used that fastens the required capillary fittings and connections to the leg of the NMR magnet. Sample recovery is typically >90%.

Manually assisted injection is a variation on manual injection. This process loads by hand the sample loop of a typical, two-position, and six-port microinjection valve. Once the loop is loaded, the valve position is switched, which puts the loop on-flow, and a micropump is simultaneously triggered to deliver the sample to the NMR flow cell. When the sample reaches the flow cell, it is held there by stopping the flow and returning the valve to its original position. The high-throughput sample loader consists of a PC-programmable micropump and appropriate valves. An initial calibration injection determines the correct stopping time for placement of the analyte in the NMR flow cell. The probe flow path requires a volume of only 15–25 µL to thoroughly rinse the flow cell for the next sample. Hence, the transit capillary length must provide this minimum solvent volume for rinsing the probe between samples. Because of the design of the CapNMR probe, a single transceiver coil is triple tuned for the deuterium lock channel as well as proton and carbon nuclei. The tune-and-match circuit has sufficient frequency isolation to allow all typical decoupling experiments to be performed in an effective manner. Since the probe does not contain the usual inner ^{1}H coil and outer ^{13}C coil, direct carbon sensitivity is particularly good when compared to a conventional 5 mm indirect detection probe [12]. The use of a microcoil for ^{13}C NMR results in high mass sensitivity for carbon and an ability to excite the full carbon bandwidth using typically ~1 W of power.

The automated flow injection configuration employs a Gilson 215 autosampler, a micropump, and the CapNMR probe. Sample handling and data acquisition are controlled from the NMR console using automated vendor sample transfer software. The tool command language was altered by instrument vendors to trigger the micropump and change the fluidic sequence to a series of customized steps. This system enables the automated analysis of 96-well plates. Analyte concentrations are ~1–10 mM; a typical sample pick-up volume is 10 µL; and DMSO is an acceptable solvent.

Capillary-based HPLC injections require interfacing the capillary electrophoresis system with the capillary HPLC/NMR system. The probe used in this configuration may be viewed as a detector for capillary-based HPLC and offers some unique capabilities unattainable at the conventional (120 µL) size scale of LC/NMR [13]. One advantage of this system is that the microcoil probe can readily recover from any solvent gradient because of the effect of diffusion on its micro-flow-cell volume is small.

4.4.2 Microcoil Tube Probes (Room Temperature)

Microcoil tube–based probes are also commercially available and offer significant practical advantages over conventional 5 mm probes. These advantages include enhanced solvent suppression, improved salt tolerance, ease of shimming, improved radiofrequency homogeneity, and reduced cost of sample production, especially for protein studies where expression yields are low. An advantage of the microcoil tube probe over the flow probe is the versatility it affords in sample handling. Unlike the flow cell, the tube enables the analyst to remove the sample from the magnet and reinsert it at a later time for further data collection. This frees the spectrometer for other studies while the initial data set is being evaluated. If additional data is required, the sample can easily be returned to the probe for additional experiments.

Biochemists have reported the use of a 1 mm triple resonance z-gradient microcoil HCN probe to determine full resonance assignment and generate a full 3D structure of a small protein [14]. The protein used was a 68 residue *Methanosarcina mazei* TRAM protein (8.7 kDa). The biological function of this protein is not understood; hence, it was expected that structure elucidation would provide insight into its function. Only 72 µg of uniformly labeled ^{13}C,^{15}N protein was used in the study. The accuracy of the overall structure, determined using the microcoil probe assessed by comparison with the 5 mm NMR sample containing ~1600 µg of doubly labeled protein, was not dramatically different. The backbone rms deviation between the mean coordinates of the ensembles of conventional and micro-coil-probe structures was 0.73 Å demonstrating that high-quality data can be obtained with

<100 µg protein samples [14]. The entire microcoil-probe data set used for resonance assignment and 3D structure determination was acquired in approximately twice the time as required using the 5 mm probe (19 d vs. 9.5 d) but on a fraction of the mass of protein typically required for NMR structure determination using a conventional 5 mm room temperature probe.

The probe used in the study consisted of a vertical two-coil design with the inner ^{1}H coil and the outer coil tuned to ^{13}C and ^{15}N. It is a top-loading probe that accommodates 1 mm tubes. Therefore, all the moving parts and and possible probe maintenance issues of flow cell probes are eliminated. The probe had a 2.5 µL NMR-active volume and a 5 µL practical-minimum-sample volume requirement.

The probe was also tested for performance as a function of protein molecular weight. It was found that comparable spectral sensitivity for two-dimensional ^{1}H,^{15}N heteronuclear single quantum correlation (HSQC) spectra for proteins up to ~20 kDa, although spectral quality, was inherent to the characteristics of the protein. Comparison of the probe performance with a conventional 5 mm room temperature probe showed a ninefold gain in mass-based sensitivity. When compared with a 5 mm cryogenically cooled probe, the practical gain in mass sensitivity was approximately threefold. Mass-based sensitivity has potential drawbacks if the sample is not well behaved at higher concentrations. For example, some proteins may tend to form aggregates at higher concentrations leading to deterioration of spectral quality.

4.5   CRYOGENICALLY COOLED PROBES

When considering conventional probe technology, the thermal noise contribution from the entire receiver circuitry is treated as constant. Therefore, the sensitivity of a probe is commonly described in terms of the quality factor parameter, otherwise known as the Q-factor, of the probe. The Q-factor is determined by the resonance frequency ω, the inductance L, and the resistance R of the entire resonant circuit [15]:

$$Q = \omega L/R \tag{4.1}$$

The S/N ratio is proportional to the square root of the Q-factor, not considering power losses in the sample. Improving the S/N ratio requires raising the Q-factor. By examining Equation 4.1, it is evident that increasing Q can be accomplished by reducing the resistance R. The inductance (L) is fixed for a given coil geometry. Hence, the Q-factor and the sensitivity of a probe are functions of the resistance of the entire resonant circuit. Because the resistance of the coil depends on the temperature of the coil, the resistance may be decreased by lowering the temperature. Thus, in the case of cryogenic probes, cooling the coil and reducing the resistance will raise the Q-factor, and consequently the sensitivity of the probe. It should also be noted that the product of the Q-factor and the filling factor is a measure of how efficiently a coil can transfer RF energy to the sample volume; hence, it is important that the distance between the coil and the sample be minimized.

The sensitivity gain obtainable by cryogenic cooling of the NMR probe was first demonstrated in 1984 as reported by Styles et al. [16] where a sensitivity improvement of ^{13}C-signal at 45.9 MHz was achieved by a factor of eight. The first probe of this kind was cooled with liquid helium and nitrogen. However, it took about 20 years for cryogenically cooled probes to become commercially available. The reason for this was due to the high engineering demands. Instead of liquid cryogens, the modern cryogenic probes use cold helium gas, which requires an efficient thermal insulation. In addition to the probe itself, the cryogenic accessory consists of a cryogenic cooling unit and a helium compressor. The cooling unit supplies the entire infrastructure for the operation of the cryogenic probe and also continuously monitors the operation of the probe. These units form a closed-loop cooling system, where helium gas is compressed in one chamber and then chilled through expansion in a second chamber. This process of heat exchange follows the ideal gas law where the temperature changes relative to pressure and volume.

The temperature is proportional to pressure and volume and inversely proportional to the number of moles and the gas constant. Cooling is achieved by means of heat exchange. The cold helium is then transferred to the probe assembly using a vacuum-insulated transfer line. The probe consists of a well-insulated Dewar system, housing both the receiver coil and preamplifier. These components are fitted in one assembly in order to keep all the RF-connections between coil and preamplifier cold, so that any other thermal noise that might be introduced to the detector system would be minimal. Typical temperatures for the receiver coil and the preamplifier in the cryogenically cooled probes are 25 and 80 K, respectively. The spectrometer is operated in a routine manner while the cooling system works separately. The cool-down and warm-up procedures are initiated by push buttons or over controlling software, and they are completed in a few hours. Figure 4.3 schematically shows the different hardware units required for the operation of a cryogenic probe.

Overall, cryogen cooling of an NMR probe has had a major impact in NMR spectroscopy. This is because improving S/N has been and continues to be a major challenge in NMR spectroscopy. The conventional solutions for S/N improvement are to increase field strength, sample concentration, or *via* signal averaging; however, the more challenging yet highly effective approach is to decrease the noise. This was accomplished by cryogenically cooling the electronics in the NMR probe to reduce thermal noise. Once the technology had been developed, the resulting probe named a "cryoprobe," or a "cold probe," went into commercial production. Because these cryogenically cooled probes operate at a temperature of 25 K, they provide a S/N improvement that is about fourfold relative to their room temperature counterpart [17]. These probes have been designed for a range of sample sizes that include capillary tubes up to 5 mm tubes and flow cells [8,18,19].

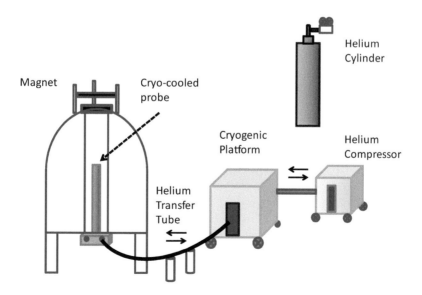

FIGURE 4.3 Illustration of the components needed to operate a cryogenically cooled probe. The helium compressor can be configured for air-cooled or water-cooled heat exchange.

4.5.1 CRYO-FLOW PROBE

As previously described, when the electronics of the NMR probe are cooled, the S/N of the resulting NMR spectra will increase [16]. Initially, tube-based 5 mm probes for solution-state NMR studies were cooled to 25 K, resulting in a fourfold increase in sensitivity [12,17]. Later, flow accessories were added to the 5 mm cryogenically cooled probe, increasing sensitivity over room temperature flow probes [18]. The addition of a cryogenically cooled flow accessory increases the versatility by enabling the conversion of a tube-based probe to a flow-through probe. The change-over operation takes only minutes and does not require warming the probe; hence, there is no change-over downtime nor loss of performance in either tube or flow operation. Cryo-flow accessories are available for spectometers ranging from 300 to 900 MHz. These probes have automatic tune-and-match capability and are comptible with sample changer accessories. The flow cells are commerically available with 30, 60, and 120 μL active NMR volume.

The design of flow accessory is optimized to avoid a filling factor loss and S/N is maximized. Because a cryogenically cooled flow probe greatly improves NMR sensitivity, it can be readily used as a high throughput analysis tool, although care must be taken to avoid flow-cell contamination. As with all flow probes used in a high throughput manner, cleaning and maintenance of the system need to be carried out to avoid contaminant carryover.

Many developments have expanded the capability of NMR flow cells. Flow probes currently incorporate improved fluidics, temperature control, and a wide range of sample-cell sizes. When combined with cryogenic cooling, substantially better sensitivity is realized. New capillary separation techniques can be supported, and mutiplex NMR that allows solution-state NMR to be truly parallel is also possible (see Chapter 8). Cryogenic flow probes can also be interfaced with an SPE system and a loop collector. These developments have greatly expanded the range of LC-NMR applications.

4.5.2 CRYO-CAPILLARY TUBE PROBE

Micro-cryo technology is commercially available with the 1.7 mm TCI MicroCryoProbe™ from Bruker Biospin. This probe represents a tremendous improvement in mass sensitivity [20]. The sample volume is 30 μL and offers more than an order of magnitude in mass sensitivity compared to the conventional 5 mm probe, and allows for highest ^{1}H sensitivity as well as enhanced performance for carbon. A comparative illustration of a 1.7 mm capillary tube is given in Figure 4.2.

The extreme sensitivity provides a powerful tool for NMR analysis of limited sample amounts. Applications are particularly targeted toward isolated materials in low abundance. These may include degradants, process impurities, extractables, natural products, peptides or small molecules, difficult-to-express proteins, and so on. This probe enables acquisition of spectra that allows difficult problems to be tackled and solved, particularly where only microgram or submicrogram quantities of material are available.

Because the sensitivity gain is 14-fold over a conventional room temperature 5 mm probe, a 200-fold reduction in experiment time can be realized. The probe is optimized for direct proton detection with indirect detection for ^{15}N and ^{13}C. Short pulses and large bandwidths for ^{15}N are ideal for small molecule structure elucidation. Incorporation of z-gradients enables signal enhancement and solvent suppression. The probe covers a temperature range of 0–80°C. In addition, the 30 μL volume can be perfectly matched to the elution volume of high-performance liquid chromatography-solid phase extraction (HPLC-SPE). Isolated samples from SPE may be sent directly to a 1.7 mm tube either in manual mode or in automation using a fraction collector.

Overall, the inherent insensitivity of NMR spectroscopy has been significantly addressed with the development of microcoils and the introduction of cryogenic cooling. While conventional NMR hardware generally requires several milligrams of sample, the micro-environment has reduced the amount to tens of micrograms. The development of these microcoil probes has had a profound impact on studies ranging from isolations of very low levels of degradants and impurities to molecules synthesized on the microgram scale for activity screening studies as well as proteins that have challenging expression and purification schemes. These probe advancements provide a tremendous boost in the successful investigation of biological NMR.

The success of combining cryo cooling with microcoil probe design relies on the concept that miniaturization of the receiver coil will produce a concomitant increase in mass sensitivity (S/N per mg of solute). This is because S/N for a given amount of compound is approximately inversely proportional to the diameter of the coil. As a result, microcoil probes are highly suited for mass-sensitive applications, where the mass of the solute is limited, yet the solute is reasonably soluble. As noted previously, the power of a room temperature microprobe was ably demonstrated using the *Methanosarcina mazei* TRAM protein. In the first published example of its kind, the Northeast Structural Genomics Consortium (NESG) determined the structure of this small (68-residue, 8.6 kDa) β-barrel protein using only 72 µg of protein. The probe used was a room temperature Bruker 1 mm TXI MicroProbe (7 µL sample volume) [10]. Although the data were acquired in approximately twice the spectrometer time required with a conventional 5 mm probe, only 1/20th the mass of protein was needed compared to a conventional 5 mm NMR sample that required 1.6 mg. The NESG consortium subsequently employed the use of the even more sensitive Bruker 1.7 mm MicroCryoProbe™ (30 µL sample volume), which combines the microcoil concept with the three- to four-fold increase in S/N afforded by cryogenically cooled coils in the probe. The very valuable application of the MicroCryoProbe™ is in the investigation of proteins that can only be obtained in minute amounts, for example, those proteins that can only be prepared using cell free and other eukaryotic expression systems where yields are low [21]. An NESG target, VpR247, with a molecular weight of 11 kDa, was solved using an 800 MHz spectrometer equipped with a 5 mm cryoprobe. The total sample volume was 300 µL and the coordinates were deposited in the protein data bank (PDB id 2KIF). The structure was also solved using a 600 MHz NMR spectrometer equipped with a 1.7 mm MicroCryoProbe™. In this case, only a 30 µL volume (1/10th the amount of protein) was used. The structure was deposited in the protein data bank (PDB id 2KIM). Hence, in both cases, the sample concentration of 0.9 mM was used and both produced nearly identical results [22]. A comparison of the published structures of VpR247 protein is shown in Figure 4.4.

Using the 1.7 mm cryoprobe at 600 MHz enabled complete resonance assignments and 3D structures of proteins ranging from 6 to 15 kDa in size, at 0.5–1 mM protein concentration. Accordingly, it is now possible to routinely determine high-quality 3D structures of small (<20 kDa) proteins by NMR using 70–300 µg samples. This opens NMR spectroscopy to many new application opportunities in Structural Biology that were previously not possible.

4.5.3 AFFORDABLE CRYOGENICALLY COOLED PROBES

A new type CryoProbe has been introduced that provides improved sensitivity approaching that of a conventional cryoprobe but at a more affordable price. This new type of broadband CryoProbe uses nitrogen-cooled RF coils and preamplifiers. The sensitivity enhancement of this probe relative to a conventional room temperature probes is reported to produce a factor of 2 to 3 for X-nuclei ranging from ^{15}N to ^{31}P.

The gain in sensitivity on the proton/fluorine channel was reported to exceed standard probe performance by a factor of 2. This probe comes equipped with a control unit and a liquid-nitrogen vessel. The advantage of this system relative to a helium cooled cryoprobe

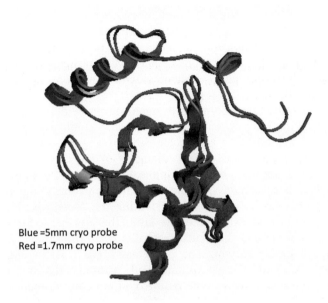

Blue =5mm cryo probe
Red =1.7mm cryo probe

FIGURE 4.4   A superimposed ribbon representation of the backbone structure of NESG target VpR247 (11 kDa) is shown. The structure in blue was solved using an 800 MHz NMR spectrometer equipped with a 5 mm cryoprobe (300 μL volume, PDB id 2KIF). The structure in red was solved using a 600 MHz NMR spectrometer equipped with a 1.7 mm MicroCryoProbe™ (30 μL volume, PDB id 2KIM). The sample concentration of 0.9 mM was used for both studies.

is that no additional infrastructure is required in preparing a location for the system. The probe may also be obtained with an optional automatic tune-and-match accessory. Currently, this probe is available for NMR systems ranging from 400 to 600 MHz. The probe does however require weekly liquid N_2 fills to replenish the cryo-cooling system, thereby adding additional cost to the cryogen maintenance.

As a routine workhorse, the nitrogen-cooled probe offers affordable sensitivity, a notable increase in sample throughput, and broadband technology covering nuclei from ^{15}N to ^{31}P. Ideal for small molecule routine labs in academia, pharmaceutical, and chemical industry, it allows full automation with minimum operating and maintenance costs and long service intervals.

4.6   PROBE COIL GEOMETRIES

Probe construction for NMR spectroscopy involves a high level of skill and precision. The optimization of NMR probes for commercial use requires advanced knowledge and expertise in the areas of physics, electronics, materials science, and mechanical engineering. While conventional probes face design and construction challenges, the microcoil probes present even greater level of difficulty. This is because obtaining optimal sensitivity and resolution requires incorporation of the following attributes:

1. Selecting appropriate coil geometry.
2. Minimizing circuit losses and magnetic susceptibility-induced line broadening.
3. Ensuring the generated B_1 field is homogeneous over the entire sample volume.

The typical coil geometry used in standard solution NMR probes is saddle type. This coil configuration generates a very homogeneous magnetic field orthogonal to the external B_o field. For miniaturization and small NMR coils, however, other coil geometries have been employed due to fabrication challenges combined with the need to optimize sensitivity and resolution. These coil geometries include solenoidal, planar helical, and transmission type (micro-slot and spripline) [23]. Figure 4.5 shows the various types of NMR coil configurations. Additional discussion on microcoil probes is given in Chapter 8.

4.7 PROBE SENSITIVITY COMPARISON

The sensitivity of NMR probes can be defined relative to concentration sensitivity, S_{conc}, or mass sensitivity, S_{mass}. The equation describing concentration sensitivity is given below:

$$S_{conc} = \frac{S/N}{C}(\mu M^{-1}) \tag{4.2}$$

With the decreasing size of the NMR probe coil, the S/N per unit mass increases. Because of this, if the mass of the sample is limited and high concentrations are possible (i.e., high solubility), then the S/N that can be obtained with a small coil will be greater than the S/N obtained using a larger coil. The concentration relationship with coil size and S/N, however, is different. If high concentrations are not possible, due to solubility constraints as may be experienced in biological samples such as proteins, and the total mass (e.g., mg of protein) is not limited, then larger coils will give rise to a higher S/N. For cases where mass and concentration are both limited, the optimum size of the coil is dictated by the relative values of these two variables.

Mass-limited samples of small organic molecules are often delivered as a dried powder or oil. Under such conditions, the volume and concentration of sample prepared for NMR analysis can be controlled by the spectroscopist, provided the analyte retains suitable solubility. It becomes useful then to be able to measure sensitivity in a manner that helps guide the spectroscopist to make the appropriate choice of an NMR probe to maximize S/N when sample is mass limited. In contrast to concentration sensitivity, the mass sensitivity, S_{mass}, is

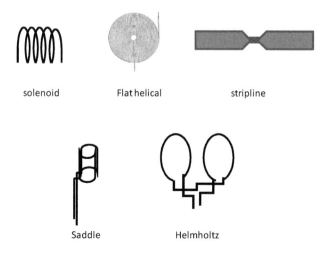

FIGURE 4.5 Nuclear Magnetic Resonance (NMR) probe coil geometries.

normalized not to sample concentration, but instead to the mass (or moles) of analyte in the observed volume (V_{obs}). Equation (4.3) will be

$$S_{mass} = \frac{S/N}{m(obs)} \, (ng^{-1}) \tag{4.3}$$

where S_{mass} is the mass sensitivity for a single scan, and m_{obs} is the mass (or moles) of sample in V_{obs} of the probe. Normalization of S/N to the amount of analyte establishes S_{mass} as a parameter that incorporates S/N as a function of number of detected nuclei. This definition of mass sensitivity (S_{mass}) is consistent with the standard definition of sensitivity that plots the slope of a curve based upon signal intensity versus sample amount observed. It should be noted, however, that when a compound is selected with a particular mass amount for measurement and data acquisition and processing parameters are selected, and presuming certain basic conditions such as tuning and matching of the circuit are completed, then S_{mass} is affected essentially by the electronics of the probe, shimmed sample homogeneity, and the ionic content of the sample. Although S_{mass} could be defined for any number of transients, it is usually determined with a single fully relaxed scan. Following optimization of acquisition conditions to maximize S_{mass} for a given sample, S_{mass} cannot be further optimized. As noted previously, multiple transients may be used in defining S_{mass}; hence, the S/N term in Equation 4.3 can be multiplied by $\sqrt{n}$, where n is the number of scans, should more than one scan be used in the evaluation.

Unlike concentration sensitivity (S_{conc}), S_{mass} enables comparison of probe performance for a sample of fixed mass. Conversely, S_{mass} is not well suited to comparing probes for a sample of fixed concentration, for example, cases where a compound may have limited solubility. Hence, use of S_{conc} and S_{mass} is dependent upon the particular sample situation (amount and solubility). However, they can both provide assistance in the choice of a probe or analytical procedure for a given set of conditions.

In addition, probe performance can be monitored over time by making periodic determinations of either S_{conc} or S_{mass} under standard conditions. Since $m_{obs} = CV_{obs}$, Equations 4.2 and 4.3 can be combined to yield Equation 4.4.

$$S_{conc} = V_{obs} \, S_{mass} \tag{4.4}$$

This expression in Equation 4.4 shows that the larger the observe volume in a probe, the greater its concentration sensitivity. This is so because for a fixed concentration, a larger volume positions more sample in the observe volume V_{obs}. Hence, the observe volume V_{obs} becomes the proportionality constant that established a relationship between mass sensitivity and concentration sensitivity. Another basic principle of NMR probe design is that $S_{mass} \sim 1/d_{coil}$. As the observe coil size decreases, transverse magnetic field strength per unit volume increases. Therefore, given a fixed mass of analyte and a consistent set of operating parameters and conditions, the NMR probe with the smallest observe coil diameter will have the largest *mass sensitivity*. This principle when combined with Equation 4.4 leads to the conclusion that a concentration-limited sample will yield the highest S/N in a probe that accommodates a relatively large volume, and conversely, a mass-limited sample is most effectively analyzed in a probe with the smallest observe coil. Therefore, if an analyte possesses high solubility and is not mass-limited, the highest S/N will be achieved by placing an analyte mass in the probe with the smallest V_{obs} and d_{coil} and highest S_{mass} and at the highest concentration allowed by sample solubility and availability. For mass sensitivity, the most challenging sample for NMR analysis is an analyte that is both concentration- and mass-limited.

Studies comparing S_{conc} and S_{mass} for a microcoil probe and a traditional 5 mm tube probe (using a $^1H\{^{13}C/^{15}N\}$ triple resonance triax), on a 600 MHz NMR spectrometer, were reported [6]. S/N was measured in D_2O from a single-scan spectrum using the anomeric proton of sucrose at 5.4 ppm. The S/N values were normalized either to the concentration or number of micromoles of material in V_{obs} to obtain the corresponding sensitivity. A Varian dsnmax algorithm was used to compute the S/N with a 200 Hz window and line broadening of 0.7 Hz. Four spectra were acquired and the S/N values were averaged. The data revealed that the microcoil probe had a 10-fold increase in mass sensitivity over the conventional 5 mm probe, but in contrast, the larger-volume probe is more concentration-sensitive by 15 folds. The data for both probes is consistent with Equation 4.4. The 10-fold mass sensitivity difference between the two probes is further confirmed by the ratio of the 90° pulse widths measured at the same gain and transmitter power. The CapNMR probe had a value of 4.6 µs, compared to the value for the 5 mm probe of 46 µs.

A second study was carried out for mass sensitivities for some common NMR probes relative to a fixed sample mass. Also included in the study was the determination of the corresponding data acquisition time required to obtain quality spectra for the same fixed mass. In practice, an analyst may choose to sacrifice a part of optimal sensitivity for a reduction in data acquisition time and the remainder for the amount of soluble sample available. For this study, a Bruker cryoflow probe was used, yielding a S_{mass} = 2688 (S/N per µmol) for the anomeric proton of sucrose, with line broadening of 1 Hz. If the S_{mass} of the CapNMR probe was recomputed using a line broadening of 1 Hz, then S_{mass} = 2400 (S/N per µmol), which was within 11% of the reported value for the cryoprobe.

The relationship between a capillary-based microcoil flow probe and a conventional tube probe can be expressed as a comparison of S/N based on relative S_{mass} values between the two different probes. If a capillary-based flow probe with a microcoil has a greater mass sensitivity than a tube probe with a larger coil, then it follows that for an identical mass in the V_{obs} of each probe, and for a given set of experimental conditions,

$$\frac{S/N(cap)}{S/N(tube)} = \frac{S(mass)(cap)}{S(mass)(tube)} \tag{4.5}$$

This expression can be modified to include two different observe volumes. Since the capillary probe has the greater mass sensitivity and smallest coil, it will also have the smaller observe volume. The ratio of the two observe volumes reflects the increase in concentration for a fixed mass in the smaller, more mass-sensitive probe, as compared to the larger probe, hence a higher concentration. If the concentration increase in dissolving a sample of fixed mass in the smaller observe volume is not compromised due to solubility issues or other limitations, then the sensitivity increase will be as stated in Equation 4.6.

$$\frac{S/N(cap)}{S/N(tube)} = \frac{S(mass)(cap)}{S(mass)(tube)} \left[\frac{C(cap)}{C(tube)} \bigg/ \frac{V(tube)}{V(cap)} \right] \tag{4.6}$$

In this equation, V_{tube} and V_{cap} represent the observe volumes for the large and small coil probes, respectively. The observe volume decrease represented by V_{tube}/V_{cap} must equal the corresponding concentration increase for a fixed mass given by C_{cap}/C_{tube} for the full S/N increase of $S_{mass,cap}/S_{mass,tube}$ to be attained. If this condition is met, the sample analysis is not limited by solubility, the concentration to volume ratios for the two probes in Equation 4.6 equals 1, and Equation 4.5 applies. Hence, Equation 4.6 allows the user to calculate the concentration increase required to achieve the maximum S/N enhancement for a given sample mass prepared for a microprobe, as compared to a larger probe. If analyte solubility or other restrictions limit the allowed concentration increase, then the final concentration actually achieved (or desired) determines the extent of S/N enhancement attained according to Equation 4.6.

It is well known that S/N is proportional to m_{obs}, and the square root of the number of scans, such that for a given probe and set of conditions, Equation 4.5 can be expanded to

$$\frac{S/N}{S/N_{(p)}} = \left[\frac{S(\text{mass})}{S(\text{mass}_{(p)})}\right]\left[\frac{m}{m_{(p)}}\right]\left[\sqrt{n/n_{(p)}}\right] \quad (4.7)$$

where m is the mass of sample in the observe volume, n is the number of acquired transients, and the subscript "p" indicates one condition, as compared to a different (non-subscripted) condition. Equation 4.7 can be interpreted as pertaining to spectral data from two probes of differing sample mass (S_{mass}), and number of transients. For the case of an equal mass in both probes, the m/m_p term in Equation 4.7 equals 1. This allows computation of the relative number of scans required to achieve the same S/N for a fixed mass placed in two different probes, each with a known S_{mass}. Equation 4.7 also indicates that for a constant S/N, n is proportional to $1/S_{mass}^2$ for a fixed mass, which indicates that an increase in S_{mass} can significantly reduce data acquisition time.

Studies have been performed comparing the sensitivity of small microcoils with the standard larger NMR probe coils. In one such study by Schlotterbeck et al. [20], the mass sensitivity of the Bruker 1 mm TXI microliter probe with a sample in a 1 mm diameter capillary tube was shown to be five times greater than a 5 mm conventional TXI probe: the sample used was sucrose in aqueous solution at 600 MHz. The small TXI probe also showed a factor of 1.7 enhancement over a 5 mm TXI cryoprobe with the sample in a 5 mm tube, and a factor of 1.3 over a 5 mm TXI cryoprobe with the sample contained in a 1 mm capillary. These studies were followed by the detailed sensitivity analysis of different probes at 600 MHz that was published by Olson et al. [10]. Although some data had to be inferred from the literature, the general results showed that the 1.5 µL CapNMR probe and 5 mm cryoprobe have very similar mass sensitivities, and that these are approximately 10 times greater than the mass sensitivity of a standard 5 mm probe. The concentration sensitivity of the CapNMR probe was found to be approximately 15 times poorer than the conventional 5 mm probe.

The concepts and data presented here illustrate how micro-NMR fills a unique role in the analysis of mass-limited and volume-limited samples. Because of the considerable reduction in required sample mass and data acquisition time (number of transients), or a combination of both, the approach is making a distinct and significant contribution to the field of NMR analysis [24]. The relationships derived here are useful in guiding the experimentalist toward the selection of optimal NMR probe and sample conditions required to obtain quality data.

REFERENCES

1. Albert, K. 2002. *LC NMR and Related Techniques*, West Sussex, England: John Wiley & Sons, Inc.
2. Martin, M.L., Martin, G., and Delpuech, J.-J. 1980. *Practical NMR Spectroscopy*, London: Heyden & Son Ltd.
3. Roth, G. 2003. Ultra high field NMR magnet design. 14–19. www2.warwick.ac.uk/fac/sci/physics/current/teach/module_home/px388/extra_material/bruker_magnets.pdf.
4. Larbalestier, D., and Ray, S. 2007. New mag lab record promises more to come. www.magnet.fsu.edu/mediacenter/news/pressreleases/2007august7.html. News Release. National High Magnetic Field Laboratory, USA.
5. Schroeder, F.C., and Gronquist, M. 2006. Extending the scope of NMR spectroscopy with microcoil probes. *Angew. Chem. Int. Ed.* 45:7122–7131.
6. Bhattacharya, A. 2010. Breaking the billion-hertz barrier. *Nature.* 463:605–606.
7. Bednorz, J.G., and Muller, K.A. 1986. Possible high T_c superconductivity in the Ba–La–Cu–O system. *Z. Phys. B. Condensed Matter.* 64:189–193.

8. Spraul, M., Freund, A.S., Nast, R.E., Withers, R.S., Maas, W.E., and Corcoran, O. 2003. Advancing NMR sensitivity for LC-NMR-MS using a cryoflow probe: application to the analysis of acetaminophen metabolites in urine. *Anal. Chem.* 75:1536–1541.

9. Russell, D.J., Hadden, C.E., Martin, G.E., Gibson, A.A., Zens, A.P., and Carolan, J.L. 2000. A comparison of inverse-detected heteronuclear NMR performance: Conventional vs cryogenic microprobe performance. *J. Nat. Prod.* 63:1047–1049.

10. Olson, D.L., Norcross, J.A., O'Neil-Johnson, M., Molitor, P.F., Detlefsen, D.J., Wilson, A.G., and Peck, T.L. 2004. Microflow NMR: concepts and capabilities. *Anal. Chem.* 76:2966–2974.

11. Gronquist, M., Meinwald, J., Eisner, T., and Schroeder, F.C. 2005. Exploring uncharted terrain in nature's structure space using capillary NMR spectroscopy: 13 steroids from 50 fireflies. *J. Am. Chem. Soc.* 127:10810–10811.

12. Hoeltzli, S.D., and Frieden, C. 1998. Refolding of [6-^{19}F] tryptophan-labeled *Escherichia coli* dihydrofolate reductase in the presence of ligand: a stopped-flow NMR spectroscopy study. *Biochemistry.* 37:387–398.

13. Hentschel, P., Krucker, M., Grynbaum, M.D., Putzbach, K., Bischoff, R., and Albert, K. 2005. Determination of regulatory phosphorylation sites in nanogram amounts of a synthetic fragment of ZAP-70 using microprobe NMR and on-line coupled capillary HPLC–NMR. *Magn. Reson. Chem.* 43:747–754.

14. Aramini, J.M., Rossi, P., Anklin, C., Xiao, R., and Montelione, G.T. 2007. Microgram-scale protein structure determination by NMR. *Nat. Methods.* 4(6):491–493.

15. Kovacsa, H., Moskaua, D., and Spraul, M. 2005. Cryogenically cooled probes—a leap in NMR technology. *Prog. Nucl. Magn. Reson. Spectrosc.* 46:131–155.

16. Styles, P., Soffe, N.F., Scott, C.A., Cragg, D.A., Row, F., White, D.J., and White, P.C.J. 1984. A high-resolution NMR probe in which the coil and preamplifier are cooled with liquid helium. *J. Magn. Reson.* 60:397–404.

17. Anderson, W.A., Brey, W.W., Brooke, A.L., Cole, B., Delin, K.A., Fuks, L.F., Hill, H.D.W., Johanson, M.E., Kotsubo, V.Y., Nast, R., Withers, R.S., and Wong, W.H. 1995. High-sensitivity NMR spectroscopy probe using superconductive coils. *Bull. Magn. Reson.* 17:98–102.

18. Corcoran, O., and Spraul, M. 2003. LC-NMR-MS in drug discovery. *Drug Discov. Today.* 8:624–631.

19. Exarchou, V., Godejohann, M., Van Beek, T.A., Gerothanassis, I.P., and Vervoort, J. 2003. LC-UV-solid-phase extraction-NMR-MS combined with a cryogenic flow probe and its application to the identification of compounds present in Greek oregano. *Anal. Chem.* 75:6288–6294.

20. Schlotterbeck, G., Ross, A., Hochstrasser, R., Senn, H., Kühn, T., Marek, D., and Schett, O. 2002. High-resolution capillary tube NMR. A miniaturized 5-microL high-sensitivity TXI probe for mass-limited samples, off-line LC NMR, and HT NMR. *Anal. Chem.* 74(17):4464–4471.

21. Zhao, L., Zhao, K., Hurst, R., Slater, M., Acton, T.B., Swapna, G.V.T., Shastri, R., Kornhaber, G. J., and Montelione, G.T. 2010. Engineering of a wheat germ expression system to provide compatibility with a high throughput pET-based cloning platform. *J. Struct. Funct. Genomics.* 11:201–209.

22. Aramini, J.M., Tubbs, J.L., Kanugula, S., Rossi, P., Ertekin, A., Maglaqui, M., Hamilton, K., Ciccosanti, C.T., Jiang, M., Xiao, R., Soong, T.T., Rost, B., Acton, T.B., Everett, J.K., Pegg, A.E., Tainer, J.A., and Montelione, G.T. 2010. Structural basis of O^6-alkylguanine recognition by a bacterial alkyltransferase-like DNA repair protein. *J. Biol. Chem.* 285:13736–13741.

23. Fratila, R.M., and Velders, A.H. 2011. Small-volume nuclear magnetic resonance spectroscopy. *Annu. Rev. Anal. Chem.* 4:227–249.

24. Webb, A.G. 2005. Microcoil nuclear magnetic resonance spectroscopy. *J. Pharmaceut. Biomed. Anal.* 38:892–903.

5 NMR-Associated Isolation Technologies

Hyphenated NMR instruments have evolved to include a number of automated operation modes for sample collection and detection. These LC-NMR modes consist of but are not limited to continuous flow, stop flow, loop collection, and solid-phase extraction [1]. As noted previously, liquid chromatography (LC) peaks of interest for the NMR transfer require the use of a non-NMR detector. Detection systems such as UV, DAD, and MS are compatible with the gradient LC methods that are required to keep the eluting peaks as sharp as possible. For flow methods, this is critical in preserving NMR sensitivity.

In the initial stages of LC-NMR development, continuous-flow NMR was the primary collection mode. With continuous-flow LC-NMR, the sample from the chromatographic column is guided through a flow cell into a dedicated NMR flow probe. The time for one measurement is determined by the chromatographic flow rate and is typically between 8 and 20 s. Active flow-cell volumes usually range from 30 to 120 μL. This approach requires the use of deuterated solvents and solvent suppression techniques (see Chapter 6) to mediate the dynamic range effects where the huge solvent peaks mask the peaks from the compound of interest.

Continuous-flow experiments work most effectively for cases where significant amounts of analyte can be isolated from a single chromatographic injection. This approach is best suited for compounds that have stability issues and cannot be stored for significant periods of time. While there are cases where continuous-flow LC-NMR is preferred, this work has been discussed extensively in the initial introduction of LC-NMR (Chapter 1). Hence, this chapter will focus on the robust collection modes of stop flow, loop collection, and SPE that enable full structure characterization. In addition, DI-NMR (direct injection NMR) and FIA NMR (flow injection NMR) along with the emerging CE isolation in conjunction with micro-flow detection will be explored.

5.1 STOP FLOW

Signal-to-noise limitations are a major challenge in many LC-NMR studies. Although NMR signal to noise can be improved by signal averaging, this requires the flow to be stopped with the chromatographic peak of interest in the flow cell for a substantial period of time. During the stopped period, multiple scans of the chromatographic peak's NMR spectrum may be carried out to achieve an acceptable signal-to-noise level.

With stop-flow analysis, the chromatographically separated sample is analyzed under static conditions. A compound separated by HPLC is sent directly to the NMR flow cell and is stopped in the NMR flow probe for as long as is needed for NMR data acquisition. Stop flow requires the calibration of the time required for the sample to travel from the detector of the HPLC to the NMR flow cell. Timing depends upon the flow rate and the length of the tubing connecting the HPLC with the NMR spectrometer. Because the chromatographic peak of interest is stopped in the flow cell, 2D homonuclear and heteronuclear NMR correlation experiments [2] can be obtained. The time the compound resides in the flow cell is under complete control of the spectroscopist.

There are several ways to acquire stopped-flow data. One approach involves stopping the chromatographic peaks of interest in the NMR flow probe as they elute from the chromatography column. Alternatively, the LC pump may be programmed to "time slice" through a chromatographic peak, stopping every few seconds to acquire a new spectrum. This can be useful for resolving multiple components by NMR from within a peak that is not fully resolved chromatographically, or for verifying the purity of a chromatographic peak; however, with such mixtures substantial amounts of material are required to obtain interpretable spectra in a few scans.

In order to use stop flow to collect NMR data on a number of chromatographic peaks in a series of stops during the chromatographic run without incurring on-column diffusion, the NMR data for each chromatographic peak must be acquired in a short time. It is estimated that 30 min or less are required to prevent diffusion effects when more than four peaks need to be analyzed, and less than two hours for the analysis of no more than three peaks [3,4]. The use of commercially available cryoprobes or cold probes improves the sensitivity of the stop-flow mode, requiring less experiment time or reduced amounts of sample (see cryogenically cooled probes).

Stop flow is preferred when the chromatography produces reasonably well separated peaks and the compound is stable in solution for extended periods. Other methods utilizing the stopped-flow approach involve loop collection or trapping the eluted peaks onto a mini chromatographic column (solid-phase extraction cartridge). These methods are discussed next in more detail.

5.2 LOOP COLLECTOR

Unlike stop flow, loop collectors allow separation of the LC and NMR functions. With loop collection, the sample undergoes chromatographic separation, but instead of sending the sample directly to the NMR probe, the peaks corresponding to individual compounds are trapped and stored on loops comprised of capillary tubing. The separated fractions may be stored in the loops until mobile phase is used to send it to the NMR flow cell. After transfer to the probe, the valves divert the mobile phase and the HPLC pump is shut off, allowing the transferred sample to remain in the NMR probe.

A loop collector may be thought of as a fraction collector; however, instead of using collection vials or tubes, the fractions are collected in pressurized loops. Although the system can be configured to allow each loop's contents to be pumped into the NMR flow probe, the system can be removed and treated as a separate remote unit. If used off-line with a separate HPLC system, the loops may be filled and then the entire loop assembly can be transferred to the NMR system [5,6].

The loop-volume capacity can be adjusted to a desired size by adjusting the tubing length while keeping the diameter fixed. The loop system typically uses multi-port rotary valves located downstream from the LC detector; hence, the LC-detector signal can be used to activate the valves. The system can be programmed to automatically trap multiple chromatographic peaks in the loops, all under computer control (based upon the LC detector's signal output).

Loop collectors are useful when lengthy NMR data collection is required or when detectable amounts of material may be collected from a single chromatographic injection. In such cases, the isolated compound may be stored in a controlled environment without concern of decomposition from solid support or changes with respect to light or heat.

A major advantage of loop collection is that this technique avoids possible peak contamination that may be experienced with a stopped-flow collection mode. For example, if the retention time of two peaks is 0.4 min apart, under direct stopped-flow conditions, approximately 20% of the first peak may be carried over to the second peak. With loop storage trapping, however, the two peaks remain pure.

A disadvantage of loop collection in LC-NMR is matching solvent composition. If solvent composition in the loop is significantly different from the mobile phase solvent that pushes the loop's contents into the NMR probe, magnetic-susceptibility mismatches between the two solvent mixtures will result causing shimming problems and broad lines. Possible solutions involve adjusting the loop volume to be larger relative to the flow-cell volume to displace a different mobile phase. Alternatively, one may attempt to match the mobile phase with that of the solvent composition in the loop; however, this can be difficult to achieve. Another disadvantage is the apparent dilution of peaks stored in loops when switching to elution. The dead volume introduced by capillary connections exacerbates the problem. Hence, loop collection combined with flow NMR would not be ideal for trapped samples having relatively low concentrations.

5.3 SOLID-PHASE EXTRACTION (SPE)

A major development in the field of hyphenated techniques is the use of a solid-phase extraction (SPE) system as an interface between LC and NMR. SPE combined with LC provides a system that enables peak collection of an analyte on a cartridge. As a compound is eluted from an HPLC column, it may be diverted and trapped on an SPE cartridge. The peak selection is typically done either by UV detection or by evaluation of MS spectra.

The integration of SPE with flow injection NMR was first demonstrated by Griffiths and Horton [7] followed by Brinkman in 1998 [8]. Subsequently, numerous reports on the use of SPE-NMR have appeared in the literature [9–20]. For SPE, the trapping process involves adding water *via* a makeup pump to the mobile phase as it eluted from the LC column (when reverse phase HPLC is used). Elution from the cartridge can occur via a separate pump to dispense stronger solvents that can transfer the analyte from the SPE cartridge. This process is illustrated in Figure 5.1.

The SPE-NMR hardware may be viewed as similar to a loop collector with the exception that an SPE cartridge is inserted into the flow path. A makeup pump introduces water to the mobile phase as it leaves the LC column and enters the SPE cartridge. This enables retention of the analyte on the cartridge. The SPE cartridge may be interfaced with an NMR flow probe to allow the sample to be directed into a flow cell for NMR detection. A separate pump is used to transfer the deuterated solvent that elutes the analytes from the SPE cartridges through the capillary tubing directly into the NMR flow probe. In such cases, the eluting solvent is usually deuterated acetonitrile or methanol (CD_3CN or CD_3OD). The deuterated solvent provides a lock signal to prevent the frequencies from drifting during NMR data acquisition and to address a dynamic range issue resulting from large proteo solvent peaks. The deuterated solvents may be stored over a stream of dry nitrogen gas to prevent absorption of water from the atmosphere. A stream of dry nitrogen gas is also used to dry the cartridges prior to eluting into the NMR probe with deuterated solvent. This drying will further reduce the large signal contribution from proteo solvents such as residual water and acetonitrile. Although large solvent peaks can be reduced using solvent-suppression pulse sequences, critical peaks from the sample may also be suppressed, thereby compromising structural analysis. Hence, minimization of proteo solvents from the NMR sample is a good common practice.

An important consideration with flow NMR is that the sample in the flow cell maintains uniform magnetic susceptibility. To achieve this, solvents in the sample must be homogeneous. If a solvent gradient exists across a sample, the lineshape of the NMR signal will become broad. One advantage of the SPE method is that the analyte is dissolved in a single homogeneous deuterated solvent that transfers the compound as a concentrated fraction, thereby eliminating the magnetic susceptibility problem. Hence SPE-NMR can successfully

SPE in LC-NMR

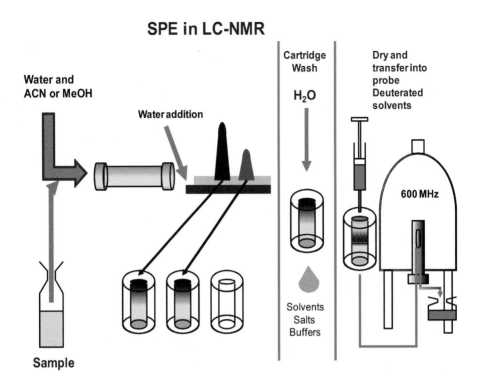

FIGURE 5.1 Illustration of the interface of liquid chromatography (LC) and solid-phase extraction (SPE) with NMR spectroscopy. Compound can be incrementally added to a cartridge. The cartridge can then be washed, dried with nitrogen gas, and eluted from the cartridge with deuterated solvent.

be used to concentrate and spectrally characterize compounds that may elute as broad chromatographic peaks with long retention times [16].

SPE also offers an advantage with respect to the integration of NMR and MS technologies. In particular, finding a solvent system that is compatible with both the gradient profile and columns used in the separation protocols for NMR and MS can be difficult. Additionally, the use of deuterated solvents, that are desired in NMR analysis, create problems in the MS analysis where the analyte is subject to hydrogen–deuterium (H–D) exchange resulting in a shift of one mass unit for every exchangeable hydrogen. With SPE, all chromatography may be performed using conditions compatible to MS analysis and without the need for deuterated solvents [21].

An additional attraction of SPE-NMR is that multiple trapping is possible. Although both the loop collector and SPE have the chromatographic step performed separately from the NMR analysis, the SPE system can trap peaks from multiple injections allowing a single compound to be built up on a specified cartridge. This enhances the concentration of material in the flow cell and can be effectively used to trap impurities or low-level metabolites. Concentration in the 0.1% range relative to the major analyte may be isolated on an SPE cartridge with proteo solvents, dried, and then eluted from the cartridge with deuterated solvents. Full spectral characterization may then be possible on the sample once sufficient concentration is obtained [9,14].

It should be noted that since an SPE cartridge may be viewed as a "mini storage column," one can consider the resulting elution step of SPE-NMR as just another form of chromatography where the sample stored on the SPE column may be eluted into a capillary tube for

NMR evaluation or into a vial. Samples from SPE-NMR studies can also be recovered from an NMR flow probe at the end of the NMR analysis by blowing it out of the probe into a vial or tube with nitrogen gas. Multiple collected samples may then be combined, dried, and taken up in an alternate deuterated solvent for further evaluation under the control of the user. Hence, low concentrations due to limited on-column loading are eliminated with SPE [10].

For samples where the quantity is limited, it is advantageous to increase the concentration by limiting the amount of solvent in which the compound is dissolved. Hence, the SPE elution volume may be adjusted to fill a range of NMR tube sizes. If a normal 5 mm tube is used, however, the volume cannot be less than about 500 µL without causing serious line shape problems (shimming problems) and loss of signal to noise. However, there are special tubes made by Shigemi that can be used to restrict the active volume and reduce the amount of solvent without causing line shape problems. Shigemi tubes matched to particular solvents are commercially available.

While the SPE system offers a tremendous advantage in building sample concentration, retention on specific cartridge types are compound dependent. Finding a cartridge with the appropriate solid support for a particular compound or system type needs to be performed for each new compound. For small organic compounds found in pharmaceutical research, C18 and general purpose (GP) resin cartridges are a good starting point. Often, very low levels of degradants and impurities may be evaluated indirectly through examination of the retention properties of parent compounds with similar structural features. Method-development cartridge trays are commercially available for this purpose. Some typical SPE-cartridge support systems used in method development studies are given in Table 5.1.

In summary, the major advantages of SPE-NMR are as follows:

- Chromatographic separation can be done with inexpensive non-deuterated solvents or with additives that are not compatible with NMR spectroscopy.
- Because no D_2O is used in the eluent, no H–D exchange occurs during the chromatographic process, which could result in elimination of critical exchangeable protons needed for structure elucidation.
- Only small amounts (approximately 300 µL) of deuterated solvents are required for the transfer.
- The complete sample can be eluted in a small volume (<30 µL) of liquid from the SPE cartridge. Because of this concentration effect, a substantial increase in sensitivity by a factor of 2 to 4 is observed, especially for broader peaks.
- By multiple collections from subsequent chromatographic separations of the same sample, the amount and concentration can be further increased, improving the sensitivity by a factor of 10 or more.
- The deuterated solvent that is used for the elution and transfer is independent of the chromatographic conditions and can be selected to improve spectral quality and make exchangeable protons observable in the NMR.
- The compatibility of integration of LC-NMR and LC-MS is enhanced with the addition of SPE.
- The SPE system may be run under automated conditions. NMR flow probes with the detection volume adapted to the small volume of the peaks eluted from the cartridges may be interfaced with the SPE system.

5.4   NON-CHROMATOGRAPHIC FLOW NMR

In the previous sections, different modes of sample collection in LC-NMR were discussed each having different capabilities and advantages. This section will cover two new types of

TABLE 5.1
SPE Cartridge Selection for Method Development

Type HySphere	Size	Structure
CN–SE	7 μm	$-\!Si\!-\!(CH_2)_3\!-\!CN$
C2–SE	7 μm	$-\!Si\!-\!(CH_2)\!-\!CH_3$
C8–EC–SE	8 μm	$-\!Si\!-\!(CH_2)_7\!-\!CH_3$
C18-HD	7 μm	$-\!Si\!-\!(CH_2)_{17}\!-\!CH_3$
Resin GP	10–12 μm	
Resin SH	25–35 μm	
MM cation	10–12 μm	Polymeric-based mixed-mode exchangers
MM anion	25–35 μm	Polymeric-based mixed-mode exchangers

non-chromatographic flow NMR. They are FIA-NMR (flow injection analysis NMR) and DI-NMR (direct injection NMR).

As the need for improved efficiency and productivity increases, there is an increased demand for automated high-resolution NMR spectroscopy that can move the samples to be analyzed in and out of the NMR probe in a fast and reliable manner. Tube-based NMR relies on mechanical robots to move traditional precision glass samples' tubes in and out of the magnet under automation. Such systems are not as rapid as may be required for high-volume throughput; hence, there is an ever-increasing need for improved speed and reliability at a reduced cost.

Flow injection analysis NMR (FIA-NMR) is a method that may be viewed as LC-NMR without a chromatography column. One way to classify LC-NMR is to view HPLC as simply a sample delivery system for an NMR spectrometer. In LC-NMR, one begins with a mixture of compounds and the liquid chromatography step separates the compounds before they are passed into the NMR flow cell. Alternatively, it is possible to analyze the samples that are intact and already separated, without the chromatography column. If the chromatography column is removed and everything else remains unchanged (i.e., the pump, injector, mobile phase, connective tubing, flow probe, and UV detector), a sample can be injected into the system and transported by the "mobile" phase into the NMR probe. Stopped-flow LC-NMR may then be implemented to retain the sample in the NMR probe. The pump can be triggered off (which stops the mobile phase, and hence the sample transport) in two different ways. The first involves using a signal from the UV detector to stop the sample in the flow cell after a calibrated delay where the UV detector registers a peak maximum as in LC-NMR. The second transport method involves simply calibrating the delay time (from the injector port to the probe) as a function of the rate of solvent flow. When the delay times are properly calibrated, the maximum concentration of the injected sample plug can be stopped reproducibly inside the NMR flow cell. This concept was initially termed "columnless LC-NMR" but later named "flow injection analysis NMR" (FIA-NMR). The sample delivery system was first demonstrated in 1997 by Keifer and coworkers [22–25]. A typical delay time (from the injector port to the probe) was calibrated at 10–20 s. FIA was found to be a fast and easy way to automate the movement of samples in and out of an NMR probe.

A more detailed (complete) description of FIA-NMR was published in 2003 [26]. FIA-NMR was shown to be especially useful for repetitive analyses. Sample volumes of 20–200 µL could be injected, with very good reproducibility on an analytical scale. The relative standard deviation of the peak integrals was 1.38% for 15 duplicate injections; it was 0.38% for 15 repetitions of the same static sample left in the probe. Although a minimum-detectable sample size was not determined, the study did show that 20 µg sample-injection amounts could be easily detected with 1 min acquisition time. Data were acquired using a mobile phase of $CH_3CN–D_2O$ (50:50) to transport the samples into the probe. WET solvent suppression was used so non-deuterated acetonitrile could be used. The FIA analysis showed that the "mobile phases" do not need to be deuterated when WET is used, although it has been generally found that when using capillary systems, the cost of using deuterated solvent is minimal. Hence, the improved spectra justify the cost.

One goal for FIA-NMR was the ability to perform analyses rapidly (high throughput). There are three requirements that need to be taken into account to accomplish this. First, the samples need to be rapidly transported from the injector port into the NMR probe. Second, good sensitivity from the NMR probe is required to allow the total NMR acquisition time to be short. And third, the sample needs to be removed from the probe with a rinse cycle to remove all traces of the sample before the next sample is introduced. The FIA-NMR system easily meets the second requirement by being able to detect 20 µg (injected) sample quantities [26]. One study, however, suggested that up to 140–840 µg of injected material might be needed for routine evaluation [27]. In contrast, when a capillary NMR flow probe (1–5 µL detection volume) is used for FIA-NMR instead of a "standard" flow probe (typically a 60 µL detection volume), the detection limits can be reduced significantly (easily to the range of 10 µg) [28].

The quality with which FIA-NMR meets the first requirement (moving the sample from the injector port into the NMR probe quickly) depends on the solvent flow properties [26]. For pumping speeds of 4.0 mL/min, the sample can be mechanically moved into the probe in 6–8 s (depending on the length of the capillary transfer line); however, once inside the probe it may take approximately 2 min for the NMR signal to fully stabilize. In addition, the NMR peak heights may be less reproducible. In contrast, when pumping speeds are reduced to 0.25 mL/min, the mechanical transfer takes much longer (~1.5 min), and an additional stabilization time of ~1.5 min is needed. Hence, a total of 3 min from injection to the start of acquisition is required. The advantage of slower flow rates is that the peak heights are most reproducible under these conditions. Data from intermediate flow rates verify these trends: the fastest flow rates (4.0 mL/min) allow faster initiation of NMR data acquisition (even though the stabilization time is then the longest, the reduced pumping time more than compensates for that delay). However, the slowest flow rates provide the most reproducible (quantitative) NMR data. For cases where quantification is not needed, one could start the NMR acquisition earlier, but it still requires 1–2 min to transfer and stabilize the sample. This is still time-saving over mechanical tube-based sample changers, which change samples at a slower rate. Also, they require additional time to then lock and shim the sample – activities which take no additional time in FIA-NMR [26]. In comparison with LC-NMR, however, it has been observed that the time needed to stabilize an injected-plug sample in FIA-NMR is longer than the stabilization time needed for an LC-NMR sample. This presumably is because an LC-NMR sample has a Gaussian distribution of concentration at the trailing and leading ends of the sample plug, whereas an injected plug sample has a more discontinuous interface of mobile phase at the sample edges.

The requirement that one removes the sample from the probe as quickly as possible is the most variable parameter in FIA-NMR. It depends in part on how much carryover from sample-to-sample can be tolerated. Carryover is a critical concern in FIA-NMR unlike a mechanical tube-based sample changer where this is absolutely 0%. To achieve approximately 0.1% carry-over, it was found that the process required 0.5 min at a flow rate of 4.0 mL/min (2 mL total solvent consumed), and 4.0 min at 0.25 mL/min (1 mL total solvent consumed). Although flush out is faster at 4.0 mL/min, flushing at 0.25 mL/min uses less solvent that may be more import-ant if a deuterated mobile phase is used. Deuterated solvent is recommended when it is possible to avoid NMR solvent suppression and achieve a generally cleaner spectrum [26]. Many add-itional reports on the use of flow injection NMR [29–31] have appeared following the original report in 1997 [23]. Many of these applications fall under the catagory of DI-NMR.

5.5   DIRECT INJECTION NMR (DI-NMR)

DI-NMR is sometimes viewed as a simplified version of FIA-NMR. However, in practice, the two sample delivery systems are mechanically very different. In DI-NMR, the pump is simplified, and the mobile phase is removed along with any detectors. This gives rise to the simplest possible flow-NMR system. DI-NMR injects the sample directly into the NMR flow probe, and the whole injection process is then driven by automation. In fact, the NMR flow probe may be viewed as a sample loop for the injector port of an automated injector. Because DI-NMR does not require a mobile phase, only the solvent in which the sample is dissolved is used. Therefore, an additional reservoir of solvent is used to either rinse the NMR flow cell and transfer lines or hydraulically push the sample into the flow cell, or both. The first demonstrations of DI-NMR were published starting in 1997 by Keifer et al. [22–25], where the technique was compared and contrasted with FIA-NMR. This was followed by a full description and evaluation of the technology [32]. The driving force in developing DI-NMR was the desire to do automated NMR faster and better. Experience with LC-NMR and FIA-NMR had shown that flow NMR had several advan-tages over traditional tube-based automation in terms of speed and economy. In flow NMR, no additional time was needed to lock or shim each sample (assuming the solvent composition was kept constant). Deuterated solvents were not required because of the effi-ciency of WET solvent suppression, and because locking the sample was not necessary due to programmed frequency registration (Scout-Scan) [32]. A major need for high-throughput NMR resulted from combinatorial chemistry and parallel syntheses libraries, since a high volume of samples were being either stored or synthesized in microtiter plates, and the automated liquid handlers could inject aliquots from those microtiter plates into standard injector ports. Direct injection also eliminated the breakage or cleaning of costly glass sample NMR tubes. Combinatorial chemistry provided an ideal justification for the devel-opment of DI-NMR because such synthetic endeavors produce thousands of samples, all dissolved in similar concentrations, normally in a common (uniform) solvent, all stored in microtiter plates. Usually, such sets of compounds had similar structural scaffolds; hence, all that was needed was a ^{1}H-NMR spectrum to identify a probably known structure or to verify purity. With such studies, the spectrum had to be obtained in a rapid manner and at a low cost. Because it was necessary to repeat the same NMR experiment thousands of times with minimal human intervention, a major improvement in the data collection process became imperative. DI-NMR was a system that could address the need for rapid and efficient scanning at low cost.

A report on the application of DI-NMR [32] analyzed combinatorial-chemistry samples stored in (septum-covered) 96-well microtiter plates, all dissolved in either DMSO-d_6 or DMSO-h_6, at a concentration of either 25 or 2.5 μM. The volume of injected sample solution was 350 μL. A Gilson 215 liquid handler was interfaced to the NMR spectrometer; under NMR control, one sample was withdrawn from a well in the microtiter plate and injected via an injector port and a transfer line into the NMR flow cell (in the probe in the magnet) using the syringe pump on the Gilson. Under instrument-controlled software, the Gilson then triggered the NMR to acquire an NMR spectrum (a 32-scan 1 min acquisition). When the NMR acquisition was complete, the NMR sent a signal back to the liquids handler, which withdrew the sample (using a synchronized source of air backpressure) and returned it to the original well of the microtiter plate (or a designated location). The syringe pump then was used to rinse the NMR flow cell by a similar "out-and-back" motion with an aliquot of clean solvent. The original report showed that the data acquired on samples dissolved in DMSO-h_6 can be almost as good as the data acquired on the samples dissolved in DMSO-d_6. It also showed several ways to display and analyze the NMR data acquired on a library of compounds stored in a 96-well plate. Discussion of the various possible ways to get quality-control information for the fluidic injection process involved using either the Scout-Scan data or signals from an internal standard [32].

Another significant way that DI-NMR differs from both LC-NMR and FIA-NMR is that the flow cell is empty of solvent when the sample is injected. This allows the DI-NMR user to acquire the highest possible sensitivity on the sample and also allows the sample to be recovered in an unaltered (undiluted) state. This does, however, require the DI-NMR user to inject at least a certain minimum sample volume in order to acquire an acceptable NMR spectrum. In common practice, this minimum sample volume is about 150 μL for standard "60 μL observe volume" flow cells, but it can be as low as 2 μL for a ".5 μL observe volume" capillary flow cell [33]. If this minimum volume (which depends on the volume of the flow cell) is not met, the flow cell will contain air, which will destroy the NMR lineshape. As there is no solvent mixing in DI-NMR, the volume difference cannot be made up with a "push" solvent. The different fluid behavior of DI-NMR as compared with FIA-NMR influences many performance aspects of the analysis, such as carryover, detection efficiency, recovery efficiency, and speed. The major disadvantage of DI-NMR (not encountered with FIA-NMR) is the possibility that the tubing in the system could become clogged with particulate matter if unfiltered samples are injected.

For both FIA-NMR and DI-NMR, sample contamination from carryover or clogged capillaries represents particular disadvantages in a controlled industrial environment where sample identity and purity need to be documented for regulatory purposes. Hence, these technologies would be more amenable for implementation in an industrial or academic research environment.

5.5.1 Applications of DI-NMR

Following the introduction of DI-NMR, NMR vendors (Bruker and Varian/Agilent) proceeded to commercialization of the technique. This resulted in numerous reports of DI-NMR applications. These applications could be classified as one of three general categories, all of which utilize repetitive analyses. The studies typically employed use of one-dimensional (1D) ^{1}H NMR spectra; however, cases of two dimensional applications were also reported.

The categories were as follows:

1. The analysis of libraries of relatively pure organic compounds. These may include synthetically prepared libraries or commercially purchased screening libraries.
2. The analysis of biofluids (urine, plasma, etc.) – an application that ultimately leads to the analysis of samples for clinical diagnoses.
3. The study of biomolecules – most commonly to use ^{1}H^{15}N HSQC to analyze combinations of ligands and ^{15}N-labeled receptors (proteins) to either study binding properties or to screen ligands for binding affinities.

Although the DI-NMR technology was adopted and heavily used in private industry (primarily the pharmaceutical industry), much of the work was proprietary and not subject to publication. Therefore, the extensive and widspresd use of this technology is not accurately reflected from literature reports. Some reported applications include Spraul et al. [34], who used DI-NMR to screen rat urine. Others used DI-NMR to study ligand binding to a biomolecule [35] or to analyze combinatorial-chemistry libraries [36]. Additional demonstrations of the utility of DI-NMR to analyze compound libraries were published by Combs and coworkers [37], Lewis and coworkers [38], and numerous studies with natural product libraries were also reported [39–42].

A key example of DI-NMR was reported by Eldridge and coworkers [39]. These researchers described a high-throughput application involving the production, analysis, and characterization of libraries of natural products. The study was conducted for the purpose of accelerating the drug discovery process for high-throughput screening in the pharmaceutical and biotechnology industries. Compound libraries were generated using automated flash chromatography, solid-phase extraction, filtration, and high-throughput preparative liquid chromatography. The libraries were prepared in 96- or 384-well plates and consisted of purified fractions with approximately one to five compounds per well. Libraries were analyzed by a high-throughput parallel liquid chromatography-evaporative light-scattering detection-mass spectrometry system to determine the molecular weight, number, and quantity of compounds in a fraction. The compounds were subsequently subjected to biological screening and active fractions were further purified at the microgram level to confirm individual activity. Purified compounds were elucidated by NMR spectroscopy utilizing a microcoil probe. A syringe pump operating at 5 µL/min pushed the sample to the microcoil probe in 2.5 min and parked the sample inside the probe. After the acquisition was completed, the sample was collected in recovery vials. Sample loading was done in 3 µL with 5–50 µg quantities. One dimensional ^{1}H NMR spectra were generated on isolated compounds of 5–50 µg.

Studies by Kautz and coworkers described an automated system for loading samples into a microcoil NMR probe using a segmented flow analysis system [33]. This approach yielded a twofold increase in the throughput of DI-NMR and FIA-NMR methods. Sample utilization was reduced by three folds. Sample volumes of 2 µL (10–30 mM, ~10µg) were drawn from a 96-well microtiter plate by a sample handler, and then pumped to a 0.5 µL microcoil NMR probe. These samples were set up as a queue of closely spaced "plugs" containing sample in DMSO-d$_6$ separated by an immiscible fluorocarbon fluid. Individual sample plugs were detected by their NMR signal and automatically positioned for stopped-flow data acquisition. The sample in the NMR coil could be changed within 35 s by advancing the queue. The fluorocarbon liquid wetted the wall of the Teflon transfer line and prevented the DMSO samples from contacting the capillary wall. The fluorocarbon coating reduced the sample losses to below 5%. A wash plug of solvent between samples reduced the sample-to-sample carryover by <1%. Significantly, the samples did not disperse into the carrier liquid during loading or during acquisitions of several days for trace analysis, thereby allowing extended 2D studies to be conducted. For automated high-throughput analysis using a 16 s acquisition time, spectra were recorded at a rate of 1.5 min/sample, and the total deuterated solvent consumption was <0.5 mL per 96-well plate [33].

Additional publications on the use of DI-NMR for the analysis of biomolecules and the screening of binding were published [30,31,43,44]. Stockman et al. reported using DI-NMR in a pharmaceutical lead-like protein screening application. In this study, samples were prepared for analysis by adding sufficient protein to each well of the 96-well library plate to give a 1:1 (protein:ligand) ratio at a concentration of approximately 50 µM. The total volume in each well was 350 µL in order to provide a 300 µL injection volume. Homogeneous sample dispersion throughout the well was facilitated by agitating the plate on a flatbed shaker. Screening allowed a good 1D ^{1}H NMR spectrum to be acquired in about 10 min. This concentration of target and small molecule required identified ligands to have affinities on the order of ~200 µM or tighter. Once the screening plate was prepared, a Gilson liquid sample handler transferred samples from 96-well plates into the flow injection probe. The system could be programmed to return the samples back into either the original 96-well plate or a new plate. Once the sample was in the magnet, changes in chemical shifts, relaxation properties, or diffusion properties were recorded. The reported NMR screening assay consisted of two 1D relaxation-edited ^{1}H NMR spectra: one spectrum was collected on the ligand mixture in the presence of protein, and the second, control spectrum was collected on the ligand mixture in the absence of protein. Ligands were identified as binding to a target based upon a reduction in signal when compared to a relaxation-edited spectrum collected in the absence of protein.

Demonstrations of the use of DI-NMR to study biofluids were published by several different investigators [45,46]. Potts and coworkers [46] reported on the ability of DI-NMR to measure metabolic responses to toxic insult as expressed in altered urine composition. Interpretation of NMR data relied upon comparison with a database of proton NMR spectra of urine collected from both control and treated animals. Pattern recognition techniques, such as principal component analysis (PCA), were used to establish whether the spectral data cluster according to a dose response. Studies were also conducted to determine the impact that NMR-related variables might impart on the data. The NMR variable study focused on solvent suppression methods, as well as instrument-to-instrument variability, including field strength. The magnitude of the NMR-induced variability was assessed in the presence of an established response to the nephrotoxin bromoethanamine. The results showed that toxin-induced changes were larger and easily distinguished from those caused by using different solvent suppression methods and field strengths.

DI-NMR has also been used to analyze foods such as olive oils and other food oils [47,48]. Even beverages such as beer have been analyzed with flow injection NMR [29]. Lachenmeier et al. have used high-resolution nuclear magnetic resonance for the quality control and authenticity assessment of beer in official food control. Measurements were performed using a 400 MHz NMR spectrometer with flow injection technology to change samples. The beer samples were degassed and buffer pH 5.6 in D_2O was added containing 0.1% TSP for referencing. Differences in the spectral profiles of beers varying in type and origin were studied by PCA, evaluating the spectral characteristic fingerprint. The high throughput afforded by the flow injection system allowed a comprehensive database of beer spectra for PCA classification to be established efficiently. Beers made with barley malt could be distinguished from those made with wheat malt. Beers were clustered based upon their brewing sites and quality. Using the partial least squares (PLS) method to correlate NMR spectra, the ethanol and lactic acid content were established. The summary of all of this work proved that DI-NMR is a useful technology for automatic NMR evaluation. Extension of this technology to food chemistry illustrates the range of applications.

5.5.2   COMPARISONS

FIA-NMR and DI-NMR have both proven to be useful, but they have different strengths and weakness when compared with each other as well as when compared with other flow-NMR techniques like LC-NMR. For example, LC-NMR is appropriate whenever chromatographic separations are needed, but it tends to be time-consuming. DI-NMR is appropriate if maximum sensitivity is required, or if the sample needs to be recovered, or if there is a sufficient sample volume, and if the sample contains no sediment. FIA-NMR is appropriate when carryover needs to be minimized or if the sample does not need to be recovered intact, and if a wider range of smaller sample volumes must be analyzed. Clearly, each technique has its own set of advantages and disadvantages and should be appropriately used depending upon the application [26].

5.6   CAPILLARY ELECTROPHORESIS AND NMR

As noted previously, the integration of capillary electrophoresis and NMR, termed CE-NMR, is an important separation technology whose integration with NMR is still in early stages of development and optimization [49–52]. Because of the low-mass sample requirements and high separation efficiencies as well as rapid separations, CE appears to be an attractive means of separating micro amounts of compounds. When combined with NMR, CE-NMR offers the possibility of combining the powerful separation capability of CE and the superior detection and structure elucidation of NMR.

5.6.1   MODES OF ELECTROPHORESIS-NMR AND EFFECTS ON NMR SPECTRAL PROPERTIES

With NMR, magnetic objects must be kept at a safe distance from the magnet depending on the magnetic field strength and the shielding properties of the magnet. However, initial studies showed that NMR microcoil probes can be coupled to a capillary CE system with no major modification to CE instrumentation. In fact, for CE-NMR studies, the solenoidal microcoil was either wrapped around the sample capillary or the capillary was housed inside a saddle coil. For solenoidal coils, however, magnetic susceptibility effects will degrade line shapes and S/N when thinner-walled capillaries are used as the coil comes into closer contact with the sample.

The initial non-commercial solenoidal coils used in CE-NMR had an axis orthogonal to the static magnetic field (B_o) and were reported to form the most mass-sensitive RF (radio frequency) coils. However, in this configuration, the electrophoretic current induces a second local magnetic field gradient that perturbs the magnetic field homogeneity. The strength of this induced magnetic field is related to the electrophoretic current and the radial distance from the center of the capillary. Any magnetic field inhomogeneity resulting from a CE effect is difficult to restore by shimming procedures normally employed in NMR to restore magnetic field homogeneity. The effect of induced magnetic field gradient leads to broader NMR lines, loss of scalar coupling, and poor S/N. The scalar coupling of resonance signals has been shown to degrade with high-field magnetic gradients created with increasing current [53]. A significant loss of S/N and increase in resonance linewidth were also found to occur at higher voltages. For example, a voltage increase from 0 to 8 kV was found to increase linewidths from 1.5 to 15 Hz and decrease S/N from 147 to 19 [54]. A solution to address the current-induced thermal and magnetic field gradient effects involved employing a form of stopped-flow experiment. This technique has the advantage of allowing both one-dimensional and two-dimensional NMR experiments to be performed. An alternate technique which acquires NMR spectra under voltage dormant conditions was also reported [54]. In this method, once the analyte reaches the NMR microcoil observe region, the voltage is periodically interrupted and 1 min high resolution NMR spectra with good S/N are acquired for every 15 s of applied voltage. A drawback of this approach, however, is that the termination of voltage can adversely affect the separation efficiency and reduce peak resolution due to diffusion over the longer period of the experiment.

A novel CE-NMR instrument with a dual-microcoil probe to record continuous flow NMR data under stopped-flow conditions has been reported [53]. The dual coil probe was constructed by wrapping two separate coils around two separate outlet capillaries. The outlet capillaries were connected to a separation capillary through a capillary splitter. In this study, CE-NMR experiments were performed with one outlet capillary while keeping the second capillary as an open circuit. A series of NMR spectra were obtained by alternating electrophoretic flow between the two coils. The severe line broadening induced by current was eliminated, because with the described instrumentation, only the NMR spectra of the sample in the floating capillary at 0.0 µA were obtained.

The best NMR sensitivity for analytes was reported using a saddle-coil probe when the coil axis is oriented parallel to the external magnetic field. In this case, the electrophoretic current-induced magnetic field does not contribute to magnetic field inhomogeneity along the z-axis. Hence, high-resolution NMR spectra with similar S/N was obtained even when varying electrophoretic currents over a range from applied voltage 0 to 20 kV, 36 µA [49]. A light chemical shift change was observed that was attributed to Joule heating at higher voltages. Hence, with saddle NMR probes, the absence of current-induced magnetic fields permits on-flow CE-NMR experiments to be performed unencumbered by severe spectral line broadening effects.

A significant challenge with CE-NMR is that the rate of migration of each analyte band in CE depends upon intrinsic electrophoretic mobility; hence, each band moves past the detector at a different rate. These differential migration rates can influence the NMR signal intensity and linewidth of each compound. By using high surface-area-to-volume ratio in capillary electrophoresis (i.d. ≤100 µm), Joule heat may be effectively dissipated, thereby allowing voltages as high as 30 kV to be used. The high voltages were important because they enabled faster separations. Because NMR is an insensitive technology compared to other detectors, CE-NMR often requires capillaries with a diameter of 75–200 µm to enhance the NMR active volume and hence signal intensity. However, an increase in the internal diameter will decrease the surface area to volume ratio. For example, increasing the i.d. from 75 to 200 µm would result in a 50% decrease in surface-area-to-volume ratio [55]. The inability to dissipate heat effectively with larger capillaries can lead to Joule heating, which will cause chemical shift changes particularly with respect to exchangeable protons [56, 57]. This in turn complicates the interpretation of NMR data.

There is a complex dependence of flow rate, electric field, and current on the observed NMR signal intensities. High flow rates reduce the effective T_1, allowing faster pulse repetition and improved sensitivity by replacing the sample volume for each acquisition so that the signal is not saturated. However, this can also increase the observed line widths. In addition, the electrophoretic current induces an internal magnetic field, and the resulting field gradient may perturb the NMR measurement. To add to the complications, the high electric field can partially align large molecules, and Joule heating affects the diffusion rates and electrophoretic mobilities.

Hence, from a practical standpoint in order to improve the utility of this approach further, the areas needing further refinements must address poor concentration, limited dynamic range and hence sensitivity, and the broad line widths.

By optimizing coil geometry and the use of uniformly wound microcoils, the line broadening due to susceptibility variations in the local region of the sample may be minimized. However, due to the limited residence time of a nucleus in the NMR microcoil, spin-spin relaxation, T_2, is reduced, and the minimum line width is increased to just over 1 Hz under typical electrophoretic flow conditions. Hence, a major obstacle to improving the sensitivity of the NMR detection is the limited time available to observe the sample since a typical sample residence time is less than 10 s. Only a few scans may be obtained in the NMR data acquisition due to the relatively long T_1s of the sample. One way to compensate for this is to decrease the analyte migration rate by decreasing the voltage across the observation window. Hence, the analyte residence can be increased and more NMR transients can be acquired.

A number of online CE injection methods have been reported that concentrate the sample as it is introduced to the capillary [58–60]. These methods extend the useful sample concentration range. To improve the concentration LOD (limits of detection), the noise introduced by the rf coil needs to be greatly reduced. This may be achieved by using lower resistance coil materials in probe construction. Continuous refinements in probe design are expected to decrease the LOD of the CE-NMR technique.

5.6.2   NMR Observe Volume

Analyte peaks in CE typically contain low nanoliter volumes. High-resolution CE electropherograms and NMR spectra can be obtained using nanoliter observe volumes; however, the obtainable NMR sensitivity often precludes the use of such small NMR active volumes. Nonetheless, CE-NMR experiments have been successfully performed with microcoil observe volumes as small as 5 nL [54, 61].

High filling factor is one of the advantages of using microcoils directly wrapped on a CE capillary. However, this configuration excludes the ability to change the observe volume. This problem was addressed with the construction of a "sleeve probe" in which the solenoidal coil is wrapped around a polyimide sleeve that can accommodate different sizes of CE capillaries [62]. High-resolution spectra with linewidths of 1–2 Hz were obtained with this new probe configuration.

As noted previously, a saddle-coil NMR probe enables CE-NMR data to be recorded in a continuous-flow mode unencumbered by magnetic susceptibility induced line broadening effects. However, difficulty arises in the construction of microprobes with saddle coils and they typically tend to be larger than solenoids, although a 2.5 μL saddle-coil probe for static samples has been reported [63]. Insets for larger-volume saddle coils are, however, available and the effective flow-cell volumes tend to range from approximately 250 to 400 nL [49,64].

The use of capillary isotachophoresis (cITP) as a separation techonology has been noted previously; however, this technology also serves to concentrate charged analytes prior to NMR detection, which is an effective strategy for improving the detection limits of microcoil probes [65]. cITP is an electrophoretic sample-stacking technique that separates and concentrates charged analytes even in the presence of a large excess of neutral sample matrix [66]. In cITP, analytes are separated according to their electrophoretic mobilities by applying a high electric field (10–30 kV) across a capillary containing a discontinuous buffer system composed of a leading electrolyte (LE) and a trailing electrolyte (TE). Sample stacking arises from the insertion of the sample matrix behind an LE, which has a higher electrophoretic mobility than the sample constituents, and ahead of a TE, which has a lower electrophoretic mobility. Upon electric field application, individual components of the sample matrix sort into bands whose relative positions depend on their electrophoretic mobility values. Owing to the placement of the sample between the LE and TE, analyte bands remain in contact after focusing and travel at constant velocity through the separation channel, adjusting their concentration in proportion to that of the LE. This collection mode approach requires careful selection of LE and TE to optimize the separation conditions and results in the concentration of analytes by 2–3 orders of magnitude (see Figure 5.2). The focusing characteristic of cITP enables online concentration which in turn improves the effective sample concentration and hence the sensitivity of NMR detection [49].

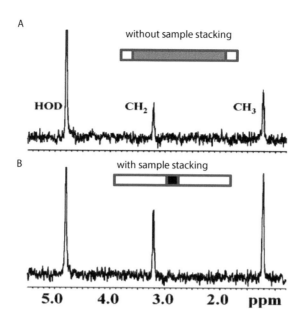

FIGURE 5.2 ^{1}H NMR spectra of TEAB (tetraethylammonium bromide). (A) Capillary filled with 5 mM TEAB without sample stacking (S/N of peak at 1.2 ppm: 13). (B) 8 µl of 200 µM TEAB injected; spectrum after sample stacking by cITP (S/N of peak at 1.2 ppm: 30). (Reproduced with permission from reference [62]. Copyright 2001 American Chemical Society.)

It should be noted that the primary reported cITP NMR results have been studies involving cationic analytes. The major challenges in conducting anionic cITP NMR experiments are the development of suitable buffer systems and overcoming electroosmotic flow (EOF), which opposes the migration of anions and degrades the ability of cITP to separate and focus the analyte bands. To overcome this limitation, commercially available zero-EOF capillaries have been employed. Application studies of cITP directly coupled with online microcoil NMR detection for the separation and analysis of nanomole quantities of compounds are discussed in Chapter 7.

5.6.3 CAPILLARY ELECTROCHROMATOGRAPHY (CEC)-NMR

Capillary electrochromatography (CEC) is a hybrid separation technique that combines the separation efficiency of CE and the selectivity of LC. CEC also provides more sample loading capacity in comparison to CE. With these features, CEC has become an attractive analytical tool to separate complex mixtures. The increased loading capacity is an important advantage to improve NMR sensitivity.

CEC-NMR is a hybrid between CE-NMR and cLC-NMR (capillary LC-NMR). A typical instrumentation schematic used for CEC-NMR measurements is shown in Figure 5.3. In CEC, the separation is performed using capillary LC columns with electro-osmotic flow instead of pressure flow. However, low pressures (>30 bar) have been used to prevent air bubble formation at frits and to achieve faster separation [67]. Integrated instrumentation to perform CEC-NMR has been designed and shown to be a promising tool to analyze complex mixtures [49]. This technology has been used in continuous-flow and stopped-flow modes to analyze drug metabolites in human urine [50,51]. CEC-NMR was also used to analyze a mixture containing caffeine, acetaminophen, and acetylsalicylic acid [68]. In this reported study, however, the peaks were poorly separated using isocratic conditions (2 mM borate,

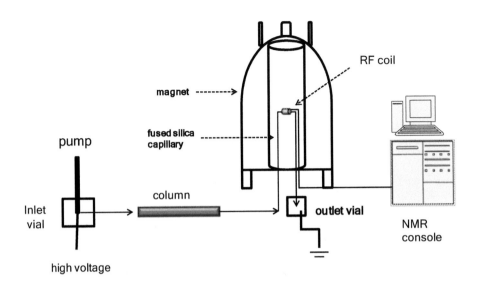

FIGURE 5.3 Schematic for CEC-NMR showing the arrangement of the separation system and NMR detection cell.

80% D_2O, and 20% CD_3CN). Apparently, the large injection volumes used to improve sensitivity caused the separation of analytes to deteriorate. Separation was improved with a solvent gradient 0–30% CD_3CN.

When pressure was applied during CEC to decrease the separation time [67], the total analysis time was reduced by a factor of 10 compared to non-pressurized CEC. When compared to capillary LC-NMR operating at 16 bar, a mixture was separated in 110 min. With pCEC-NMR at 16 bar and 20kV, the mixture was separated in 13 min. However, better resolution was achieved with capilary LC. Increasing the pressure of pCEC did not improve the migration rate but did result in a decrease in separation efficiency.

Like CE-NMR, CEC-NMR allows compound separation on a micro scale. A drawback of CEC-NMR, however, is that the frit used to contain the packing material creates a region of electrical resistance and can lead to local temperature increase. CEC-NMR is still relatively new and more studies are required to improve the separation capability and reproducibility of this technology.

5.6.4   CE-NMR IN PRACTICE

The first hyphenation of CE and NMR used 800 mM arginine, and the CE inlet and outlet vials were within the bore of the NMR magnet [69]. It was reported that separation of arginine, cysteine, and glycine had detection limits of 50 ng under static modes [70]. To record NMR spectra in 16 s, a relatively high injection volume (20 nL) had to be used which reduced the separation efficiency. The periodic stopped-flow technique excluded the electrophoretic current-induced effects associated with the horizontally positioned solenoidal coil, and was used to analyze a mixture of arginine and triethylamine (TEA) [70]. Improved concentration LODs for arginine with similar mass sensitivities (57 ng; 330 pmol) and TEA (9 ng; 88 pmol) were realized with field amplified stacking. Because of the sample stacking and stopped-flow measurement, the sensitivity was increased by two- to fourfolds without compromising the separation efficiency. Separation efficiencies of 50,000 were reported for arginine with periodic stopped flow.

Continuous-flow CE-NMR with a saddle coil was used to separate a mixture of lysine and histidine in phosphate buffer [49]. In this report, a detection limit of 336 ng (2.3 nmol) was recorded for lysine. CE-NMR with a saddle-coil probe was also used to analyze drug metabolites. The ability to separate and identify drug metabolites and bio-transformed products in biological fluids represents a considerable challenge. CE-NMR was able to successfully analyze of the major metabolites of paracetamol in human urine [50,51]. Two major metabolites, paracetamol glucuronide and paracetamol sulfate conjugates, as well as endogenous material (hippurate) were separated and identified with CE-NMR. The experiment was performed with a 400 nL NMR detection cell at 600 MHz, and used an injected volume of 8 nL for an 80 μm i.d. column (effective length 1 m). Comparison of chemical shifts confirmed the identification of the metabolites. The estimated amount that could be detected in this study with an S/N of 3:1 was ~10 ng.

Often, the separation process can be improved by optimizing the buffer pH. Continuous-flow CE-NMR was used to analyze two commonly found food additives, caffeine, and aspartame at two pH values [68]. The separation was achieved using a glycine buffer at pH 10 in less than 70 min. However, complete separation was not realized and some NMR signals overlapped with the glycine resonances. Changing the buffer to formate (pH 5) removed the spectral overlap but reversed the order of migration and increased migration time. However, at pH 5, both caffeine and aspartame migrate closer to each other. Thus, as with all chromatographic endeavors, the experimental conditions are a compromise between separation efficiency and NMR spectral quality. Injections of larger analyte volumes and concentrations greater than the buffer concentration to improve NMR sensitivity usually degrade the separation efficiency of CE.

As described above, online concentration methods can enhance the sensitivity by concentrating the sample during the separation process [54]. Capillary isotachophoresis can concentrate either positively or negatively charged analytes [71]. During the cITP process, the analytes are stacked between the leading electrolyte and the trailing electrolyte according to their electrophoretic mobilities. Using cITP, concentration sensitivity enhancements of two orders of magnitude have been reported for online nanoliter volume NMR detection [62]. The concentration features of cITP-NMR have been shown to increase S/N by a factor of 2 compared to analyses without preconcentration. The overall concentration enhancement is ~1000. Furthermore, the sample observation efficiency has also been increased from 0.5% to 50% since cITP stacking enables the recording of high S/N NMR in a shorter period of time. Stopped-flow COrrelation SpectroscopY (COSY) has been reported in as little as 22 min. The application of cITP-NMR is attractive for separating and identifying trace drug metabolites and synthetic organic products. In a reported study [72], 200 μM (1.9 nmol) atenolol, a beta blocker used for treatment of cardiovascular disease, was successfully isolated in the presence of 200 mM sucrose. At the pH of analysis, atenolol was positively charged and therefore could effectively be separated from sucrose. Spectral comparisons estimated the focused band of atenolol to be about 40 mM. Hence, cITP significantly improved the concentration sensitivity of most mass-sensitive NMR microcoils by allowing microliter samples to be concentrated and measured using 10–100 μL NMR observe volumes. The study demonstrated that cITC-NMR can be used to analyze trace materials at the 0.1% level in the presence of excess uncharged species.

Overall capillary techniques are useful for analyzing mass-limited samples as required in pharmaceutical analysis, biomedical research, and environmental analysis. Sample amounts as small as picomoles have been analyzed with capillary NMR, and this technology is gaining wider acceptance in the pharmaceutical industry. A major drawback to the implementation of separation and concentration technologies such as CE, CEC, and cITP is their lack of commercial availability. Advancements that would improve, sensitivity, ease of use, maintenance, robustness, and general utility would render this technology a more viable investment.

REFERENCES

1. Spraul, M. 2006. Developments in NMR hyphenation for pharmaceutical industry. In *Modern Magnetic Resonance*, ed. G.A. Webb, 1203–1210. The Netherlands: Springer.

2. Smallcombe, S.H., Patt, S.L., and Keifer, P.A. 1995. WET solvent suppression and its applications to LC NMR and high-resolution NMR spectroscopy. *J. Magn. Reson. A.* 117:295–303.

3. Novak, P., Cindrić, M., Tepeš, P., Dragojević, S., Ilijas, M., and Mihaljević, K. 2005. Identification of impurities in acarbose by using an integrated liquid chromatography-nuclear magnetic resonance and liquid chromatography-mass spectrometry approach. *J. Sep. Sci.* 28:1442–1447.

4. Sidelmann, U.G., Gavaghan, C., Carless, H.A.J., Farrant, R.D., Lindon, J.C., Wilson, I.D., and Nicholson, J.K. 1995. Identification of the positional isomers of 2-fluorobenzoic acid 1-O-acyl glucuronide by directly coupled HPLC-NMR. *Anal. Chem.* 67:3401–3404.

5. Tseng, L.H., Braumann, U., Godejohann, M., Lee, S.S., and Albert, K. 2000. Structure identification of aporphine alkaloids by on-line coupling of HPLC-NMR with loop storage. *J. Chin. Chem. Soc.* 47:1231–1236.

6. Exarchou, V., Krucker, M., Van Beek, T.A., Vervoort, J., Gerothanassis, I.P., and Albert, K. 2005. LC-NMR coupling technology: recent advancements and applications in natural products analysis. *Magn. Reson. Chem.* 43:681–687.

7. Griffiths, L., and Horton, R. 1998. Optimization of LC-NMR. III. increased signal-to-noise ratio through column trapping. *Magn. Reson. Chem.* 36:104–109.

8. de Koning, J.A., Hogenboom, A.C., Lacker, T., Strohschein, S., Albert, K., and Brinkman, U.A.T. 1998. Online trace enrichment in hyphenated liquid chromatography-nuclear magnetic resonance spectroscopy. *J. Chromatogr. A.* 813:55–61.

9. Exarchou, V., Godejohann, M., Van Beek, T.A., Gerothanassis, I.P., and Vervoort, J. 2003. LC-UV-solid-phase extraction-NMR-MS combined with a cryogenic flow probe and its application to the identification of compounds present in Greek oregano. *Anal. Chem.* 75:6288–6294.

10. Godejohann, M., Tseng, L.H., Braumann, U., Fuchser, J., and Spraul, M. 2004. Characterization of a paracetamol metabolite using on-line LC-SPE-NMR-MS and a cryogenic NMR probe. *J. Chromatogr. A.* 1058:191–196.

11. Nyberg, N.T., Baumann, H., and Kenne, L. 2001. Application of solid-phase extraction coupled to an NMR flow-probe in the analysis of HPLC fractions. *Magn. Reson. Chem.* 39:236–240.

12. Nyberg, N.T., Baumann, H., and Kenne, L. 2003. Solid-phase extraction NMR studies of chromatographic fractions of saponins from Quillaja saponaria. *Anal. Chem.* 75:268–274.

13. Clarkson, C., Strk, D., Hansen, S.H., and Jaroszewski, J.W. 2005. Hyphenation of solid-phase extraction with liquid chromatography and nuclear magnetic resonance: application of HPLC-DAD-SPE-NMR to identification of constituents of Kanahia laniflora. *Anal. Chem.* 77:3547–3553.

14. Lambert, M., Staerk, D., Hansen, S.H., and Jaroszewski, J.W. 2005. HPLC-SPE-NMR hyphenation in natural products research: optimization of analysis of croton membranaceus extract. *Magn. Reson. Chem.* 43:771–775.

15. Lewis, R.J., Bernstein, M.A., Duncan, S.J., and Sleigh, C.J. 2005. A comparison of capillary-scale LC-NMR with alternative techniques: spectroscopic and practical considerations. *Magn. Reson. Chem.* 43:783–789.

16. Sandvoss, M., Bardsley, B., Beck, T.L., Lee-Smith, E., North, S.E., Moore, P.J., Edwards, A.J., and Smith, R.J. 2005. HPLC-SPE-NMR in pharmaceutical development: capabilities and applications. *Magn. Reson. Chem.* 43:762–770.

17. Seger, C., Godejohann, M., Tseng, L.H., Spraul, M., Girtler, A., Sturm, S., and Stuppner, H. 2005. LC-DAD-MS/SPE-NMR hyphenation. A tool for the analysis of pharmaceutically used plant extracts: identification of isobaric iridoid glycoside regioisomers from harpagophytum procumbens. *Anal. Chem.* 77:878–885.

18. Alexander, A.J., Xu, F., and Bernard, C. 2006. The design of a multi-dimensional LC-SPE-NMR system (LC(2)-SPE-NMR) for complex mixture analysis. *Magn. Reson. Chem.* 44:1–6.

19. Wilson, S.R., Maleroed, H., Petersen, D., Simic, N., Bobu, M.M., Rise, F., Lundanes, E., and Greibrokk, T. 2006. Controlling LC-SPE-NMR systems. *J. Sep. Sci.* 29:582–589.

20. Simpson, A.J., Tseng, L.H., Simpson, M.J., Spraul, M., Braumann, U., Kingery, W.L., Kelleher, B.P., and Hayes, M.H.B. 2004. The application of LC-NMR and LC-SPE-NMR to compositional studies of natural organic matter. *Analyst.* 129:1216–1222.

21. Schlotterbeck, G., and Ceccarelli, S.M. 2009. LC-SPE-NMR-MS: a total analysis system for bioanalysis. *Bioanalysis.* 1(3):549–559.

22. Keifer, P.A. 2000. NMR spectroscopy in drug discovery: tools for combinatorial chemistry, natural products, and metabolism research. In *Progress in Drug Research*, ed. E. Jucker, 137–211. 55, Basel: Birkhauser Verlag.

23. Keifer, P.A. 1997. High-resolution NMR techniques for solid-phase synthesis and combinatorial chemistry. *Drug Discov. Today.* 2:468–478.

24. Keifer, P.A. 1998. New methods for obtaining high-resolution NMR spectra of solid-phase synthesis resins, natural products, and solution-state combinatorial chemistry libraries. *Drugs of the Future.* 23:301–319.

25. Keifer, P.A. 2002. The NMR "tool kit" for compound characterization. In *Integrated Drug Discovery Technologies*, eds. H.-Y. Mei and A.W. Czarnik, 451–485. New York: Marcel Dekker Inc.

26. Keifer, P.A. 2003. Flow injection analysis NMR (FIA- NMR): A novel flow NMR technique that complements LC- NMR and direct injection NMR (DI- NMR). *Magn. Reson. Chem.* 41:509–516.

27. Lenz, E., Taylor, S., Collins, C., Wilson, I.D., Louden, D., and Handley, A. 2002. Flow injection analysis with multiple on-line spectroscopic analysis (UV, IR, 1H- NMR and MS). *J. Pharm. Biomed. Anal.* 27:191–200.

28. Bailey, N.J.C., and Marshall, I.R. 2005. Development of ultrahigh-throughput NMR spectroscopic analysis utilizing capillary flow NMR technology. *Anal. Chem.* 77:3947–3953.

29. Lachenmeier, D.W., Frank, W., Humpfer, E., Schaefer, H., Keller, S., Moertter, M., and Spraul, M. 2005. Quality control of beer using high-resolution nuclear magnetic resonance spectroscopy and multivariate analysis. *Eur. Food Res. Technol.* 220:215–221.

30. Tisne, C., and Dardel, F. 2002. Optimisation of a peptide library for screening specific RNA ligands by flow-injection NMR. *Comb. Chem. High Throughput Screen.* 5:523–529.

31. Stockman, B.J., Farley, K.A., and Angwin, D.T. 2001. Screening of compound libraries for protein binding using flow-injection nuclear magnetic resonance spectroscopy. *Methods Enzymol.* 338:230–246.

32. Keifer, P.A., Smallcombe, S.H., Williams, E.H., Salomon, K.E., Mendez, G., Belletire, J.L., and Moore, C.D. 2000. Direct-injection NMR (DI- NMR): a flow NMR technique for the analysis of combinatorial chemistry libraries. *J. Comb. Chem.* 2:151–171.

33. Kautz, R.A., Goetzinge, W.K., and Karger, B.L. 2005. High-throughput microcoil NMR of compound libraries using zero-dispersion segmented flow analysis. *J. Comb. Chem.* 7:14–20.

34. Spraul, M., Hofmann, M., Ackermann, M., Shockcor, J.P., Lindon, J.C., Nicholls, A.W., Nicholson, J.K., Damment, S.J.P., and Haselden, J.N. 1997. Flow injection proton nuclear magnetic resonance spectroscopy combined with pattern recognition methods: implications for rapid structural studies and high throughput biochemical screening. *Anal. Commun.* 34:339–341.

35. Gmeiner, W.H., Cui, W., Konerding, D.E., Keifer, P.A., Sharma, S.K., Soto, A.M., Marky, L.A., and Lown, J.W. 1999. Shape-selective recognition of a model Okazaki fragment by geometrically-constrained bis-distamycins. *J. Biomol. Struct. Dyn.* 17:507–518.

36. Hamper, B.C., Snyderman, D.M., Owen, T.J., Scates, A.M., Owsley, D.C., Kesselring, A.S., and Chott, R.C. 1999. High-throughput ^{1}H NMR and HPLC characterization of a 96-member substituted methylene malonamic acid library. *J. Comb. Chem.* 1:140–150.

37. Combs, A., and Rafalski, M. 2000. N-Arylation of sulfonamides on solid supports. *J. Comb. Chem.* 2:29–32.

38. Lewis, K., Phelps, D., and Sefler, A. 2000. Automated high-throughput quantification of combinatorial arrays. *Am. Pharm. Rev.* 3:63–68.

39. Eldridge, G.R., Vervoort, H.C., Lee, C.M., Cremin, P.A., Williams, C.T., Hart, S.M., Goering, M. G., O'Neil-Johnson, M., and Zeng, L. 2002. High-throughput method for the production and analysis of large natural product libraries for drug discovery. *Anal. Chem.* 74:3963–3971.

40. Kalelkar, S., Dow, E.R., Grimes, J., Clapham, M., and Hu, H. 2002. Automated analysis of proton NMR spectra from combinatorial rapid parallel synthesis using self-organizing maps. *J. Comb. Chem.* 4:622–629.

41. Leo, G.C., Krikava, A., and Caldwell, G.W. 2003. Application of flow NMR to an open-access pharmaceutical environment. *Anal. Chem.* 75:1954–1957.

42. Pierens, G.K., Palframan, M.E., Tranter, C.J., Carroll, A.R., and Quinn, R.J. 2005. A robust clustering approach for NMR spectra of natural product extracts. *Magn. Reson. Chem.* 43:359–365.

43. Stockman, B.J. 2000. Flow NMR spectroscopy in drug discovery. *Curr. Opin. Drug Disc. Dev.* 3:269–274.

44. Ross, A., Schlotterbeck, G., Klaus, W., and Senn, H. 2000. Automation of NMR measurements and data evaluation for systematically screening interactions of small molecules with target proteins. *J. Biomol. NMR.* 16:139–146.

45. Robertson, D.G., Reily, M.D., Sigler, R.E., Wells, D.F., Paterson, D.A., and Braden, T.K. 2000. Metabonomics: evaluation of nuclear magnetic resonance (NMR) and pattern recognition technology for rapid in vivo screening of liver and kidney toxicants. *Toxicol. Sci.* 57:326–337.

46. Potts, B.C.M., Deese, A.J., Stevens, G.J., Reily, M.D., Robertson, D.G., and Theiss, J. 2001. NMR of biofluids and pattern recognition: assessing the impact of NMR parameters on the principal component analysis of urine from rat and mouse. *J. Pharm. Biomed. Anal.* 26:463–476.

47. Rezzi, S., Axelson, D.E., Heberger, K., Reniero, F., Mariani, C., and Guillou, C. 2005. Classification of olive oils using high throughput flow ¹H NMR fingerprinting with principal component analysis, linear discriminant analysis and probabilistic neural networks. *Anal. Chim. Acta.* 552:13–24.

48. Rezzi, S., Spraul, M., Axelson, D.E., Heberger, K., Mariani, C., Reniero, F., and Guillou, C. 2005. Injection flow NMR as a tool for the high throughput screening of oils. *Special Publication – Royal Society of Chemistry.* 299:124–130.

49. Pusecker, K., Schewitz, J., Gfrorer, P., Tseng, L.H., Albert, K., and Bayer, E. 1998. Online coupling of capillary electrochromatography, capillary electrophoresis, and capillary HPLC with nuclear magnetic resonance spectroscopy. *Anal. Chem.* 70:3280–3285.

50. Pusecker, K., Schewitz, J., Gfrorer, P., Tseng, L.H., Albert, K., Bayer, E., Wilson, I.D., Bailey, N.J., Scarfe, G.B., Nicholson, J.K., and Lindon, J.C. 1998. On-flow identification of metabolites of paracetamol from human urine by using directly coupled CZE- NMR and CEC- NMR spectroscopy. *Anal. Commun.* 35:213–215.

51. Schewitz, J., Gfrorer, P., Pusecker, K., Tseng, L.H., Albert, K., Bayer, E., Wilson, I.D., Bailey, N. J., Scarfe, G.B., Nicholson, J.K., and Lindon, J.C. 1998. Directly coupled CZE-NMR and CEC-NMR spectroscopy for metabolite analysis: paracetamol metabolites in human urine. *Analyst.* 123:2835–2837.

52. Schewitz, J., Pusecker, K., Gfrorer, P., Gotz, U., Tseng, L.H., Albert, K., and Bayer, E. 1999. Direct coupling of capillary electrophoresis and nuclear magnetic resonance spectroscopy for the identification of a dinucleotide. *Chromatographia.* 50:333–337.

53. Wolters, A.M., Jayawickrama, D.A., Webb, A.G., and Sweedler, J.V. 2002. NMR detection with multiple solenoidal microcoils for continuous-flow capillary electrophoresis. *Anal. Chem.* 21:5550–5555.

54. Olson, D.L., Lacey, M.E., Webb, A.G., and Sweedler, J.V. 1999. Nanoliter-volume ¹H NMR detection using periodic stopped-flow capillary electrophoresis. *Anal. Chem.* 71:3070–3076.

55. Oda, R.P., and Landers, J.P. 1997. Introduction to capillary electrophoresis. In *Handbook of Capillary Electrophoresis*, ed. J.P. Landers, 1. 9, Boca Raton, FL: CRC Press.

56. Lacey, M.E., Webb, A.G., and Sweedler, J.V. 2000. Monitoring temperature changes in capillary electrophoresis with nanoliter-volume NMR thermometry. *Anal. Chem.* 72:4991–4998.

57. Lacey, M.E., Webb, A.G., and Sweedler, J.V. 2002. On-line temperature monitoring in a capillary electrochromatography frit using microcoil NMR. *Anal. Chem.* 74:4583–4587.

58. Black, R.D., Early, T.A., Roemer, P.B., Mueller, O.M., Mogrocampero, A., Turner, L.G., and Johnson, G.A. 1993. A high-temperature superconducting receiver for nuclear magnetic resonance microscopy. *Science.* 259:793–795.

59. Webb, A.G. 1997. Radiofrequency microcoils in magnetic resonance. *Prog. Nucl. Mag. Res. Spec.* 31:1–42.

60. Olson, D.L., Peck, T.L., Webb, A.G., Magin, R.L., and Sweedler, J.V. 1995. High-resolution microcoil ¹H-NMR for mass-limited, nanoliter-volume samples. *Science.* 270:1967–1970.

61. Wu, N.A., Peck, T.L., Webb, A.G., Magin, R.L., and Sweedler, J.V. 1994. ¹H- NMR spectroscopy on the nanoliter scale for static and online measurements. *Anal. Chem.* 66:3849–3857.

62. Kautz, R.A., Lacey, M.E., Wolters, A.M., Foret, F., Webb, A.G., Karger, B.L., and Sweedler, J.V. 2001. Sample concentration and separation for nanoliter-volume NMR spectroscopy using capillary isotachophoresis. *J. Am. Chem. Soc.* 123:3159–3160.

63. Schlotterbeck, G., Ross, A., Hochstrasser, R., Senn, H., Kuhn, T., Marek, D., and Schett, O. 2002. High-resolution capillary tube NMR. A miniaturized 5-μL high-sensitivity TXI probe for mass-limited samples, off-line LC NMR, and HT NMR. *Anal. Chem.* 74:4464–4471.

64. Gfrorer, P., Schewitz, J., Pusecker, K., and Bayer, E. 1999. On-line coupling of capillary separation techniques with 1H NMR. *Anal. Chem.* 71:315A–321A.

65. Wu, N., Webb, L., Peck, T.L., and Sweedler, J.V. 1995. Online NMR detection of amino acids and peptides in microbore LC. *Anal. Chem.* 67:3101–3107.

66. Albert, K., Schlotterbeck, G., Tseng, L.H., and Braumann, U. 1996. Application of online capillary high-performance liquid chromatography-nuclear magnetic resonance spectrometry coupling for the analysis of vitamin A derivatives. *J. Chromatogr. A.* 750:303–309.

67. Gfrorer, P., Tseng, L.H., Rapp, E., Albert, K., and Bayer, E. 2001. Influence of pressure upon coupling pressurized capillary electrochromatography with nuclear magnetic resonance spectroscopy. *Anal. Chem.* 73:3234–3238.

68. Gfrorer, P., Schewitz, J., Pusecker, K., Tseng, L.H., Albert, K., and Bayer, E. 1999. Gradient elution capillary electrochromatography and hyphenation with nuclear magnetic resonance. *Electrophoresis.* 20:3–8.

69. Wu, N., Peck, T.L., Webb, A.G., Magin, R.L., and Sweedler, J.V. 1994. Nanoliter volume sample cells for '^{1}H-NMR:application to on-line detection in capillary electrophoresis. *J. Am. Chem. Soc.* 116:7929–7930.

70. Wu, N.A., Peck, T.L., Webb, A.G., Magin, R.L., and Sweedler, J.V. 1994. ^{1}H-NMR spectroscopy on the nanoliter scale for static and online measurements. *Anal. Chem.* 66:3849–3857.

71. Wanders, B.J., and Everaerts, F.M. 1994. Isotachophoresis in capillary electrophoresis. In *Handbook of Capillary Electrophoresis*, ed. J.P. Landers, 111–128. Boca Raton, FL: CRC Press.

72. Wolters, A.M., Jayawickrama, D.A., Larive, C.K., and Sweedler, J.V. 2002. Capillary isotachophoresis/NMR: extension to trace impurity analysis and improved instrumental coupling. *Anal. Chem.* 74:2306–2313.

6 NMR Experiments

6.1 SOLVENT SUPPRESSION

Liquid chromatography-nuclear magnetic resonance (LC-NMR) spectra typically contain mobile-phase deuterated solvents that have significant amounts of proteo solvent. Because these solvent signals are large relative to the peaks of interest, the solvent signals need to be suppressed. Solvent suppression is required to address dynamic range issues that prevent observation of the resonances arising from the compounds of interest. A number of solvent suppression techniques were invented for this purpose. These solvent suppression techniques include the standard presaturation [1] as well as binomial sequences such as $1\bar{3}3\bar{1}$ [2,3] $1\bar{1}$ or $1\bar{2}1$ [4]. The binomial sequence $1\bar{3}3\bar{1}$ is a very efficient sequence for obtaining proton NMR spectra of dilute solutes in H_2O. This sequence has been shown to suppress an intense water resonance by more than three orders of magnitude. The $1\bar{3}3\bar{1}$ sequence consists of short, strong radio frequency (RF) pulses designed to be insensitive to many of the imperfections of the NMR spectrometers. This sequence enables high-quality spectra to be acquired without the need for fine adjustment of pulse lengths, phase shifts, delays, and transmitter frequency. The numbers in the $1\bar{3}3\bar{1}$ sequence give the relative pulse lengths, the overbars denote a 180° phase-shifted pulse, and equal delays exist between the pulses.

The first reports of $1\bar{3}3\bar{1}$ showed it outperformed the other binomial sequences. With the $1\bar{3}3\bar{1}$ sequence, suppression of the pulse and receiver phases is inverted on alternate scans; hence, any solvent excitation arising from slightly imperfect 180° phase shifts is canceled. The phases of all four pulses and the receiver are then incremented together in 90° steps. In executing this solvent suppression routine, the transmitter is set at the exact resonance frequency of the solvent and chosen to excite the desired spectral region. The reason that $1\bar{3}3\bar{1}$ was more successful relative to the other binomial sequences was that it is less sensitive to the effects of nonideally shaped pulses. Thus, if the relative pulse angles in $1\bar{3}3\bar{1}$ are not exactly 3:1, the symmetry of the sequence ensures that there is still a null, whereas errors in the other binomial sequences would result in several fold less solvent suppression.

In the mid-1990s, the WET (water suppression enhanced through T_1 effects) solvent-suppression method was created by Smallcombe et al. [5]. This pulse sequence uses shaped, selective pulses, and pulsed magnetic field gradients to suppress one or more solvent signals. This method yields very excellent signal suppression and selectivity. It had been found to be highly effective when using solvents such as 90% H_2O/10% D_2O or solvents with multiple resonances. A modified version of WET (WETDC) that is commonly used employs low-power decoupling of ^{13}C satellites during WET and acquisition. WETDC is shown in Figure 6.1. The WET signal-suppression technique was rapid enough to work for both continuous-flow and stopped-flow samples. This experiment was amenable for use with samples obtained from reversed-phase LC-NMR studies and can easily be adjusted to suppress multiple solvent signals in the NMR spectrum.

The invention of robust solvent suppression methods is of critical importance for LC-NMR structure-elucidation studies. This is because multiple proteo solvents (typically acetonitrile and water) can broaden resonance peaks and complicate spectral displays, especially when using steep solvent gradients during the course of the chromatographic run. Such conditions will compromise the observation of small analyte signals relative to those from the large solvent peaks. The WET pulse sequence, however, can suppress these multiple solvent lines with only a single RF channel by using shaped pulses [6]. In addition, WET suppresses the ^{13}C satellites,

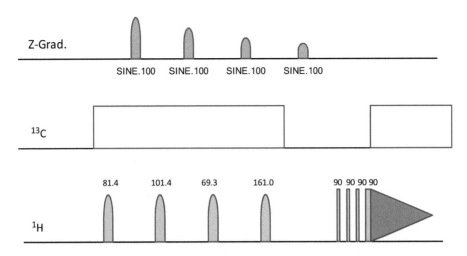

FIGURE 6.1 The NMR pulse sequence for WET solvent suppression modified with low power ^{13}C decoupling during WET and acquisition. The tip angles of each of the ^{1}H pulses are shown above each shaped pulse. The ^{13}C decoupling pulses remove the ^{13}C-satellites of the suppressed signals. The composite read pulse is shown as four 90° pulses having phases of X, Y, $-X$, and $-Y$, respectively.

and allows one to view any unsuppressed resonances that reside under these ^{13}C satellites. WET has proven itself to be a robust and forgiving technique that requires no tedious adjustments or extensive optimization. This technique has been found to be considerably more frequency selective than other solvent suppression techniques.

6.2 STRUCTURE ELUCIDATION EXPERIMENTS

For NMR structural characterization, two-dimensional (2D) correlation experiments are of special value because they connect signals through chemical bonds. Examples of these correlation experiments are COrrelation SpectroscopY (COSY) [7] and TOtal Correlation Spectroscopy (TOCSY) [8]. COSY is a 2D NMR technique that gives correlations between J-coupled signals by incrementing the delay between two 90° proton pulses. The resulting 2D spectrum is generally displayed as a contour plot, which may be viewed as a topographical map. The COSY contour map provides a cross-section (slice) of a 3D image of an NMR spectrum. When displaying a COSY spectrum, the one-dimensional (1D) proton spectrum is traced on the diagonal of the plot. Any peaks that are not on the diagonal represent cross peaks or correlation peaks that are primarily a result of ^3J-coupling. Thus, by simply tracing a rectangle using the diagonal and cross peaks as vertices it is possible to determine which protons are coupled to each other. Standard COSY experiments require phase cycling to remove unwanted signals and thus can be quite time consuming. This can be largely circumvented using gradient-selected COSY (gCOSY), which utilizes pulsed field gradients to destroy unwanted z-magnetization and hence their associated signals (axial peaks). Quality gCOSY spectra can be acquired with one scan in as little as 5 min. A sample gCOSY spectrum is shown in Figure 6.2.

^{1}H–^{1}H TOCSY experiment (also known as HOHAHA – HOmonuclear HArtmann HAhn) enables the grouping of the proton signals based on their coupling networks. This experiment is useful in establishing connectivities in complex overlapping multiplets (where different protons have very similar chemical shifts) or where there is extensive second-order coupling. The TOCSY spectrum yields through bond correlations via spin-spin coupling

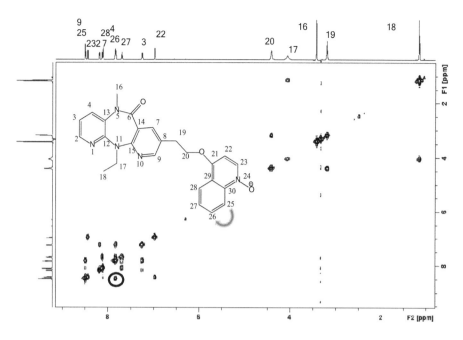

FIGURE 6.2 A gCOSY spectrum obtained on a 600-MHz NMR spectrometer equipped with a 1.7-mm micro-capillary cryo probe. The sample contained 10 μg of 5,11-dihydro-11-ethyl-5 methyl-8-(2-(1-oxido-4-quinolinyl)ethyl-6H-dipyrido(3,2-B:2′,3′-E)(1,4)diazepin-6-one in 30 μl DMSO-d_6. Spectra were acquired with 2K data points, 160 increments, and ns = 4. Total experiment time was 23 min.

correlations; however, unlike the COSY experiment, the couplings are observed throughout the coupling network. In addition, the intensity of the TOCSY cross-peak is not related to the number of bonds connecting the protons. For example, a five-bond correlation may or may not be stronger than a three-bond correlation. The TOCSY experiment is a 2D technique that employs a spin lock during the mixing time of the sequence, for which chemical shifts are invariant but *J*-couplings evolve. The technique is used to correlate spins in the same *J*-coupled spin system. A 1D selective version of the TOCSY is also available [9]. This version uses shaped pulses to select a single spin producing 1D spectrum, where all resonances in the same spin system as the selected spin are displayed. The intensity of the coupled resonances depends on the duration of the mixing time as well as on the magnitude of the coupling between spins. Mixing times are commonly adjusted from 10 to 100 ms. Short mixing times will produce correlations to large *J*-coupled spins, that is spins in the immediate vicinity and produce a COSY-like spectrum. As the mixing time is increased, correlations to smaller *J*-coupled spins will be observed. The TOCSY experiment proves especially useful in the analysis of crowded spectra where correlations from a single-resolved proton may be used to trace the coupling network. This experiment is often used in the analysis of peptides, proteins, and oligosaccharides where molecules are composed of discrete subunits or spin systems (i.e., amino acids or saccharide units). While this experiment is very useful in determining long-range spin connectivity in complex structures, to determine the order in which bonds are connected, a COSY spectrum is preferable.

The gradient-enhanced gTOCSY pulse sequence, where the spin-lock is a composite pulse between 20 and 200 ms, with a pulse power sufficient to cover the spectral width is a commonly used version of the experiment. Although long spin-lock (mixing time) holds the magnetic vector in the *x–y* plane, allowing correlations over large coupling networks, the length of the spin-lock needs to be carefully adjusted. If the spin-lock is too long, the sample will be heated, causing signal distortion in the spectrum. In addition, it can damage the electronics of the spectrometer. In practice, the attenuation should be set so that the 90° pulse width will be less than 1/(4 SW) (SW is the spectral width in Hz).

Other 2D experiments such as Nuclear Overhauser Effect SpectroscopY (NOESY) [10] and Rotating frame Overhauser Effect SpectroscopY (ROESY) [11] provide information on protons that are connected through space to establish molecular conformations. A more detailed description of the NOESY and ROESY experiments and their application and interpretation in structure elucidation are given in Section 6.3. These experiments play a critical role in identifying the chemical structure.

6.2.1 HETERONUCLEAR CORRELATION EXPERIMENTS (HSQC, HMQC, AND HMBC)

Spectra such as HMQC (heteronuclear multiple-quantum correlation spectroscopy), HSQC (heteronuclear single-quantum correlation spectroscopy), and HMBC (heteronuclear multiple-bond correlation spectroscopy) are indirect/inverse NMR experiments that correlate ^{1}H with an X nucleus, where ^{1}H is the observe nucleus [12,13]. For small molecules, homonuclear 2D techniques such as COSY and NOESY are often sufficient for structure identification. However, in cases when there is extensive overlap in the 2D homonuclear spectrum, the heteronuclear correlation experiments can be used to provide definitive assignment.

HMQC is an experiment that allows one to associate directly bonded XH resonances, where X is a nucleus other than proton. It can employ X-nucleus decoupling and hence gives rise to RF heating. The experiment requires reasonably well-calibrated pulses and many steady-state scans. Protons directly bonded to X nuclei (e.g., ^{13}C or ^{15}N, etc.) will produce cross-peaks.

HSQC provides the same information as HMQC. The difference between the two techniques is that during the evolution time of an HMQC pulse sequence, both the ^{1}H and X magnetization are allowed to evolve, whereas in an HSQC, only X magnetization is allowed to evolve. This means that an HMQC experiment is affected by homonuclear proton J-coupling during the evolution period while an HSQC is not, since there is no proton magnetization during the evolution time. The homonuclear proton J-coupling manifests itself in the HMQC spectrum as broader resonances for ^{1}H–^{13}C correlations relative to the HSQC. Hence spectral resolution is better in the HSQC compared to the HMQC. A disadvantage of the HSQC experiment is that the resolution of X nuclei such as ^{13}C is highly dependent upon well-calibrated pulses. Since there are many more pulses in an HSQC compared to an HMQC, the HSQC is more susceptible to losses in signal-to-noise ratio (S/N) due to poor probe tuning or poor pulse calibration.

Experiments such as HSQC and HMQC allow facile assignment of carbon resonances without the need for a 1D ^{13}C spectrum. Hence, once proton resonances are determined, their associated carbon resonances may be easily assigned. This becomes a critical experiment when only tens of micrograms of compound are available preventing timely acquisition of direct observe carbon spectra. These HSQC experiments aid in the assignment of aliphatic protons that suffer from severe overlap and cannot be unambiguously identified in homonuclear 2D experiments. A primary example is in the case of nonequivalent methylene protons that may be readily assigned in the HSQC spectrum through their correlation to the same carbon atom.

The advent of pulsed field gradients has had a major impact on the speed of 2D spectral acquisition. Incorporation of gradients in 2D NMR pulse sequences such as HSQC has enabled the selection of the desired coherence pathway and suppressed the ^{12}C–H signal by gradient scrambling of the coherence. This produces excellent suppression of ^{12}C–H artifacts in a single scan. In heteronuclear 2D carbon-based experiments, ^{13}C–H is the observed signal and the receiver gain can be dramatically increased, thereby increasing selectivity and signal intensity. Because there is no need for phase cycling to cancel proton signals from ^{12}C–H, a 2D HSQC may be acquired with only one scan per FID (free induction decay). The result is very short experiment times, provided that there is an adequate concentration of sample. When using a conventional 5-mm probe, the amount of compound required is ≥10 mg in 600-µL solvent for small organic molecules. When using a microcapillary cryoprobe, the amount of material decreases 1000-fold with 20-fold less solvent. Improvement in spectral quality, particularly for more dilute samples, can be accomplished by increasing the number of scans by multiples of 2. An example of a gHSQC experiment is shown in Figure 6.3.

HMBC is a 2D nuclear magnetic resonance experiment that is one of the most powerful contemporary tools for the elucidation of molecular structure, function, and dynamics. HMBC relies on long-range internuclear correlation, which provides information about carbons bonded to protons separated by two to four bonds. In HMBC studies, magnetization transfer occurs between two different types of nuclei. Like HSQC or HMQC, these nuclei may include ^{13}C and ^{1}H, ^{15}N and ^{1}H, and so on. Full ^{13}C assignments may be realized using these experiments even for samples where 1D ^{13}C spectra cannot be obtained due to poor S/N.

In the HMBC pulse sequence, the first 90° pulse on carbon-13 (C-13) serves as a low-pass filter that suppresses one-bond correlation and passes the smaller coupling. This pulse creates multiple-quantum coherence for the one-bond coupling, which is removed from the spectra by alternating the phase of the C-13 pulse. The second 90° pulse on C-13 creates multiple-quantum coherence for the long-range couplings. After the evolution time t_1, the magnetization is converted back into detectable single-quantum proton magnetization. Since carbon decoupling is not used in this sequence, the protons will display heteronuclear couplings.

The HMBC is relatively insensitive as compared to HMQC, because multiple-bond correlations are less efficient than one-bond correlations. Typical one-bond coupling constants are around 145 Hz, whereas multiple-bond coupling constants fall in the range of 2–15 Hz.

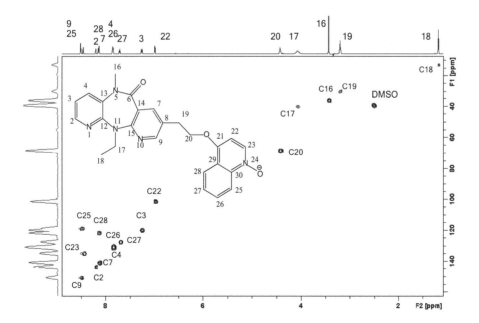

FIGURE 6.3 A gHSQC spectrum obtained on a 600-MHz NMR spectrometer equipped with a 1.7-mm microcapillary cryoprobe. The sample contained 10 µg of 5,11-dihydro-11-ethyl-5 methyl-8-(2-(1-oxido-4-quinolinyl)ethyl-6H-dipyrido(3,2-B:2′,3′-E)(1,4)diazepin-6-one in 30-µL DMSO-d$_6$. Spectra were acquired with 1K data points, 128 increments, and ns = 64. The total experiment time was 5 hours.

For the HMBC pulse sequence, experimental parameters can be adjusted to detect relatively large coupling constants (7–15 Hz) or smaller couplings (2–7 Hz).

The ^{1}H,^{13}C-HMBC technique can detect quaternary carbons coupled to protons and like HSQC and HMQC is especially useful when direct observation of C-13 spectra is not possible to obtain in a timely manner due to low sample concentration. This very useful sequence provides information about the backbone skeletal connectivities of a molecule. Many 1D proton spectra can be complex when signals overlap heavily. Using a heteronuclear second dimension helps to separate and simplify the assignment process. It has primary application in carbohydrate chemistry as a sequence analysis tool that provides unique information concerning connectivities across glycosidic linkages. Another critical area for using HMBC is in peptide-protein structural studies especially when applied to a ^{15}N-labeled protein. It is possible with this technique to get connectivities between the nitrogen and the CHα proton of the amino acid of the next residue, although as protein size increases, three-dimensional NMR experiments become necessary.

6.2.2 WET SOLVENT SUPPRESSION AND 2D NMR

By developing the WET solvent suppression technique and applying it to 2D homonuclear NMR experiments, the quality of NMR spectra obtained from continuous-flow or stop-flow experiments has vastly improved. Consequently, the characterization of chemical structures by LC-NMR has been greatly enhanced. The WET solvent-suppression technique has been effectively applied to obtain high-quality 1D spectra in a continuous-flow and stop-flow mode. The power of applying the WET pulse sequence in LC-NMR is illustrated in Figure 6.4.

Clearly in the example shown, the use of this very efficient solvent suppression technique resulted in the ability to observe peaks from the compound of interest, although the experiment required approximately 1 hour of data acquisition time using a 30-μL room temperature flow probe. This time can be reduced 30-fold by employing a microcapillary cryoprobe.

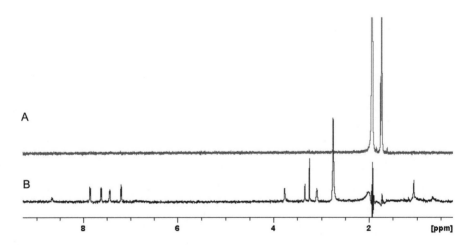

FIGURE 6.4 The spectra were obtained using a sample eluted from a C18 SPE cartridge with acetonitrile-d$_3$. (a) 1D ^{1}H spectrum obtained on a 600-MHz NMR spectrometer equipped with a 30-μL room temperature flow probe. Data were acquired with 16 scans. (b) 1D ^{1}H WETDC spectrum obtained on a 600-MHz NMR spectrometer equipped with a 30-μL room temperature flow probe. A total of 1024 scans was obtained, yielding 1 hour of experiment time.

The WET solvent suppression technique has been incorporated into 2D spectra such as WET-TOCSY, WET-COSY, WET-NOESY, and others [5]. This incorporation has enabled quality datasets to be collected and full structure characterization to be obtained on samples containing large amounts of proteo solvents relative to the amount of analyte present.

The WET pulse sequence has also been successfully incorporated into the gHSQC experiment enabling rapid acquisition of quality data. However, it should be noted that indirect detection pulse sequences such as gHMQC and gHMBC do not require solvent suppression. These experiments may be used even with samples having low analyte mass in deuterated organic solvent without the need to suppress residual proteo solvents such as water. While such heteronuclear experiments are less sensitive relative to gHSQC, when examining low mass (μg) quantities of compound, the required sensitivity can be realized by combining highly concentrated capillary samples with cryogenically cooled probes or microcapillary flow probes for NMR detection.

Overall, the WET solvent suppression pulse sequence has been shown to be highly effective in the acquisition of quality NMR data for samples dissolved in nondeuterated solvents and solvent mixtures [14,15]. Because of its robustness and versatility, the WET sequence has become a preferred solvent suppression method for LC-NMR.

6.2.3   DECOUPLING AND POWER LEVELS IN 2D NMR

In heteronuclear 2D NMR experiments, it is necessary to obtain "broadband" decoupling of protons over the range of resonances from 0 to 10 ppm. Although broadband decoupling is one of the most heavily used techniques in NMR spectroscopy, power requirements for removing spin-spin couplings at high fields are significantly increased, since the frequency bandwidth of the decoupling region is also increasing. Power dissipation throughout the NMR sample volume causes noticeable overheating and temperature gradients, leading to significant losses of the spectra quality. This problem is of particular concern in the execution of inverse spectroscopic methods when ^{1}H[X] decoupling must be applied to the low-γ nuclei. In such cases, the decoupling field strength (defined as γB_2) makes spectral acquisition problematic, since in such cases it is not possible to achieve high field strength without sample heating.

To address this issue, very efficient methods using the lowest possible power have been developed employing repeated pulses of different phase and duration at a single frequency. The result is a method known as composite pulse decoupling. A composite pulse is a sandwich of pulses that consist of $90°_x$, $-180°_{-x}$, $-270°_x$. Such pulse sandwiches may be written as 1**2**3 with 2 in bold and underlined to indicate a phase of $-x$. This pulse sandwich produces a very efficient overall 180° pulse over a wide chemical shift range. The rapid repetition of these 180° pulse sandwiches results in spin inversion ($\alpha \rightarrow \beta$, $\beta \rightarrow \alpha$) over and over again that rapidly averages the C–H J-coupling to zero. Pulse-calibration errors are ameliorated through inversion of the pulse phase from x to $-x$ at regular intervals in the sequence.

The WALTZ-16 pulse sequence is a commonly used decoupling sequence often applied in inverse heteronuclear experiments such as HMQC and HMBC [16]. This pulse sequence was a very important advance in the history of decoupling. The WALTZ sequence may be described beginning with the sequence 1**2**31**2**31**2**3. By rearranging the sequence to **2**31**2**3 1**2**31**2**3**1** and combining 31 = 4 **3**1 = **4** and adding a reversed phase, then one obtains **2**4**2**31 **2**4**2**31 **2**4**2**31 **2**4**2**31. Moving the ending 1 back to the beginning and writing 12 = 3 **1****2** = **3**, one obtains 3**4****2**31**2**4**2**3 3**4****2**31**2**4**2**3, which when repeated with opposite phase gives a pulse/phase cycle known as "WALTZ-16" because of the 123 basic building block of the sequence. The total 36 pulse block of WALTZ-16 is repeated over the entire acquisition time. The parameters that need to be set in a WALTZ-16 decoupling sequence are the RF amplitude (power) and the duration of the pulse at the designated power level. The robustness of the WALTZ-16 pulse sequence is due to the fact that it is insensitive to small errors in the 90° phase shift that are required for the implementation of the composite pulse. When accurate 90° phase shift in the decoupler channel was found to be problematic, the solution was to use a pulse train with 180° phase shifts as a means of improving the performance of the whole sequence. Owing to relative insensitivity to phase errors produced in the decoupler channel and very small residual couplings, WALTZ-16 has become a standard for inverse experiments.

GARP (Globally optimized Alternating-phase Rectangular Pulses) is another decoupling method that also makes use of composite pulses. This approach uses fractional multipliers to describe a composite pulse more precisely, hence obtaining improved inversion properties. The GARP pulse invokes a four-step RR-$\overline{RR}$ supercycle to achieve a more uniform decoupling profile.

where

$$R = 30.5\ \overline{55.2}\ 257.8\ \overline{268.3}\ 69.3\ \overline{62.2}\ 85.0\ \overline{91.8}\ 134.5\ \overline{256.1}\ 66.4\ \overline{45.9}\ 25.5$$
$$\overline{72.7}\ 119.5\ \overline{138.2}\ 258.4\ \overline{64.9}\ 70.9\ \overline{77.2}\ 98.2\ \overline{133.6}\ 255.9\ \overline{65.6}\ 53.4$$

Here, barred symbols denote phase inversion [17].

Garp has much wider decoupling bandwidths than WALTZ-16, but will introduce significantly larger residual splitting for ^{13}C spectra. Cycling sidebands can compromise spectral quality; however, these cycling sidebands can be attenuated by an order of magnitude when a progressively permuted GARP sequence is used.

Overall, WALTZ-16 is superior to GARP for cases where optimal spectral resolving power is required. However, in cases where decoupling of a wide bandwidth range is required, GARP decoupling has the advantage over WALTZ-16. Hence GARP is the method of choice for decoupling of heteronuclei having a wide chemical shift range (such as ^{13}C, ^{19}F, ^{15}N, and so on) as well as for the highest magnetic field intensities for protons. Because GARP can cover a wider bandwidth range of X nuclei, while minimizing sample heating and related effects, it is also preferred for highly temperature-sensitive samples.

Decoupling power levels are described by the expression ($\gamma_H B_2/2\pi$), which is about one-tenth of the amplitude (2500 Hz) or 1% of the power used for a single-pulse excitation of proton resonances, (25,000 Hz) for a 10-μs 90° proton pulse. The decoupler field in units of Hz may be calculated from the B_1 amplitude (db) and the 90° pulsewidth ratio (e.g., the hard pulse length divided by the decoupler pulse length) using the equation,

$$\text{Decoupler power(Hz)} = 1/[4 \times \text{decoupler pulse width}]. \tag{6.1}$$

Hence a decoupler pulse of 100 μs would be equal to 2500 Hz decoupling power.

The decoupler field strength expressed in units of Hz is proportional to the B_1 amplitude. Hence the relation

$$\begin{aligned} dB &= 10\,\text{Log[amplitude ratio]}^2 \\ &= 10[\text{Log90°pulsewidth ratio}]^2 \\ &= 10\,\text{Log[power ratio]}. \end{aligned} \tag{6.2}$$

Since power is the square of the amplitude, the power level required for decoupling is 100 times less than that of hard pulses. So if the power for a hard pulse is 60 W, the decoupling power would be 0.6 W.

The NMR instrument manufacturers (Bruker or Agilent) express power output in their software differently. In adjusting power and pulses, a useful relationship is as follows: to shorten the 90° pulse by a factor of 2 requires a −6 dB (Bruker) or +6 dB (Agilent) change in pulse power.

The relationship between power in watts versus decibels is logarithmic, such that a factor of 2 increase in watts corresponds to a −3 dB (Bruker) or +3 dB (Agilent) change in power attenuation. Spectrometer parameters often include how much RF power (in watts) is being used in the NMR experiments. This is important in transitioning between the two major instrument manufacturers (Bruker and Agilent). If one uses a power level that is too high, the NMR probe can be damaged. Although both Agilent and Bruker express the output power in decibels, on a Bruker spectrometer the maximum output power is −6 dB and the minimum power is 120 dB. On Agilent spectrometers, the maximum power is 63 dB and the minimum power is 0 dB. Because of amplifier nonlinearity, maximum power may be achieved at a lower dB settings on Bruker instruments, and the nonlinearity can be calibrated using a correction table.

If an amplifier with a maximum output power of 100 W is used and the effects of amplifier nonlinearity are neglected, the output power conversion from dB to W may be calculated using the following equations:

$$P_{\text{Agilent}} = P_{\text{max}}10^{(\text{dB}-63)/10} \tag{6.3}$$

$$P_{\text{Bruker}} = P_{\text{max}}10^{-(\text{dB}+6)/10} \tag{6.4}$$

where P_{max} is the maximum output power, hence

100 W/63 dB (Agilent) and −6 dB (Bruker)
1 W/43 dB (Agilent) and 14 dB (Bruker)
1 mW/13 dB (Agilent) and 44 dB (Bruker)

However, it is important to note that the exact output power (P_{max}) on any particular instrument must always be measured and calibrated. Although an amplifier may be rated at 100 W, this does not mean that it puts out 100 W. The output must be measured after the preamplifier and calibrated to obtain the true power that goes into the probe.

6.3 NOE EXPERIMENTS

NOESY is one of the most useful techniques in structure elucidation, as it allows one to correlate nuclei through space (distance smaller than 5 Å). Nuclear Overhauser effects (NOEs) provide spatial proximity information that allows assignment of noncoupled but spatially adjacent protons. By measuring cross-peak intensities, distance information can also be extracted.

The pulse sequence starts with a 90° pulse followed by an evolution time t_1. This delay is varied systematically to provide chemical shift information in the F1 domain. Then a 90° pulse transmits the magnetization to the z-axis, and during the following mixing period, the nonequilibria z-component will exchange magnetization through relaxation (a dipole-dipole mechanism). This exchange of magnetization is known as NOE. After some time (shorter than the relaxation time T_1), the transverse magnetization is restored and detected. If relaxation exchange (or chemical exchange) has taken place during the mixing time, cross-peaks will also be observed in the 2D spectra.

NOEs are produced by dipolar cross-relaxation between nuclei in a close spatial relationship. Because of the r^{-6} dependence of NOEs on interproton distance, cross peaks appear only if protons are less than 5 Å apart. The NOE interaction does not depend on through-bond interactions like J-coupling does, so NOEs can be seen between spatially adjacent protons on different parts of a molecule. This makes NOEs extremely useful in determining the conformation of a molecule. A limitation of NOESY spectra is that the intensities of the cross peaks depend on the relationship between molecular rotational correlation time (τ_c) and the spectrometer proton frequency (w_0). Therefore, cross peaks in NOESY spectra may change in intensity, depending upon different magnetic-field strengths and the viscosity of different solvents. It should also be noted that it is sometimes not possible to observe cross peaks in a NOESY spectrum, even when the protons are less than 5 Å apart. This usually occurs for molecules with molecular weights of ~1000 Da, where $w_0\tau_c \sim 1$. For small molecules, where $w_0\tau_c \ll 1$, cross peaks are observed in a NOESY spectrum that are opposite in sign to the diagonal peaks. For large molecules of several thousand molecular weight, where $w_0\tau_c \gg 1$, cross peaks observed in a NOESY spectrum have the same sign as diagonal peaks. For compounds that fall in the intermediate molecular weight category (i.e., NOESY cross peaks are not observable), it becomes necessary to run a ROESY experiment to observe NOEs (or ROEs).

Selection of the mixing time in the NOESY experiment is a vital parameter because NOEs require time for "buildup." The mixing time is dependent on the molecular rotational correlation time (τ_c), which is associated with molecular weight of the molecule. Mixing times comparable to 300 to 500 ms provide maximum NOE enhancements, although such measurements may not be quantitative. Also, if the 90° pulse width is incorrect or the mixing time is too short, "COSY-type" (antiphase) cross peaks may appear in the spectrum. True NOE peaks are pure absorptive signals. It should also be noted that appropriate phase cycling with the requisite number of scans is required to ensure proper detection of NOESY cross-peaks.

In small/medium size molecules, the mixing time can be selected to be a fraction of the relaxation time. For larger molecules, shorter mixing times should be used to avoid "spin-diffusion" problems.

Spin diffusion is a T_1 phenomenon, frequently observed in NMR of macromolecules. In the limit of $\omega\tau_c \gg 1$, the rate of transfer of spin energy between nuclei becomes much larger than the rate of transfer of energy to the lattice. This produces longitudinal relaxation rates of protons within the molecule, which all tend to possess the same value. Since the NOE is dependent upon T_1, in the presence of spin diffusion, this parameter will no longer be specific for proximal nuclei. In the extreme case, a homogeneous negative NOE will be observed throughout the spectrum. For experimental applications, spin diffusion may be observed in the generation of relayed NOEs in either 1D NOE difference or 2D NOESY spectroscopy. Hence when spin diffusion occurs, there is no simple relationship between the magnitude of the NOE and the internuclear distance. Spin diffusion is most frequently observed for large macromolecules such as proteins, and these issues may be addressed using short mixing times (typically 30–60 ms) (see Section 6.3.4).

In the NOESY experiment, the distinction between cross-peaks originating from the NOE effect and those originating from chemical or conformational exchange is not readily determined. However, in small molecules, having long correlation times, the phase of the peak can be used to provide such a distinction. The relationship between the sign of the phase and the magnetization-transfer phenomenon may be described as follows:

When the diagonal signal is phased "up" (small molecules),

- The NOE cross-peak is phased "down"
- The chemical exchange cross-peak is phased "up"
- The conformational exchange cross-peak is phased "up"

For large molecules (i.e., molecular weight >1000 Da), having long correlation times, the phase of the diagonal, the NOE cross-peak, and the exchange cross-peak is the same in the NOESY experiment. It is therefore impossible to distinguish NOE from chemical or conformational exchange. For such cases, the ROESY (NOE in the rotating frame) pulse sequence should be used.

^{1}H, ^{1}H ROESY (Rotating-frame Overhauser Effect SpectroscopY) is an experiment that measures homonuclear NOEs under spin-locked conditions. ROESY is applied to studies of molecules with motional correlation times (τ_c), such that $\omega\tau_c \sim 1$, where ω is the angular frequency $\omega = \gamma B$. In the ROESY experiment, the rotating-frame NOE (i.e., ROE) remains positive and increases as the molecular correlation time increases. Setting the mixing time or spin-lock period is important in the ROESY experiment, since the mixing time is when spin exchange occurs among spin-locked magnetization components of different nuclei. Different spectral density functions apply to ROESY interactions than for the NOESY transfers, hence the ROEs are always positive.

The ROESY experiment is very similar to the NOESY, in that it provides information concerning distance between nuclei. This technique is based on NOE in the rotating frame, and the ROESY pulse sequence is almost identical to the TOCSY, which leads to complications caused by the introduction of TOCSY artifacts. TOCSY artifacts occur between spins that are *J*-coupled and that are relatively close in chemical shift. They also occur for peaks that are symmetric about the central spectral frequency. The TOCSY signals are very large compared to the ROE cross-peaks; therefore, it is essential to take as many precautions as necessary to cancel those signals. To avoid TOCSY artifacts, the power used to achieve spin-lock can be reduced. Also to further reduce the chances of Hartmann-Hahn match, the spin-lock offset can be shifted to one end of the spectra (instead of the center). In practice, TOCSY cross-peaks are the major artifact in ROESY spectra; hence, to check for such TOCSY artifacts with small molecules (<1000 Da), it would be advisable to run a NOESY spectrum.

An advantage over the NOE, which can be positive (for small molecule), negative (for large molecule), or null (if the correlation time leads to a suppressed NOE), resides in the fact that the ROE (NOE in the rotating frame) is always positive; hence, the phase of the cross peak in a ROESY spectrum is diagnostic of the type of magnetization transfer that is taking place regardless of the compound's molecular weight. Alternation in sign of the ROE effect allows one to distinguish TOCSY or exchange peaks from true small ROE resonances.

The peak phase behavior from the ROESY experiment may be summarized as follows:
If the diagonal peaks are phased "up" or positive

- ROE cross-peaks will be phased "down" or negative
- TOCSY peaks will be phased "up" or positive
- Chemical exchange peaks will be phased "up" or positive
- Conformational exchange peaks will be phased "up" or positive

A diagram of the relationship between phase and molecular weight for ROESY and NOESY is given in Figure 6.5.

Another possible complication in the ROESY spectrum is the relay of ROE through TOCSY resulting in false cross peaks. This may be observed with geminal methylene peaks that give rise to TOCSY cross peaks. For example, if there is a proton in close proximity to only one of the two geminal protons, ideally there should be only one ROE cross peak from the third proton to only one of the geminal protons but not its partner. The TOCSY artifact occurs through the transfer from the ROE proton to its geminal partner, making it appear as if the third proton has an ROE to both geminal protons. Once again it is advisable to confirm such observations with a NOESY spectrum.

6.3.1 SAMPLE CONSIDERATIONS

When conducting an NOE study, the preparation of the sample can be critical for observing through space magnetization transfer. In particular, dissolved oxygen or other paramagnetic species such as Cu^{2+} can reduce or completely quench the NOE. For small molecules, it can

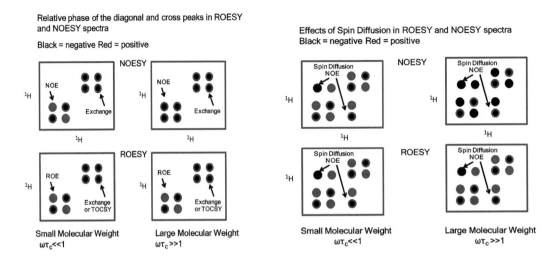

FIGURE 6.5 Diagram of NOE and ROE cross peaks indicating their phase relative to the phase of cross peaks arising from spin diffusion or exchange. For small molecules (molecular weight <1000 Da), the NOESY and ROESY cross peaks will always have a phase opposite to that of the diagonal.

be extremely important to remove dissolved oxygen. For large molecules, the removal of oxygen is not critical. A common means of removing oxygen from an NMR sample is to employ the freeze-pump-thaw method. Bubbling argon or nitrogen through the sample may not be sufficient. The freeze-pump-thaw procedure is described as follows:

1) Freeze the sample in liquid nitrogen or CO_2/acetone.
2) Evacuate the space above the solution.
3) Turn off vacuum, but keep sample isolated and allow to thaw. As it thaws, bubbling should be noticed.
4) Repeat several times (three to four times).
5) Backfill with nitrogen gas.

When finished, the sample should be sealed in some manner. Tubes (size = 5 mm) with attached stopcocks have been designed for such purposes.

Of course, when using a flow probe, such freeze-thaw procedures are not possible. With LC-NMR, however, degassed solvents are used to elute compounds from the column into the flow probe. This helps to alleviate any paramagnetic-induced relaxation issues that may arise from dissolved oxygen. Transfer of samples stored on a solid-phase extraction cartridge or a loop collection unit may also be purged of dissolved oxygen by ensuring that the deuterated solvent reservoir used to transfer the sample to the flow probe is maintained under a blanket of dry nitrogen gas.

Overall, when conducting 2D ROESY or NOESY experiments, it is highly recommended that certain procedures are followed to ensure the highest level of accuracy.

1. The temperature should be actively controlled. This is easily achieved with modern NMR spectrometers.
2. One should not spin the sample to avoid spinning side-band artifacts. (Clearly this is not an issue with flow cells or capillary tubes.)
3. Ensure that the probe is optimally tuned/matched and shimmed and that the receiver gain is optimally adjusted. High receiver gain will produce parallel diagonals in 2D

spectra. This may require some manual adjustment. The automatic receiver gain optimization does not work very well for 2D experiments, as it uses only the first increment to test the receiver gain. Often the signals for subsequent increments are larger and can saturate the receiver.

4. To ensure the appropriate phase cycling for NOESY and ROESY, the number of scans should be set as a multiple of 8.

5. Although transform of a 2D spectrum prior to its completion is possible, transform upon completion is required to obtain the full digital resolution.

6.3.2 Processing (NOESY and ROESY) 2D Spectra

Processing is a critical part of analysis for phase-sensitive 2D experiments. The Fourier transformation of 2D NOESY and ROESY require apodization in both dimensions. Apodization is the process of multiplying the FID prior to Fourier transformation by a mathematical function. The type of mathematical or "window" function that is used will be determined by the type of enhancement that is required. For example, if S/N needs to be improved, an exponential window (EM) function may be applied. EM is commonly applied to 1D NMR spectra with line broadening between 0.05 and 0.5 Hz. For 2D spectra, the two most common window functions are the sine bell, which gives resolution enhancement at the expense of sensitivity, and the cosine bell (or 90° shifted sine bell), which gives sensitivity enhancement with some line broadening. Both functions come to zero smoothly at the end of the acquired FID data, eliminating any truncation wiggles in the spectrum. The window function determines the shape of the resultant FID and the shifted sine bell (ssb) determines the shift. For NOESY and ROESY data, the window function commonly selected is a sine-squared function with ssb set to 2. The second parameter, ssb, describes the amount by which the standard sine bell (0° to 180°) is shifted at its starting point. If ssb is set to 0, the sine bell starts at the 0° point of the sine function. If ssb is set to 2, the window function starts at 90° and ends at 180°. If ssb is set to 4, the window function starts at 45° of the sine function and ends at 180°. In all cases, the function comes to zero smoothly, ending at the 180° point of the sine function.

Since both the NOESY and ROESY spectra are phase sensitive, they must be phased prior to analysis. As noted previously for ROESY and small-molecule NOESY, the NOE cross peaks have phase opposite to the main diagonal peaks. Thus, if the main peaks are phased down, then NOE cross peaks will be up. For large-molecule NOESY, NOE cross peaks have the same phase as the diagonal. Chemical exchange peaks always have the same phase as the diagonal.

In phasing a 2D spectrum, the approach is to observe and phase individual traces through the 2D spectrum. These traces consist of rows (horizontal direction, f2 dimension) and columns (vertical direction, f1 dimension). It is recommended that the rows be phased first, followed by the columns. To phase the rows, select a row containing a proton signal (preferably one that contains cross peaks). Then select another row at the opposite end of the spectrum. The strategy is to select rows such that the peaks cover a large chemical-shift range. Additional rows may be added, but usually two are sufficient to start. Once the rows are selected, use zero-order and first-order phase corrections to adjust the phase. Repeat the process with the columns. Often the columns do not require much phase correction. If the phasing is performed properly, the 1D rows and columns will have a flat baseline. If the phased spectrum has a rolling baseline, it is best to set the phase parameters to zero, Fourier transform the spectrum in both dimensions and try again.

Once the 2D spectrum is appropriately phased, it may still require baseline correction for the f2 and f1 dimensions, respectively. Baseline correction is necessary for NOESY and ROESY spectra, since it removes the horizontal and vertical tailing effects and enables clear display of the cross peaks. Baseline correction needs to be carried out after each full

2D transformation, since the corrections are performed on processed data only, and each Fourier transformation "erases" the baseline correction effects. Finally, to enable facile analysis of 2D spectra, projections of 1D spectra on the horizontal and vertical axes are recommended. This enables easy identification of the resonance correlations.

The analysis of the ROESY and NOESY data provides three types of information. The first and most frequently used information provides confirmation of assignments. The second provides through-space interactions that can give insights into the conformation or the configuration of the molecule. Finally for small molecules, the experiments can provide information on chemical or conformational exchange processes.

6.3.3   Chemical/Conformational Exchange

As noted previously, the NOE cross peaks versus chemical or conformational exchange cross peaks can have a different phase depending on the compound's molecular weight. The NOE magnetization transfer between protons on molecules with molecular weights <1000 Da will show cross peaks with opposite phase from the diagonal in the 2D NOESY spectrum. Protons on low molecular weight compounds that experience chemical exchange will show NOE cross peaks with the same sign as the diagonal. For example, protons such as NH or OH that are in rapid exchange with water or with each other will show strong cross peaks with the water resonance in NOESY spectrum that are the same phase as the diagonal.

For non-exchangeable protons in small molecules that are in conformational equilibrium, once again the NOESY spectra of such systems will show the same phase as the diagonal. NOESY spectra of small molecules provide clear distinction between the NOE versus exchange phenomenon. There are different types of conformational exchange that can be observed on the NMR timescale. They may be classified as keto-enol tautomerization, restricted rotation, and ring interconversion. The ability to observe such phenomena depends on the intrinsic properties of the molecule plus conditions such as solvent and temperature.

Keto-enol interconversion refers to a chemical equilibrium between a keto form (i.e., a ketone or an aldehyde) and an enol form (an alcohol). The enol and keto forms are termed tautomers of each other. The interconversion of the two forms involves the movement of a proton on the carbon alpha to the carbonyl to the carbonyl oxygen and the subsequent formation of a double bond. A compound containing a carbonyl group (R-CH-C=O) is often in rapid equilibrium with an enol (R-C=C-OH) tautomer. For cases where interconversion is slow on the NMR time scale, two sets of peaks in the 1D ^{1}H spectrum would be observed corresponding to the ketone and enol forms [18,19].

Restricted rotation may be described as conformational isomerism. This is a condition where molecules exist as isomers that can be interconverted by rotations about single bonds. These isomers are referred to as conformational isomers, conformers, or rotamers. Because these conformers experience restricted rotation, separate sets of resonance can be observed in an NMR spectrum. The reason that separate sets of resonances can be observed is that there exists a rotational energy barrier that needs to be overcome to convert one conformer to another. Hence the rotamers are distinct from each other. To overcome the rotational barrier, heat may be applied to the sample to introduce the activation energy required to interconvert rotamers. A commonly observed rotamer occurs in molecules where amide bonds are present. In such cases, rotation about the C–N bond is restricted due to partial double bond character. Another interconversion that may be observed in NMR is ring conformational interconversion. Usually such interconversions are rapid at room temperature on the NMR timescale. However, there are cases where significant energy barriers caused by steric strain can exist. These cases would produce separate sets of resonances in the NMR spectrum [20,21].

An example of exchange-related cross peaks in a NOESY spectrum is shown in Figure 6.6.

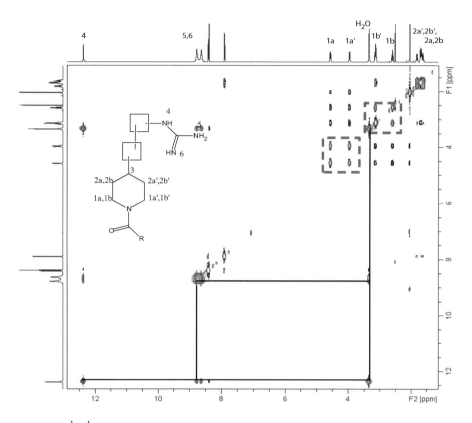

FIGURE 6.6 2D ^{1}H,^{1}H-NOESY spectrum showing chemical and conformational exchange. Cross peaks in red have the same sign as the diagonal and indicate exchange with water as well as restricted rotation about the amide bond. Hindered rotation about the amide bond gives rise to two conformers, hence, two sets of rotamer peaks. The cross peaks in black represent NOE interactions within the molecule.

6.3.4 SPIN DIFFUSION (PROBLEMATIC FOR LARGE MOLECULES)

When analyzing NOESY spectra, one must be aware of the consequences of spin diffusion. Spin diffusion occurs primarily for large molecules and for long mixing times. In NOESY spectra, spin diffusion can produce misleading cross peaks and incorrect distances. To visualize the spin diffusion phenomenon, assume there are four protons (H_1, H_2, and H_3, and H_4) and that H_1 and H_2, H_2 and H_3, and H_3 and H_4 are close (see Figure 6.7). That is, NOEs are expected between the three pairs. The expected cross peaks between protons that are close will be called direct contributions. When spin diffusion is present, indirect contributions will also be present. For example, a cross peak between H_1 and H_3 (and H_1 and H_4) will be present, even though protons H_1 and H_4 are not spatially within 5 Å distance of each other. This is because, in spin diffusion, the magnetization follows a path from H_1 to H_2 and then from H_2 to H_3 to H_4 but appears to be directly from H_1 to H_4. In NOESY spectra of large molecules, the phase of these indirect peaks is the same as for direct contributions, and the resulting cross peaks are impossible to distinguish at a single mixing time. The appearance of the H_1 to H_4 cross peak could lead the spectroscopist to erroneously conclude that protons H_1 and H_4 are close (i.e., <5 Å).

ROESY spectra suffer less from spin diffusion artifacts. With ROESY, the phase of indirect contributions may be different from direct contributions. In addition, the phase of indirect contributions alternates with the number of steps of transfer, which allows their distinction [22].

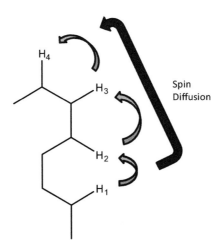

FIGURE 6.7 Diagram illustrating the relay of magnetization as a result of spin diffusion. With this condition, there is no simple relationship between the magnitude of the NOE and the inter-nuclear distance.

6.3.5   NOE, CONFORMATIONAL ANALYSIS, AND DISTANCE DETERMINATION

A NOESY spectrum may be acquired to determine the conformation by establishing a distance between two protons. When more than one conformation can exist, however, the NOE will produce inaccurate distances. For conformations that are being averaged over the time scale of the NMR experiment (the mixing time), the NOE will not reflect the average distance between the protons, but rather the average of the inverse sixth power of the distance.

If one considers a two-spin system in a conformational equilibrium, the observed NOE value will be the result of the inverse of the sixth power of a weighted average of the distances that can be achieved by two spins. The effective distance for NOE is generally expected to be less than the average distance for two spins, because the effective distance is weighted toward that of the "closest approach."

$$r_{eff} < r_{avg} \qquad (6.5)$$

For example, consider a molecule with two conformations X and Y, and the X conformation is 10% populated. If the protons are 2 Å apart in conformation X and 6 Å apart in conformation Y, then the effective distance (experimentally determined distance) might be much closer than the 6 Å separation, which is present in the dominant conformer. Thus, in such cases, the NOE will reflect the conformation where the protons are closer.

The quantitative determination of interproton distances of two spins, I and S, from NOE data is based on comparison of relative NOE intensities derived from the magnetization transfer between the two nuclei and the NOE of nuclei with a known fixed distance. For flexible small molecules (molecular weight below 500 Da) in solution, there is an expectation that the NOE-distance relationship will lack significant accuracy. This is because factors such as spin diffusion, selective polarization transfer, variation in correlation times between spins, accuracy of integration, and conformation populations can contribute to the observed NOE integrations. In spite of these limitations, a report by Jones et al. described a method for extracting interproton distances from NOE data in flexible molecules [23]. The study makes the assumption that a small molecule in solution undergoes rapid tumbling and that the initial rate approximation applies, where the NOE intensity (η_{IS})

between the two spins, I and S, is proportional to the cross-relaxation rate, σ_{IS}, between these spins and the experimental mixing time, τ_{mix} (see Equation 6.6). In addition, the cross-relaxation rate, σ_{IS}, between nuclei I and S is assumed to be proportional to the internuclear distance between spins I and S (r_{IS}^{-6}) (see Equation 6.7). For a more detailed description of the derivation of these equations and their use in determining interproton distances, the reader is referred to other sources [23,24].

$$\eta_{IS} = \delta_{IS}\tau_m \tag{6.6}$$

$$\delta_{IS} = kr_{IS}^{-6} \tag{6.7}$$

where

$$k = \left[\frac{\mu_o}{4\tau}\right] \times \frac{h^2\gamma^4}{10} \times \left[\frac{6\tau_c}{1 + 4\omega^2\tau_c^2} - \tau_c\right] \tag{6.8}$$

If the values defining k (ω – Larmor precessional frequency, τ_c – rotational correlation time, γ – magnetogyric ratio) remain constant for each spin pair in a given selective inversion experiment, the ratio of intensities of a pair of NOE signals, $\eta_{I1S}{:}\eta_{I2S}$, within the NOE experiment can thus be assumed to be proportional to the ratio of their internuclear distances (see Equation 6.9). Thus, by comparing η_{I1S} and η_{I2S} within the same selective inversion experiment, it is only necessary to know one distance, e.g., r_{I1S}, in order to calculate the second distance, r_{I2S}.

$$\frac{\eta_{I1S}}{\eta_{I2S}} = \frac{r_{I1S}^{-6}}{r_{I2S}^{-6}} \tag{6.9}$$

If an actual distance is needed, Equation 6.9 may be rearranged to give Equation 6.10, where the distance is inversely proportional to the distance to the sixth power of the integrated intensity of the NOE signal.

$$r_{ij} = r_{ref}(\eta_{ref}/\eta_{ij})^{1/6} \tag{6.10}$$

where η_{ij} is the NOE cross-peak volume and r_{ij} is the interproton distance of the two protons i and j. Given a known distance between two protons (r_{ref}) and its NOE volume (η_{ref}), a distance can be calculated from another NOE volume.

When there are multiple conformations describing a flexible molecular system, the matter of internuclear distance determination becomes more challenging. A general outline of the treatment of flexible small molecules using NOE experiments for both conformational and population analysis can be found in the work of Neuhaus and Williamson [24]. Nonetheless, it has been shown that accurate interproton distances from NOE data will enable accurate modeling of conformer populations in solution. In addition, with the appropriate number of NOE constraints, these NOE-derived distances could even improve the accuracy of modeling.

Considering molecules with high flexibility generate multiple conformations in solution that rapidly interconvert on the NMR timescale, then this conformational flexibility will produce ensemble averaging of the observed NOEs. Hence observed NOEs will be the result of an average of interproton distance from each contributing conformer. The ensemble-averaged NOEs may then be analyzed by making some assumptions that the molecule will occupy distinct low-energy conformations with particular populations in the solution. The averaged NOE-derived distances can be correlated with computationally generated low-energy conformers.

This approach has been critically evaluated with respect to the challenges of fitting multi-conformer, multi-isomeric models that require large numbers of NOEs to extract the best fit [25]. However, it was postulated and shown that with more accurate NOE-derived distances, geometry and conformation populations may likewise be accurately determined.

A study was described by Jones et al., in which 4-propylanaline was examined. Four low-energy conformers obtained from a B3LYP/6-31G* conformational search yielded equilibrium populations of 24, 25, 25, and 26%, respectively. NOE intensities were obtained from 1D NOESY experiments. The NOE intensities were corrected for chemical equivalence/symmetry by dividing the intensity signals η_{IS} by n_I/n_S, where n_I and n_S are the number of chemically equivalent spins giving rise to signals I and S.

Experimentally determined NOE intensities were converted to distances that were scaled against a well-defined distance in the molecule. Care was exercised to ensure that the reference distance was not compromised by strong coupling artifacts.

Interproton distances were calculated for each of the four B3LYP/6-31G* optimized conformers. The measured distances compared to the calculated distances with the largest % error less than 4%. Overall, this study showed that accurate NOE-derived distances can confirm the relative populations of contributing conformers in small flexible molecules with reasonable certainty.

For this relationship to be valid, a strict experimental protocol must be followed. First, the mixing time must be relatively short, so that the linear approximation is valid and spin diffusion is avoided. For small molecules, the mixing time must be less than several hundred milliseconds. For large molecules in general, the mixing time must be less than 100 ms. Whether spin diffusion leads to an apparent increase or decrease in distance depends on the details of the molecular geometry. Linear geometries lead to shorter apparent distances while nonlinear geometries may lead to longer distances. To help ensure that the mixing time is within the linear region, a buildup curve is performed. A buildup curve is a series of NOE spectra taken at different mixing times. If one is within the linear region, then the NOE will linearly increase with mixing time. A second requirement for quantitative work is that the relaxation delay must be long enough to allow reasonable recovery of the magnetization between scans. The proper relaxation delay must be three times T_1, hence a T_1 determination may be necessary.

For larger flexible systems with multiple conformations such as peptides and proteins, NOE data may be used in conjunction with distance geometry, allowed dihedral angles, simulated annealing, and energy-minimization techniques to produce an ensemble of averaged conformations. The backbone conformations for large molecules such as proteins may be achieved with a high degree of accuracy in particular for regions with well-defined secondary structure.

For flexible portions of a protein (i.e., loop regions or terminal regions), there is a lower degree of conformational definition. The relative proportions of the various conformations contributing to the ensemble of structures have been studied by numerous investigators [26] to find a good match with the experimentally measured NOEs. Unfortunately, accurate determination of populations in these analyses has been limited by the inherently low accuracy of the NOE-derived restraints used, resulting in a broad range of conformer populations fitting the observed NOE restraints.

For more rigid organic molecules, macromolecules, or for proteins with well defined secondary and tertiary backbone structure, it is generally sufficient to classify NOE peak intensities as strong, medium, and weak and make qualitative deductions about relative distances.

6.3.6   ROESY – QUANTITATIVE DISTANCE DETERMINATION

Attempts to use a ROESY spectrum to obtain quantitative distance information can present complications. One issue involves the fact that the cross-peak intensities have an offset dependence relative to the central transmitter frequency. Because of this, ROE cross peaks are less intense for resonances furthest from the center of the spectrum, regardless of spatial distance. For example, if one assumes that the center of the NMR spectrum is 5 ppm, then a cross peak between protons at 1 and 2 ppm will have lower intensity than between protons at 4 and 6 ppm, even if they have the same interproton distance. This dependence has been well characterized by Ammalahti et al. [27].

An additional complication with quantitation of ROESY spectra is that TOCSY transfer may occur and cancel or partially cancel ROESY cross peaks, leading to inaccuracy in peak integrations. This obviously has deleterious effects on distance determination. This is a particular problem for the reference ROE, for which a J-coupled methylene pair is often chosen.

In spite of these drawbacks, it was still reported possible to calculate distances from corrected ROESY intensities. If the offset (ω_0) of the spin-locked pulse is positioned at the far lower side of the spectrum and a moderate spin-lock field strength is applied, the distortion in the intensities of cross and diagonal peaks caused by their offset difference with the spin-locked pulse can be adjusted to obtain quantitative data.

For a correlation between spins at ω_i and ω_j frequencies, the correction factor, c_{ij}, in Equation 6.11 may be used.

$$r_{ij} = r_{ref} \left(a_{ref} c_{ref} / a_{ij} c_{ij} \right)^{1/6} \tag{6.11}$$

where

$$c_{ij} = 1 / (\sin^2\theta_i \sin^2\theta_j)$$

and

$$\tan\theta_I = \gamma B_1 / (\omega_I - \omega_0)$$

where ($\omega_I - \omega_0$) is the difference between the chemical shift of the peak frequency (in Hz) and the center frequency (in Hz) and γB_1 is the spin-lock power. Volume corrections of up to a factor of 4, in off resonance cases, may be required.

Overall, although inter-proton distances have been reported to be successfully determined from compounds that experience conformational equilibrium or from NOE integrations obtained from 2D ROESY experiments, care must be taken in setting the experimental conditions, and rigorous calculations and correction factors must be applied.

REFERENCES

1. Hoult, D.I. 1976. Solvent peak saturation with single phase and quadrature fourier transformation. *J. Magn. Reson.* 21(2):337–347.
2. Hore, P.J. 1983. A new method for water suppression in the proton NMR spectra of aqueous solution. *J. Magn. Reson.* 54(3):539–542.
3. Albert, K., Nieder, M., Bayer, E., and Spraul, M. 1985. Continuous-flow nuclear magnetic resonance. *J. Chromatogr.* 346:17–24.
4. Laude, Jr., D.A., and Wilkins, C.L. 1987. Reverse phase high performance liquid chromatography/nuclear magnetic resonance spectrometry in protonated solvents. *Anal. Chem.* 59:546–551.
5. Smallcombe, S.H., Patt, S.L., and Keifer, P.A. 1995. WET solvent suppression and its applications to LC NMR and high-resolution NMR spectroscopy. *J. Magn. Reson. A.* 117(2):295–303.
6. Patt, S.L. 1992. Single and multiple-frequency-shifted laminar pulses. *J. Magn. Reson.* 96(1):94–102.
7. Aue, W.P., Bartholdi, E., and Ernst, R.R. 1976. Two-dimensional spectroscopy. Application to nuclear magnetic resonance. *J. Chem. Phys.* 64:2229–2246.
8. Braunschweiler, L., and Ernst, R.R. 1983. Coherence transfer by isotropic mixing: application to proton correlation spectroscopy. *J. Magn. Reson.* 53(3):521–528.

9. Shockcor, J.P., Crouch, R.C., Martin, G.E., Cherif, A., Luo, J.-K., and Castle, R.N. 1990. Disentangling proton connectivity networks in highly overlapped ^{1}H-NMR spectra of polynuclear aromatics using 1D-HOHAHA. *J. Heterocycl. Chem.* 27:455–458.

10. Jeener, J., Meier, B.H., Bachmann, P., and Ernst, R.R. 1979. Investigation of exchange processes by two-dimensional NMR spectroscopy. *J. Chem. Phys.* 71:4546–4553.

11. Bax, A., and Davis, D.G. 1985. Practical aspects of two-dimensional transverse NOE spectroscopy. *J. Magn. Res.* 63(1):207–213.

12. Bodenhausen, G., and Ruben, D.J. 1980. Natural abundance nitrogen-15 NMR by enhanced heteronuclear spectroscopy. *Chem. Phys. Lett.* 69:185–189.

13. Bax, A., and Summers, M.F. 1986. Proton and carbon-13 assignments from sensitivity-enhanced detection of heteronuclear multiple-bond connectivity by 2D multiple quantum NMR. *J. Am. Chem. Soc.* 108:2093–2094.

14. Keifer, P.A., Smallcombe, S.H., Williams, E.H., Salomon, K.E., Mendez, G., Belletire, J.L., and Moore, C.D. 2000. Direct-injection NMR (DI-NMR): a flow NMR technique for the analysis of combinatorial chemistry libraries. *J. Comb. Chem.* 2:151–171.

15. Hoye, T.R., Eklov, B.M., Ryba, T.D., Voloshin, M., and Yao, L.J. 2004. NMR (no-deuterium proton NMR) spectroscopy: a simple yet powerful method for analyzing reaction and reagent solutions. *Org. Lett.* 6:953–956.

16. Sklenar, V., and Starcuk, Z. 1985. Composite pulse sequences with variable performance. *J. Magn. Reson.* 62(1):113–122.

17. Shaka, A.J., Barker, P., and Freeman, R. 1985. Computer-optimized decoupling scheme for wideband applications and low-level operation. *J. Magn. Reson.* 64(3):547–552.

18. Sung, K., Wu, R.-R., and Sun, S.-Y. 2002. Keto-enol tautomerism of b-ketoamides and characterization of a sterically crowded a-amido-b-ketoamide. *J. Phys. Org. Chem.* 15(11):775–781.

19. Derogis, P.B.M.C., Martins, F.T., de Souza, T.C., de C. Moreira, M.E., Souza Filho, J.D., Doriguetto, A.C., de Souza, K.R.D., Velosoa, M.P., and Dos Santosa, M.H. 2008. Complete assignment of the 1Hand 13C NMR spectra of garciniaphenone and keto-enol equilibrium statements for prenylated benzophenones. *Magn. Reson. Chem.* 46:278–282.

20. Katritzky, A.R., Akhmedov, N.G., Myshakin, E.M., Verma, A.K., and Hall, C.D. 2005. Low-temperature ^{1}H and ^{13}C NMR spectra of *N*-substituted 1,2,3,4-tetrahydropyrazino [1,2-*a*]indoles. *Magn. Reson. Chem.* 43(5):351–358.

21. Akhmedov, N.G., Myshakin, E.M., and Hall, C.D. 2004. Dynamic NMR and *ab initio* studies of exchange between rotamers of derivatives of octahydrofuro[3,4-*f*]isoquinoline-7(1*H*)-carboxylate and tetrahydro-2,5,6(1*H*)-isoquinolinetricarboxylate. *Magn. Reson. Chem.* 42:39–48.

22. Bax, A., Sklenar, V., and Summers, M.F. 1986. Direct identification of relayed nuclear overhauser effects. *J. Magn. Reson.* 70(2):327–331.

23. Jones, C.R., Butts, C.P., and Harvey, J.N. 2011. Accuracy in determining interproton distances using nuclear overhauser effect data from a flexible molecule, beilstein. *J. Org. Chem.* 7:145–150.

24. Neuhaus, D., and Williamson, M.P. 2000. *The Nuclear Overhauser Effect in Structural and Conformational Analysis*, 2nd edition. New York: John Wiley & Sons, Inc.

25. Kozerski, L., Krajewski, P., Pupek, K., Blackwell, P.G., and Williamson, M.P. 1997. Towards stereochemical and conformational assignment in flexible molecules using NOEs and molecular modeling. *J. Chem. Soc., Perkin Trans.* 2:1811–1818.

26. Salmon, L., Bouvignies, G., Markwick, P., and Blackledge, M. 2011. Nuclear magnetic resonance provides a quantitative description of protein conformational flexibility on physiologically important time scales. *Biochemistry.* 50:2735–2747.

27. Ammalahti, E., Bardet, M., Molko, D., and Cadet, J. 1996. Evaluation of distances from ROESY experiments with the intensity-ratio method. *J. Magn. Reson. Series A.* 122(2):230–232.

7 Applications

There are a vast number of applications for liquid chromatography-nuclear magnetic reson-ance (LC-NMR) in the pharmaceutical sciences. In pharmaceutical development alone, these include, but are not limited to, isolation and characterization of drug impurities resulting from drug-stability tests, synthetic intermediates, drug production, and formula-tion reactivity. LC-NMR has been used to identify structures of degradation products, unstable molecules, compounds formed in situ, those that cannot be easily isolated, or that may be sensitive to light, oxygen, heat, or time. The combination of LC and NMR can be used to evaluate the components of crude natural product extracts and components of paral-lel synthesis. In addition, applications in drug metabolism and in the analysis of biofluids, for example, metabolites in urine or plasma, have provided valuable information in the areas of toxicological safety and efficacy. Recent applications combining LC with NMR spectroscopy to address problems in key areas of interest in both the pharmaceutical industry and academia are discussed below.

7.1 DEGRADATION PRODUCTS

Because drug products may undergo physicochemical degradation during manufacturing and storage, it is imperative that the inherent stability characteristics of a product be evaluated to ensure that safety is not compromised by the presence of toxic impurities [1, 2]. In particular, the isolation and structure elucidation of degradation products becomes critical in regulatory documents where the assessment of safety must be established. Potential toxic or mutagenic degradation products often need to be identified and unambiguously characterized.

Although mass spectrometry (MS) technologies, in conjunction with high-performance liquid chromatography (HPLC), enable rapid, online structural elucidation of pharmaceutical degradation products [3–5], the results may not always provide conclusive or unambiguous results. To address this issue, NMR spectroscopy may be introduced as a highly definitive tech-nology for degradation product analysis.

In view of the sensitivity issues associated with NMR spectroscopy, degradant compounds of interest need to be isolated in sufficient quantities to enable structural characterization. While compounds can be isolated by preparative LC and analyzed by conventional solution-state NMR, often considerable losses of material are realized during the isolation process, and the use of a microprobe or cryoprobe becomes necessary. However, when hyphenated techniques such as LC-NMR and solid-phase extraction (LC-SPE-NMR) are used in con-junction with micro flow-probe or cryoprobe analysis, problems with low concentrations and poor signal-to-noise ratio (S/N) can be addressed. For cases where only microgram quantities of analyte can be isolated, use of a cryoprobe may be necessary [6–8].

A recent report by Wu et al. [9] described the isolation and structural characterization of a compound formed from solid {4-(4-chloro-3-fluorophenyl)-2-[4-(methyloxy) phenyl]-1, 3-thia-zol-5-yl} acetic acid **1** upon photoirradiation (Figure 7.1). This molecule is a pharmaceutically active compound previously under development for the treatment of overactive bladder [10]. LC-MS/MS analysis and accurate mass determinations were used to identify the major photo-degradation product as 4-chloro-N-(4-methoxybenzoyl)-3-fluorobenzamide **2**.

Preparative chromatography was used to isolate the degradation product of interest for NMR analysis using an Agilent 1100 preparative LC system interfaced with an automatic frac-tion collector (Agilent Technologies). An isocratic method was developed for the preparative

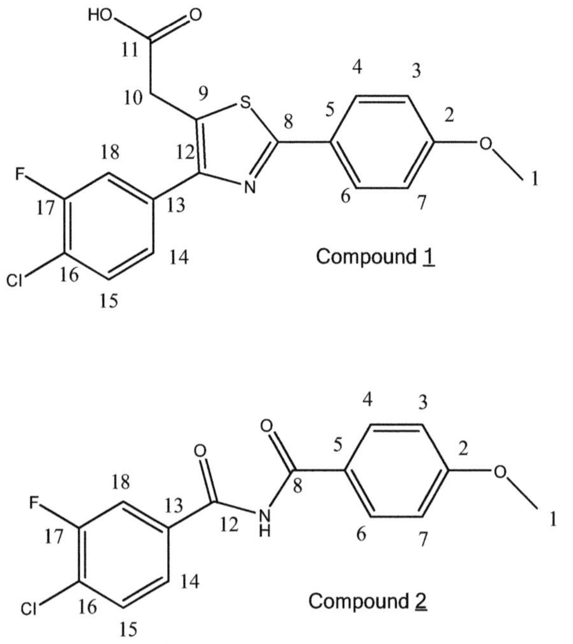

FIGURE 7.1 Photo-oxygenation of solid {4-(4-chloro-3-fluorophenyl)-2-[4-(methyloxy) phenyl]-1, 3-thiazol-5-yl} acetic acid **1** upon photoirradiation yields a major photo-degradation product, 4-chloro-*N*-(4-methoxybenzoyl)-3-fluorobenzamide **2** [9].

isolation of **2** using 45% and 55% mobile phase A of 0.1% formic acid in water and mobile phase B of 0.1% formic acid in acetonitrile (ACN), respectively. A Zorbax SB-Phenyl (150 mm × 9.4 mm, 5-μm particle size) column was selected and the flow rate was 4.7 mL/min for the preparative chromatography. The photo-degradation product had a retention time of 15 min using the isocratic LC method.

A loading study enabled maximum injection mass to be determined resulting in a concentration of approximately 5.5 mg/mL and a total of 85 runs, each using 700-μL injections. Each injection required a 32-min run. ACN was removed from the sample using a rotary evaporator and the sample was subsequently lyophilized to remove water. Approximately 2 mg of compound was produced with a purity of about 97% (estimated by both LC and solution-state NMR). Both one-dimensional (1D) and two-dimensional (2D) NMR experiments were obtained, enabling full-structure characterization [3–5]. Spectra of parent compound **1** and degradant **2** are shown in Figure 7.2.

It was proposed that the solid thiazole **1** underwent photooxygenation by reacting with singlet oxygen upon photoirradiation, yielding an unstable endoperoxide. The unstable intermediate subsequently underwent rearrangement to yield degradation product **2**. These studies enabled the researchers to understand the structure of significant photodegradation products of thiazoles, allowing for better decision-making and control during the drug-development process.

Another example of the isolation and NMR spectroscopic characterization of a degradation product involved an active pharmaceutical ingredient (API) that had a carboxylic acid moiety and existed as an ethylene diamine salt and an ethanol solvate [11]. In a closed container at room temperature over several weeks, this compound formed an impurity at about 0.1% level. To enable full-structure characterization, the compound had to be chromato-graphically isolated in sufficient quantities for NMR analysis. This was accomplished using an Atlantis

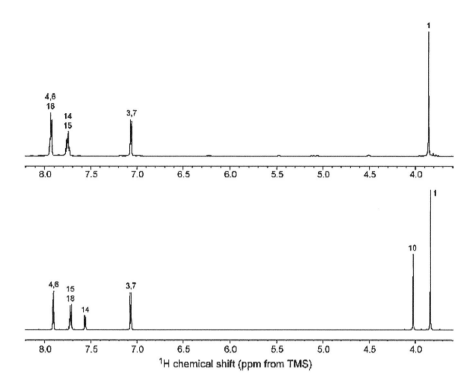

FIGURE 7.2 Comparison of the ^{1}H-NMR spectra in DMSO-d_6 solution of the parent compound **1** (top) and the preparatively isolated photo-degradation product compound **2** (bottom). Numbered peaks are in reference to the structures shown in Figure 7.1. (Reprinted from Wu, Hong, and Vogt [9]. Copyright 2007, with permission from Elsevier.)

T3 C18 column. Separation was achieved using a mobile-phase solvent gradient comprising ACN and water with 0.05% trifluoroacetic acid (TFA). The chromatogram of the API and degradant is shown in Figure 7.3.

To enable capture of material in sufficient quantities for NMR analysis, trapping of the compound was carried out using an SPE C8 cartridge. A total of 60 collections (combined from 18, 20, and 22 collections) were added together and taken to dryness using a Biotage microDryer. The sample was then dissolved in 30-μL DMSO-d_6, and spectra were collected on the 1.7-mm capillary cryoprobe. A 1D spectrum of the sample is shown in Figure 7.4.

The spectrum shows a sample relatively free of sample impurities, yet the residual hydrocarbons that bleed from the cartridge are still visible in about the same signal intensity as the sample peaks of interest. However, the 1D WET-^{1}H spectrum in DMSO-d_6 showed the presence of exchangeable protons that were not observed for the sample in ACN. Observation of the exchangeable protons was critical in establishing structural connectivity. Full-structure characterization was achieved using a combination of wet-COSY, presat-ROESY, wet-HSQC, and HMBC data. The initial proposed structural modification of the degradant shown in Figure 7.5 was supported by ROESY data, which established a through-space connectivity between exchangeable proton H19 and methylene protons H2. This addition of the ethylene diamine moiety to the carboxylic acid was confirmed in the HMBC spectrum, which showed clear through-bond correlations between the methylene protons, H20 and H2, and carbonyl carbon C1 (Figure 7.5).

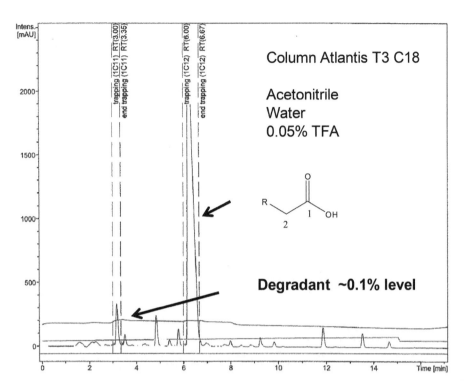

FIGURE 7.3 UV chromatogram showing the separation of the degradant from the API. An Atlantis T3 C18 analytical column was used. The mobile phase consisted of ACN and water with addition of 0.05% TFA as a modifier. The relative level of degradant to API was about 0.1% [11].

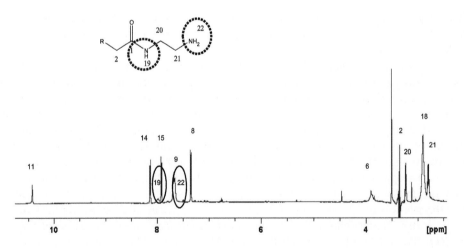

FIGURE 7.4 A 1D ^{1}H WETDC spectrum acquired on a 600-MHz spectrometer operating at 600.2 MHz for ^{1}H and equipped with a 1.7-mm capillary cryoprobe. The sample was dissolved in 30-μL DMSO-d$_6$. Exchangeable protons NH-19 and NH-22 were observed enabling full-structure characterization of the compound [11].

HMBC /HSQC (overlay red)

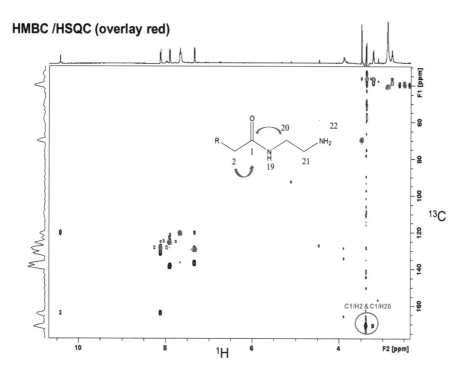

FIGURE 7.5 A ^{1}H-^{13}C WET-HSQC spectrum (red) is overlaid with a ^{1}H-^{13}C HMBC spectrum. Both spectra were acquired on a 600-MHz NMR spectrometer operating at 600.2 MHz for ^{1}H and at 150.92 MHz for ^{13}C. The spectrometer was equipped with a 1.7-mm capillary cryoprobe. Long range through-bond couplings between H2 and C1, and H20 and C1 establish the amide bond formation of the ethylene diamine portion of the molecule to parent molecule [11].

These data unequivocally established the structure in which ethylene diamine adds to the carboxylic acid of the parent compound, forming an amide bond. The structure was further corroborated by MS data.

It should also be noted that the final amount of compound in the NMR tube was determined by HPLC analysis to be about 8 μg of material. The HMBC spectrum required about six days of data acquisition, even with the mass sensitivity of the cryoprobe. Hence, although it was possible to obtain data on as little as 8 μg of material using the capillary cryoprobe, significant data acquisition time was required.

A second degradation study involved an oxidation product that required enrichment to obtain material in sufficient quantities for NMR analysis [11]. To obtain an enriched sample of the degradation product, the API was taken up in ACN: methanol (50:50) with the addition of rose bengal (1.5 mg/mL). Exposure to UV/Vis (1 ICH [250 W/m^2/21 h]) produced the enriched sample. The degradant was isolated using preparative LC. A C8 column was used with a mobile phase of 75%ACN/25%MeOH and 0.025% formic acid in water. A total of 22 collections were made.

Mass spectrometry was carried out showing that the oxidized product contained three additional oxygen atoms and two additional exchangeable hydrogen atoms. This information was important in designing a relevant hypothesis for the structure of the oxidized product.

A 1D ^{1}H spectrum of the compound was subsequently obtained in DMSO. The spectrum revealed the sample to be a mixture of species. This spectrum was compared with the parent API, which showed the oxidized impurity had numerous similarities to the API as well as some very pronounced differences (see Figure 7.6).

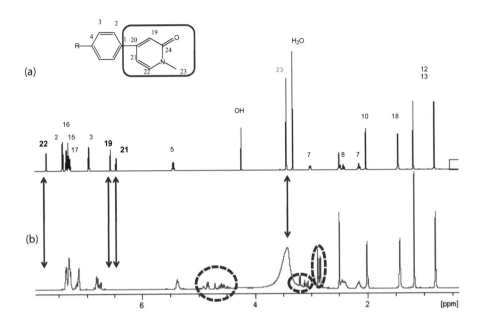

FIGURE 7.6 A 1D ^{1}H spectra (a) API and (b) oxidation product acquired on a 600-MHz spectrometer operating at 600.2 MHz for ^{1}H and equipped with a 1.7-mm capillary cryoprobe. The sample was prepared by dissolving API in ACN /water solution with a small amount of rose bengal to enrich the degradation product via singlet oxygen mechanism. Sample was then placed in light chamber and stored until 100% conversion was observed. Isolation of the oxidized products was carried out using SPE. Arrows indicate missing resonances relative to the API, while the circled peaks show introduction of new resonances [11].

Initial examination of the 1D ^{1}H data revealed that all proton resonances associated with the N-methyl pyridone ring had disappeared, while clusters of new resonances appeared in the aliphatic region. This information was consistent with oxidation of the pyridone ring leaving the remainder of the structure relatively unaffected. Integrations of clusters of protons were consistent with the existence of a cluster of diastereomers or regioisomers. This was evident from correlations in the COSY and ROSEY spectra, but was clearly illustrated in the HSQC spectrum, which showed clusters of distinct proton resonances associated with clusters of distinct carbons. If one considers that the degradant needed to contain two additional exchangeable protons and three additional oxygens, a possible structure could be devised that fit the NMR data (Figure 7.7).

Closer examination of one of the proton clusters showed correlation to a set of carbon resonances. This cluster of resonances was assigned to an H21–C21 correlation and cluster of proton resonances that integrated as one proton. Examination of the HSQC spectrum revealed the existence of at least eight different species (see Figure 7.8).

Since the proposed oxidized product introduced four new chiral centers, this condition could account for multiple diastereomers and, hence, the multiple observed ^{1}H and ^{13}C NMR resonances for the proposed structure. However, seven other regioisomers can be drawn that could also satisfy the NMR data (Figure 7.9). Using knowledge-based ^{13}C chemical shift predictions, the experimentally determined chemical shifts were found to most closely match the circled structure in Figure 7.9. More rigorous chemical shift predictions using low-energy conformations and density functional theory (DFT) calculations would be required to provide more accurate predictions [12].

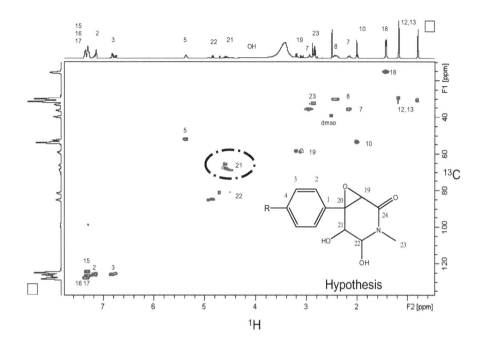

FIGURE 7.7 A ^{1}H-^{13}C WET-HSQC spectrum acquired on a 600-MHz NMR spectrometer operating at 600.2 MHz for ^{1}H and at 150.92 MHz for ^{13}C. The spectrometer was equipped with a 1.7-mm capillary cryoprobe. The spectrum was acquired using 1k and 256 points in F2 and F1 dimensions, respectively, and 64 scans. The spectrum shows clusters of proton and carbon resonances, indicating multiple species [11].

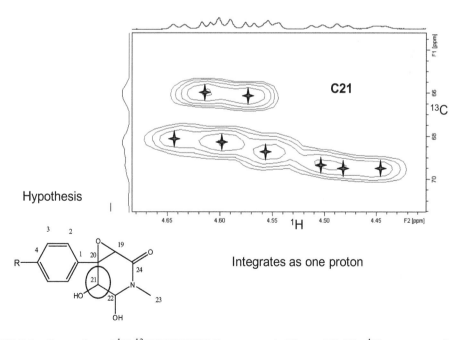

FIGURE 7.8 Expansion of ^{1}H-^{13}C WET-HSQC spectrum in Figure 7.7. The ^{1}H resonances from this cluster integrate as a single proton and show correlation peaks that indicate at least eight different species [11].

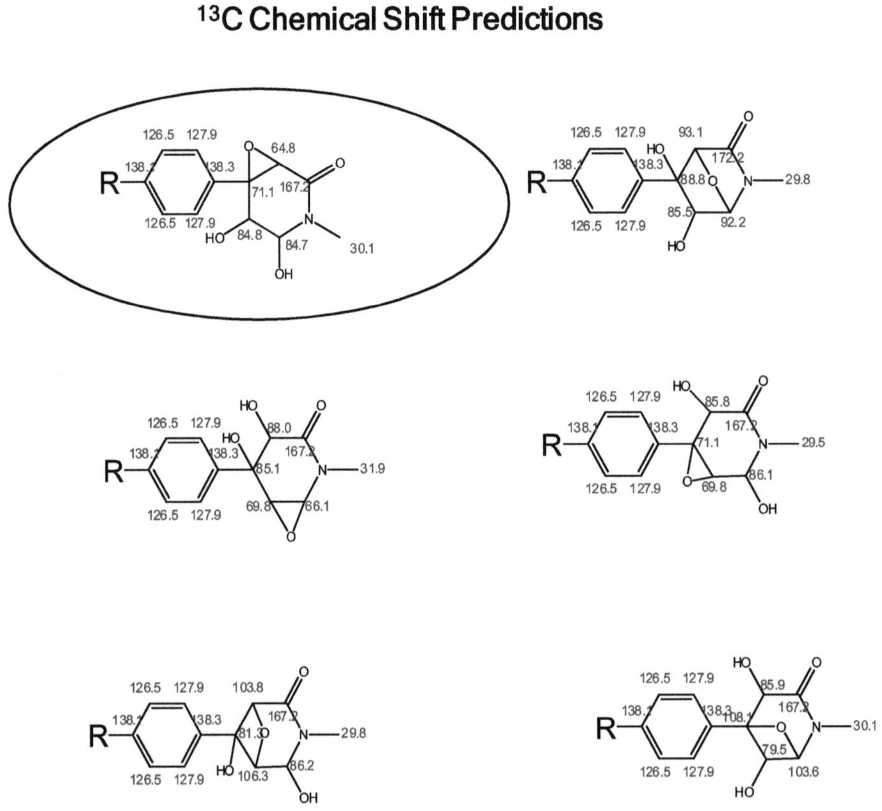

^{13}C Chemical Shift Predictions

FIGURE 7.9   Knowledge-based chemical shift predictions (ChemBioDraw Ultra 11.0) favor circled compound; however, other possibilities cannot be ruled out based upon NMR data [11].

7.2   IMPURITIES

LC-NMR has been applied to the isolation and structural characterization of impurities [13, 14]. As with degradants, structure determination of impurities is of critical importance for drug-development candidates to satisfy the strong criteria imposed by regulatory agencies.

Novak et al. [15] have reported on the application of LC-NMR for rapid identification and characterization of an unknown impurity in icofungipen **3**. The impurity compound **4** (Figure 7.10) was formed during the preparation of icofungipen, a novel, orally bioavailable antifungal drug, under development for the treatment of *Candida* infections [16–21].

Using LC and stop-flow isolation, 1D ^{1}H-NMR and 2D ^{1}H^{1}H-COSY experiments were acquired, which provided information about the structure of **4**. Analysis of the ^{1}H LC-NMR spectrum revealed the presence of both aliphatic and aromatic moieties (Figure 7.11) [22]. Integration of the proton intensities revealed a total of 17 nonexchangeable protons. Comparison between the ^{1}H LC-NMR spectra of **4** and icofungipen showed the presence of a side chain with aliphatic and aromatic structural units in addition to the cyclopentane ring and olefinic spins. A 2D DQCOSY (double quantum filtered COSY) LC-NMR spectrum produced correlation peaks diagnostic of three spin systems belonging to the benzene ring, methylenecyclopentane, and – CH CH–CH$_2$ – moieties. A vicinal proton-proton coupling constant of 15.9 Hz between the two olefinic protons was consistent with the *trans*-configuration of the double bond. Coupling constants involving protons at the chiral centers

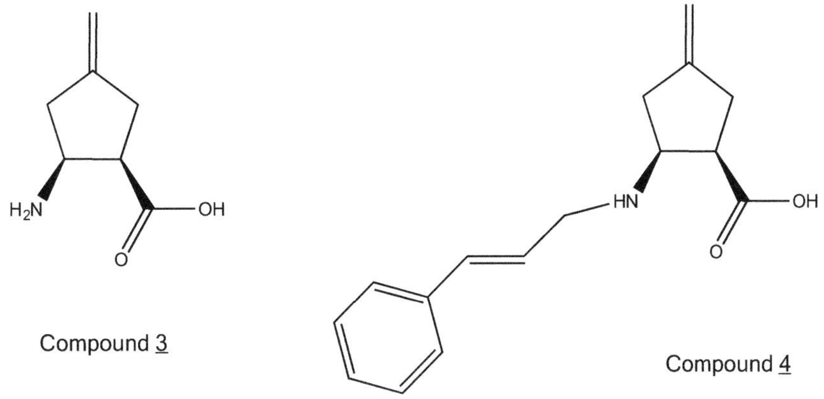

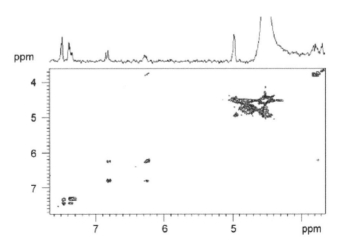

Compound $\underline{3}$

Compound $\underline{4}$

FIGURE 7.10   Structures of icofungipen $\underline{3}$ and its impurity $\underline{4}$.

of the cyclopentane ring in icofungipen and compound $\underline{4}$ revealed a similar structure consistent with that part of the molecule remaining unchanged.

In order to confirm the proposed structure, the investigators synthesized the impurity (**compound 4**). The structure was subsequently elucidated using NMR spectroscopy and MS/ MS spectrometry. A combination of 1D and 2D homo- and heteronuclear NMR experiments (APT (attached proton test), DQCOSY, HSQC, and HMBC) enabled full characterization of $\underline{4}$. The COSY spectrum of the synthesized compound $\underline{4}$ was in agreement with the COSY from LC-NMR. HSQC spectra provided protonated carbon atom assignments, while the HMBC spectra enabled assignment of the quaternary carbons. Likewise, the MS/MS spectrum showed the same fragmentation pattern observed in the LC-MS/MS spectrum.

While unambiguous structure determination required synthetic confirmation, in this example, the LC-NMR data provided a starting point for synthetic follow-up. Alternate approaches to similar problems can include SPE combined with cryoprobe technology. In such cases, structural confirmation may be achieved without the need for synthetic intervention.

A general review by Rinaldi [23] highlights the use of LC-SPE-NMR and a 30-μL cryo flow probe. This study describes the isolation and NMR spectroscopic characterization of lowlevel

FIGURE 7.11   The aromatic region of 2D DQFCOSY LC–NMR spectrum of compound $\underline{4}$ showing the main correlation peaks. (Reprinted from Novak et al. [22]. Copyright 2009, with permission from Elsevier.)

impurities. The author reported efficient trapping of impurity mixtures on C8/C18 SPE cartridges, although it was noted that those compounds that elute with less than 35% organic solvent were too polar to efficiently trap on the C18 stationary phase. Approximately 50 µg of one of the impurities was isolated and transferred to a cryo flow probe enabling full structural characterization (Figure 7.12). The use of LC-SPE integrated with

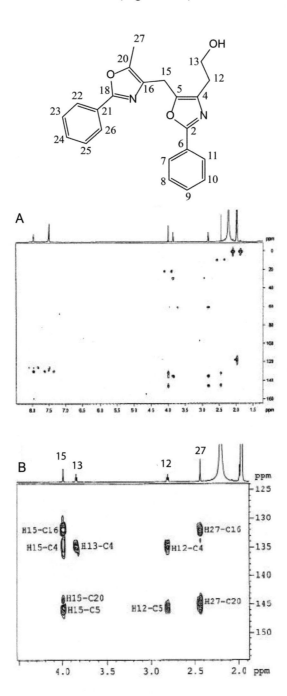

FIGURE 7.12 The HMBC spectrum of an impurity eluting at 17.7 min in acetonitrile-d_3, enabling full characterization of the compound. (a) The spectrum was acquired using 4k and 128 points in F2 and F1 dimensions, respectively, and 32 scans. (b) Expanded region of the HMBC spectrum shows proton–carbon correlated assignments. (Figure reproduced from Rinaldi [23]. Copyright rests with the publisher.)

cryoprobe technology was deemed highly efficient in the analysis of low-level components of complex mixtures that are typically encountered in pharmaceutical development.

Analysis of trace impurities by NMR may even be achieved in samples that contain significant amounts of other impurities. One example of a trace impurity obtained from an API synthesis was a compound that had a molecular weight indicating the presence of a bromine atom instead of fluorine relative to the molecular weight of the API [11]. Based upon synthetic considerations, two possible structures were proposed. Key structural features of the API and two possible impurity structures are given in Figure 7.13.

An enriched sample of the isolated impurity was separated from other minor impurities using a Waters Atlantis T3 C18 column using a mobile-phase solvent gradient of methanol and formic acid/water with the column temperature at 40°C. LC-SPE-NMR was employed, and the sample was trapped on a general purpose (GP) resin SPE cartridge from two 40-μL injections of enriched sample. Following elution and concentration to dryness, the sample was dissolved in 30-μL DMSO-d_6. A 1D proton spectrum taken using a 1.7-mm cryoprobe is shown in Figure 7.14. The spectrum shows a major component with numbered peaks corresponding to the compound of interest and a plethora of lesser impurity peaks marked with a plus sign. Even in the presence of numerous impurity peaks, it was possible to establish correlations for the impurity compound in COSY and ROESY spectra. The initial chemical shift assignments using COSY and ROESY data were consistent with the structure of impurity **1** in Figure 7.13.

Full structural analysis using ROESY, TOCSY, HSQC, and HMBC confirmed the structure as impurity **1**, where the fluorine in the API is replaced by a bromine atom. Through-bond ^{1}H and ^{13}C correlations in the HMBC spectrum, between H11 and C18, H19 and C18, and H1 and C18 as well as correlations between H8 and C10, shown in Figure 7.15, are consistent with the structure.

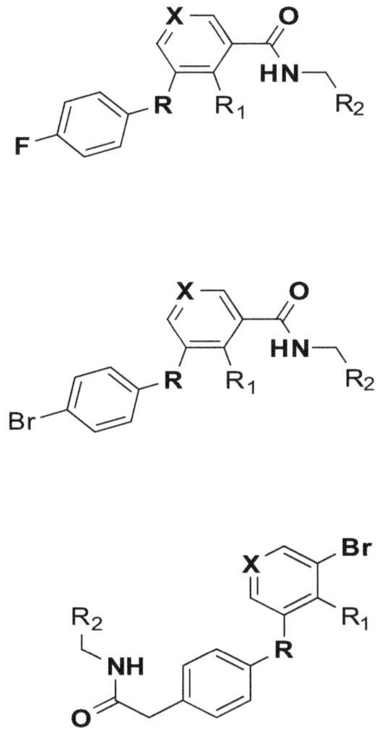

FIGURE 7.13 Chemical structures of an API and two possible impurities having identical mass. Mass spectrometry was unable to unambiguously determine the structure of the impurity [11].

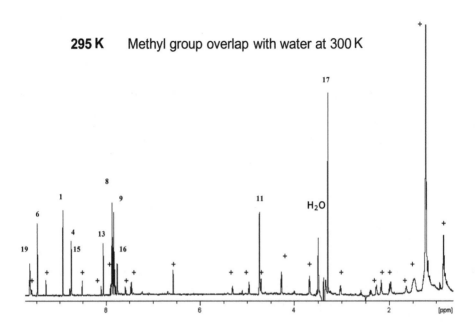

FIGURE 7.14 A 1D ^{1}H WETDC (a pulse program that executes ^{13}C decoupling during WET and acquisition of the signal) spectrum acquired on a 600-MHz spectrometer operating at 600.2 MHz for ^{1}H and equipped with a 1.7-mm capillary cryoprobe. The sample was dissolved in 30-μL DMSO-d$_6$. Numbered peaks correspond to the compound of interest. The "+" marks indicated the impurities in the sample. The intensity of largest peak in the sample was less than the peak corresponding to the residual hydrocarbon carryover from the chromatography [11].

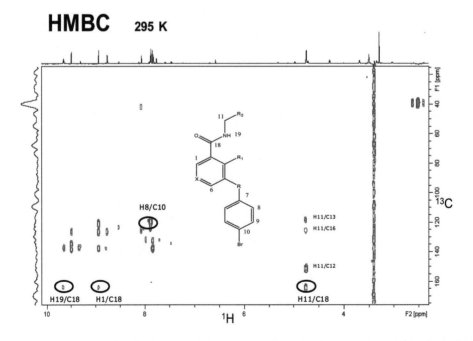

FIGURE 7.15 The HMBC spectrum of the isolated impurity in DMSO-d$_6$ acquired on a 600-MHz spectrometer enabling full characterization of the compound. The spectrum was acquired using 2k and 128 points in F2 and F1 dimensions, respectively, and 1216 scans. Correlations between H1/C18 and H11/C18 establish the structural scaffold and enable unambiguous assignment [11].

Furthermore, the upfield chemical shift for C10 (~120 ppm) is consistent with the replacement of the fluorine atom by a bromine atom. Because replacement of a fluorine atom with a bromine atom was not synthetically feasible, it is more likely that the fluorinated starting material possessed a trace amount of bromonated compound that was carried through the reaction scheme.

Another impurity characterization using LC-SPE-NMR involved enrichment of a compound that was produced in amounts less than 0.1% [11]. The API and the impurity eluted approximately 18 s apart. Mass spectrometry data showed that the impurity had the same molecular weight as the API. Initial preparative isolation was able to reduce the amount of API and provide a mixture of the two closely eluting compounds. The sample was then transferred to the LC-NMR unit using analytical column Agilent Eclipse XDB-C8 with a mobile phase of 0.025% formic acid/ACN and 0.025% formic acid/H_2O. The column temperature was maintained at 45° C. The impurity was trapped using a GP resin SPE cartridge (see Figure 7.16).

The process was repeated three times, and a total of 80 injections (25 µL) were reduced to dryness and dissolved in DMSO-d_6. NMR spectra were acquired on the 1.7-mm capillary cryoprobe, and comparison was made with the impurity and the API. The results showed significant changes in the aliphatic region of the molecule with the aromatic region remaining virtually unchanged. Based upon COSY and ROESY data, a structure of the impurity was proposed. The structural change is shown in Figure 7.17.

This structure was later confirmed by an HMBC spectrum with correlations between H21 and C24, H23 and C24, and OH (25) and C24 (Figure 7.18a). The structure of the impurity introduced a new chiral center into the molecule; however, the impurity corresponded to only one of the two possible diastereomers.

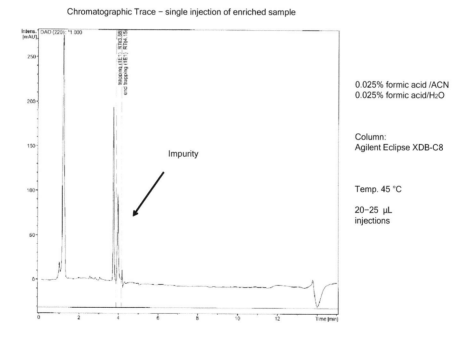

FIGURE 7.16   UV chromatogram showing the separation of an enriched impurity. An Eclipse XDB C8 analytical column was used in the separation. The mobile phase consisted of ACN /0.025% formic acid and water/0.025% formic acid. The relative level of impurity compared to API was <0.1%. Closely eluting peaks could be isolated in sufficient quantities for full characterization by NMR spectroscopy [11].

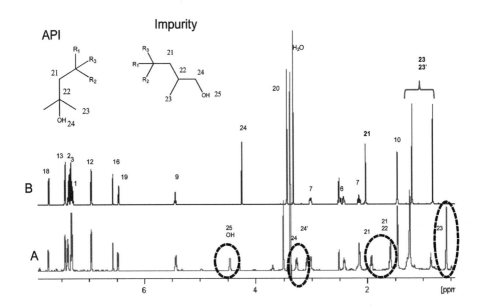

FIGURE 7.17 (A) 1D ^{1}H WETDC spectrum of the isolated impurity acquired on a 600-MHz spectrometer operating at 600.2 MHz for ^{1}H and equipped with a 1.7-mm capillary cryoprobe. Circled peaks show the new resonances consistent with changes occurring in the aliphatic region of the molecule. (B) 1D ^{1}H-NMR spectrum of the isolated impurity acquired on a 600-MHz spectrometer operating at 600.2 MHz for ^{1}H and equipped with a 1.7-mm capillary cryoprobe. Both samples were dissolved in ~30-μL DMSO-d_6. Proposed differences in the structure of the impurity relative to the API are shown [11].

An independent synthesis of the impurity molecule was carried out, producing a mixture of the two possible diastereomers (Figure 7.18b). One of the two compounds had a ^{1}H proton NMR spectrum identical to the isolated impurity. Because this particular structure was found to possess no potential toxic or mutagenic properties, no further structural investigation was carried out to determine absolute stereochemistry.

7.3 TRACE ANALYSIS

The coupling of LC with solid-phase extraction and NMR spectroscopy has been applied to structure elucidation studies of natural product extracts, drug metabolites, and pharmaceutical impurities formed in trace amounts [24, 25]. By combining LC with SPE, low levels of compound mixtures may be enriched to allow NMR detection. The LC-SPE combination provides sensitivity enhancements in NMR by factors of 3 to 4 over conventional LC-NMR analysis of a mixture. When LC-SPE is used in conjunction with an NMR cryogenic probe [26, 27], the detection sensitivity further increases by about a factor of 4. This provides sensitivity sufficient for full NMR structural characterization on <50 μg of material.

Although the combination of LC-SPE with NMR is very powerful in characterizing low levels of compound, the approach requires repeated LC runs and multiple trappings to obtain sufficient NMR sensitivity, even with a cryogenic probe [24, 26]. Large-scale preparative or semi-preparative LC off-line may be used to isolate low-level analytes for NMR analysis [28, 29]. However, in practice, such an approach can be time consuming and lacking efficiency.

To address this issue, Xu and Alexander [30] reported the design of a system that combines semipreparative LC and SPE HiCRAM (high capacity retention and mixing). This approach reportedly expanded the isolation capacity by at least fivefold. With this system,

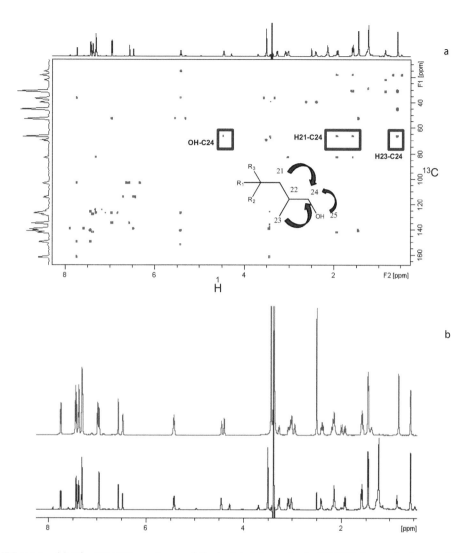

FIGURE 7.18   (a) The HMBC spectrum of the isolated impurity in DMSO-d_6 acquired on a 600-MHz spectrometer enabling full characterization of the compound. The spectrum was acquired using 4k and 256 points in F2 and F1 dimensions, respectively, and 128 scans. Correlations between H23/C24, H21/C24, and OH/C24 establish the structural scaffold and enabled unambiguous assignment. (b) Comparison of isolated impurity with synthesized mixture of the two possible diastereomers. The spectral comparison showed that the isolated impurity (bottom spectrum) overlayed identically with one of the two synthesized diastereomers (top spectrum), thereby confirming the structure of the impurity [11].

^{1}H–^{13}C heteronuclear experiments could be carried out on the isolated components from a mixture of compounds present at the 1% level using a single LC-SPE cycle. For analytes at the 0.02% level, replicate LC isolations were needed with multiple SPE trappings.

The study involving replicate LC isolations and SPE trappings was reported for buspirone at 0.02%. This component was enriched by pooling peak cuts from three replicate LC analyses into the HiCRAM unit. Both ^{1}H 1D (64 scans, 3 min) and 2D WETTNTOCSY (a pulse program that combines WET solvent suppression with TOtal Correlation SpectroscopY) (96 scans, 6 h) spectra, were obtained from ~6 µg (Figure 7.19). Alternatively, the eluent may be collected in an

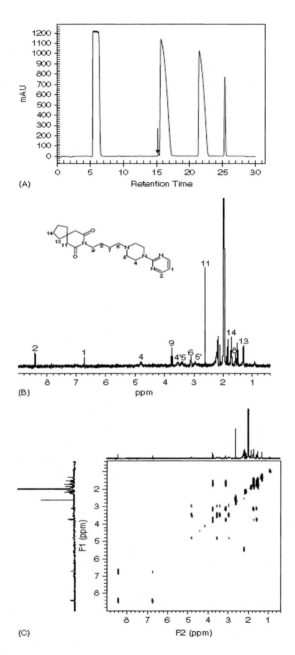

FIGURE 7.19 LC-SPE-NMR study of a trace level (0.02%, 1 μg/mL) component (buspirone). (A) Semipreparative LC chromatogram obtained at a flow rate of 5 mL/min; buspirone concentration (1 μg/mL at 15.3 min). (B) 1D ^{1}H spectrum (64 scans, 3 min) for the analyte of interest isolated in one LC run. Solvent suppression was applied to the water signal. (C) 2D WETTNTOCSY (96 scans, ~6 h) with solvent peaks suppressed. (Reprinted from Xu and Alexander [30]. Copyright 2005, with permission from John Wiley & Sons Inc.)

Eppendorf tube, evaporated to dryness, the residue dissolved in a deuterated solvent, and the solution introduced into a microcoil probe. The use of microcoil technology in trace analysis provides mass sensitivity as well as superior signal-to-solvent ratio relative to that obtained using a 5-mm sample tube [30]. Because analyte amounts are at the low microgram level, solvent dynamic range problems may be present even with a 5-mm NMR cryogenic probe. In such cases, a microcoil flow probe or a cryocapillary probe, which involves the use of <20 μL and 30 μL of solvent, respectively (as opposed to 600 μL in a 5-mm NMR tube) are the preferred probes.

Overall, the coupling of analytical LC-SPE with NMR has significantly reclaimed the sensitivity loss inherent in LC-NMR. When combined with microcoil flow or microcryo technology, the demands for large quantities of material have been substantially addressed.

7.4 ANALYSIS OF MIXTURES

The use of LC-NMR to characterize low-level (<1%), low-concentration (<100 μg/mL) components in a complex mixture is a highly challenging problem in the pharmaceutical industry. The ability to use NMR for structural elucidation of trace amounts of compound can add significant value in metabolite analysis, identifying impurities in drug synthesis scale-up or route optimization, drug stability studies, and the characterization of impurities exceeding regulatory limits (>0.1%) [31–33]. For such studies, analyte isolation and enrichment are the rate-limiting steps. This is because analytical chromatography is interfaced with the NMR spectrometer, where significant amounts of material are required for spectral evaluation. Even when semiprep columns and SPE are integrated, sufficient material still requires multiple collections and cryo- or microcapillary NMR technology.

While the use of semipreparative chromatography coupled to NMR (through SPE) for low-level component analysis [34] may enhance concentration of the sample, this process has an inherently greater propensity for peak tailing and/or peak fronting [35]. In particular, mass overload from the major component can distort the peak shape of minor components. This is true for drug impurity analysis, where the active pharmaceutical ingredient (API) is normally present in 1000-fold excess. In addition, the need to use large injection volumes for poorly soluble APIs can also adversely affect peak width owing to volume overload [35, 36].

Automated multidimensional chromatography [37, 38], achieved through column switching [39], can be a useful strategy for the isolation and analysis of complex mixtures. Reported applications include sample enrichment and target analysis of components in various complex matrices by LC-MS [40, 41].

The coupling of NMR spectroscopy with an online sample preparation system using column-switching HPLC has been reported [39]. However, with NMR, even when similar phases are employed, the two separations needed to ensure that conditions for sample focusing are maintained when the fraction of interest is cut onto the second-stage column [35, 37, 38]. Alternatively, sample preparation by off-line multistep HPLC was more flexible, in that the sample solvent could be optimized for each chromatographic step. This traditional approach, however, is time consuming, expensive, and potentially subject to analyte contamination.

Alexander et al. [43] reported the design and development of a system on the basis of automated 2D HPLC, with the incorporation of SPE and NMR technologyhis system, termed LC2-SPE-NMR, allowed for maximal mass loading in the preparative LC dimension and retained nearly optimal chromatographic resolution. SPE was used to isolate the impurity of interest. The first SPE trap–elute cycle was used to reduce the analyte volume cut from the first-stage column to about 200 μL of organic solvent [43]. This approach allowed a wider range of dilution options to be employed to achieve effective sample focusing when the analyte was injected onto the second-stage column.

LC2-SPE-NMR was applied to study a mixture of buspirone and propranolol, where buspirone was the API and propranolol was the minor component at the 0.1% level [43]. The preparative LC (ρLC) separation (10 × 150 mm column) was carried out under conditions that maximize the column loading and minimize the run time (1-mL injection of a 10-mg/mL sample in water followed by a rapid isocratic elution). Under these conditions, the resolution on the semipreparative column was poor for the analyte of interest (propranolol). However, once cut from the ρLC column, it was well separated from buspirone when rechromatographed in the second LC dimension.

As can be seen, the resulting 1D and 2D ^{1}H-^{1}H TOCSY spectra (Figure 7.20), obtained from approximately 10 μg of analyte, showed sufficient quality for use in structural analysis. Recovery of analyte was estimated at about 90%. It was reported, however, that in other cases

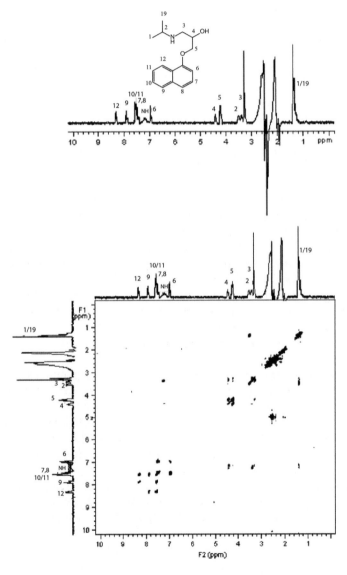

FIGURE 7.20   1D ^{1}H (solvent suppression top: 64 scans, 5 min) and 2D ^{1}H-^{1}H long-range correlation experiments (WETTNTOCSY, bottom; 64 scans, 16 h) of the final isolated analyte of interest (propranolol). (Reprinted from Alexander, Xu, and Bernard [43]. Copyright 2006, with permission from John Wiley & Sons Inc.)

the recovery can be less than optimal. The most significant factor was the variability associated with the efficiency of analyte retention/elution in the SPE trapping step. This issue is also complicated by the fact that NMR spectral quality may be influenced by the intrinsic solubility, aggregation propensity, and molecular dynamics of the analyte in the eluting solvent (sometimes containing acidic or basic modifiers such as TFA). A combination of these factors may contribute to compromised NMR lineshape. Such limitations are generally inherent in all SPE operations and can sometimes be minimized by judicious choice of solvents (e.g., methanol vs. ACN, binary vs. pure solvents) and a better understanding of analyte-absorbent interactions [44].

7.5 FORMULATION ADDUCT

The formulation of an API with excipients, as a blend or a compressed tablet, can lead to the reactivity of the active ingredient with the other components in the mixture. Investigation and identification of such reactive products is a critical part of the development process to ensure the safety and integrity of the drug product. Often such chemicals have unusual structures or mechanisms of formation; hence, it becomes important to isolate and structurally characterize the reacted species. As with degradation products, the potential toxic or mutagenic behavior of these unknown compounds needs to be identified and unambiguously characterized to assess the next steps in the formulation design and development.

An example that highlights the importance and complexity of investigations of formulation reaction products was reported for faldaprevir [45]. Faldaprevir (Figure 7.21a) is a hepatitis C virus protease inhibitor that blocks the action of an enzyme called "NS3/4A serine protease" in the hepatitis C virus [46, 47]. One of the excipients in the faldaprevir formulation was vitamin E-TPGS (Figure 7.21b), which is a known component used to improve the bioavailability of drugs by enhancing absorption and permeability [48]. Vitamin E-TPGS is also a source of natural (+)-α-Tocopherol (Figure 7.21c), which has been reported to possess advantages for drug delivery, including extending the half-life of the drug in plasma and enhancing the cellular uptake of the drug.

While conducting stability studies of an oral formulation containing faldaprevir and vitamin E-TPGS, investigators found the presence of an unknown impurity. The impurity was found to form at relative humidity of 60–75% RH stability, while stored over a 36-month time frame.

Initial LC/MS studies showed that the impurity mass was consistent with the structural components of faldaprevir and (±)-α-Tocopherol minus two mass units; however, the point of attachment and structural arrangement was not discernible. LC/MS UV analysis showed the impurity of interest eluting at 37.61 min. The m/z value of the protonated molecule was 1297.61944, with a molecular formula of $(C_{69}H_{98}N_6O_{11}SBr)^+$ and mass error of 0.172 ppm.

The full structure elucidation of the unknown adduct impurity (Figure 7.22) was solved employing a combination of LC-SPE, MS, and NMR technologies. LC-SPE was used to isolate, purify, and concentrate the impurity for characterization by NMR spectroscopy. The chromatographic method was optimized such that the time per injection was reduced and better separation of the unknown impurity from vitamin E-TPGS was achieved. The LC method was also optimized SPE trapping. Evaluation and optimization of the SPE trapping and elution was important to maximize yield of the desired material. The SPE method development found the Cyano SPE cartridge to provide good trapping efficiency, yielding more than 75% compound recovery.

Because vitamin E-TPGS is highly nonpolar and is the most abundant formulation component, multiple highly concentrated sample injections coated the column with vitamin E-TPGS. Hence, after multiple injections, the SPE-concentrated sample contained ~0.5:1 ratio of vitamin E-TPGS:impurity. An additional chromatographic purification using a new column was used to purify a suitable sample for NMR characterization.

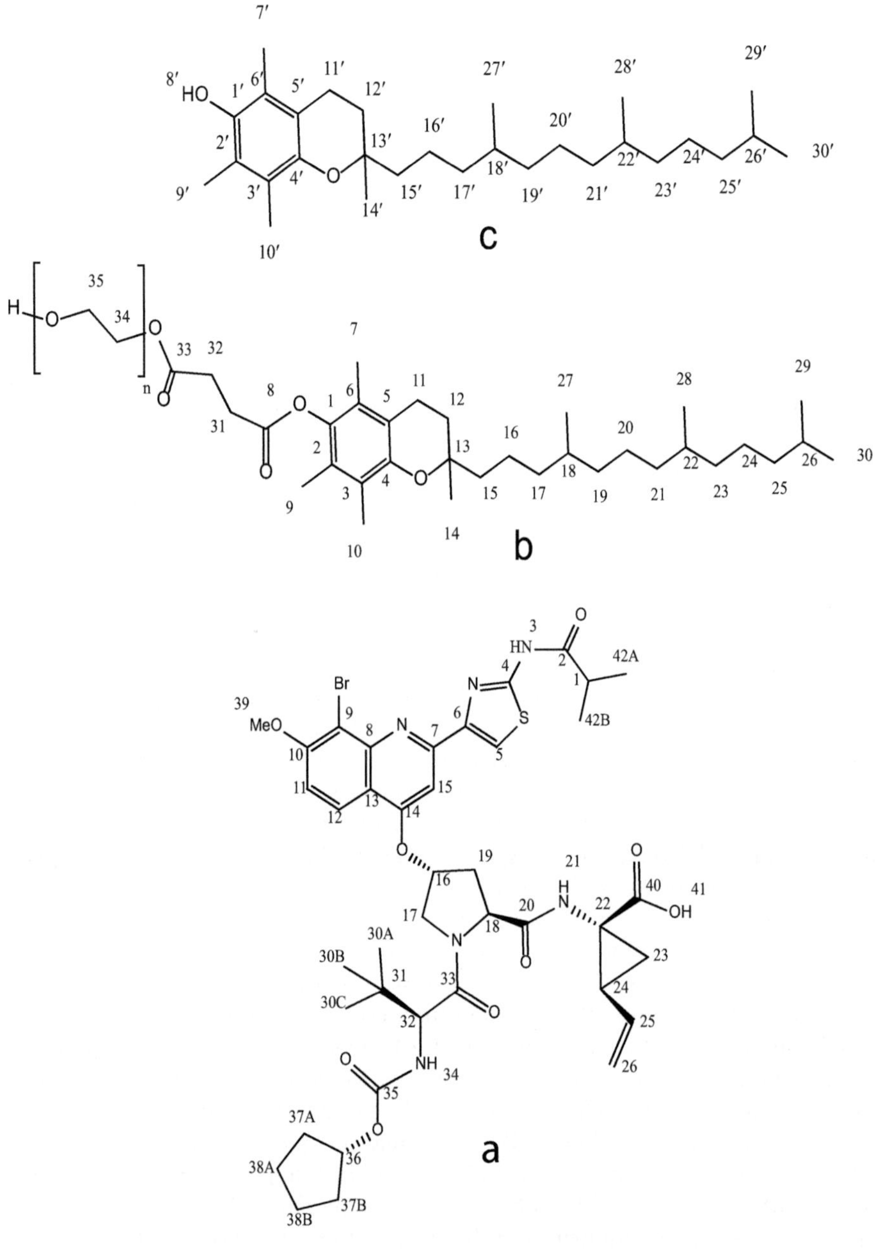

FIGURE 7.21 Chemical structures and numbering of (a) faldaprevir, (b) vitamin E-TPGS, and (c) (±)-α-Tocopherol. (Reprinted (adapted) with permission from Gonnella *et al.* [45]. Copyright 2016, with permission from John Wiley & Sons Inc.)

NMR analysis was carried out in ACN-d$_3$ and DMSO-d$_6$. The 1D spectrum had compromised integrals due to peak overlap, line broadening, impurities, and baseline distortion induced by solvent suppression. Assignments were obtained using WET-COSY, WET-TOCSY, and 2D presat-ROESY. WET-gHSQC and ^{1}H,^{13}C-gHMBC. The WET-gHSQC spectra were used to identify carbon atoms corresponding to CH, CH2, or CH3 groups in

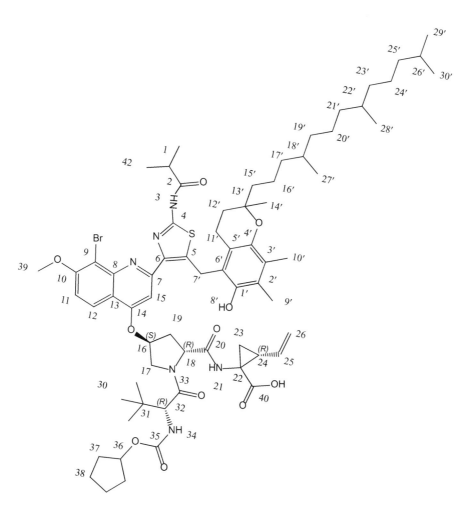

FIGURE 7.22 Numbered chemical structure of faldaprevir-(±)-α-Tocopherol adduct that is consistent with MS data and with the NMR data. Primed numbers correspond to the (±)-α-Tocopherol portion of the molecule. (Reprinted (adapted) with permission from Gonnella et al. [45]. Copyright 2016, with permission from John Wiley & Sons Inc.)

the molecule. Indirect detection obtained from ^{1}H,^{13}C gHMBC spectra established two and three bond connectivities to define the carbon skeleton.

The NMR evaluation of the unknown adduct isolated from the oral formulation showed significant amounts of vitamin E-TPGS causing problems with spectral interpretation due to severe spectral overlap. To obtain a pure sample without vitamin E-TPGS as a spectral complication, the faldaprevir-(+)-α-Tocopherol adduct was synthesized using faldaprevir and (±)-α-Tocopherol (vitamin E). NMR spectra of the synthesized adduct showed good agreement with that of the isolated impurity. The WET-gHSQC spectrum did not show a proton-carbon correlation at position 5 on the faldaprevir segment, suggesting position 5 as a point of attachment on the faldaprevir portion of the adduct.

Chemical shift assignments (^{1}H and ^{13}C) of the adduct obtained in ACN-d$_3$, however, produced NMR data, where key correlations needed to establish through-bond connectivities required to unequivocally define the adduct were not observed. The NMR sample of the synthetic adduct was subsequently transferred to DMSO-d$_6$. The 1D ^{1}H WETDC 1D spectrum of the isolated adduct in DMSO-d$_6$ with assignments displayed is shown in Figure 7.23. Presat-ROESY

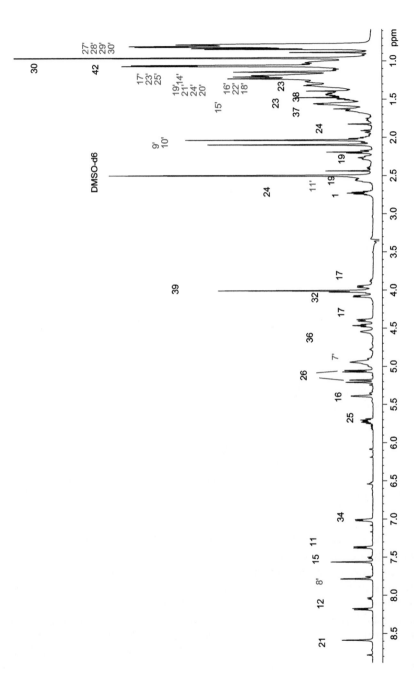

FIGURE 7.23 ^{1}H-NMR (WETDC) spectrum (600.2 MHz) of synthesized faldaprevir-(±)-α-Tocopherol adduct in dimethylsulfoxide-d$_6$. Numbered assignments correspond with the faldaprevir portion of the molecule. Numbered assignments (primed) correspond with the α-Tocopherol portion of the molecule. (Reprinted (adapted) with permission from Gonnella et al. [45]. Copyright 2016, with permission from John Wiley & Sons Inc.)

and qHMBC spectra obtained in DMSO-d_6 (Figures 7.24 and 7.25, respectively) showed key ROESY correlations and gHMBC correlations, not found in ACN, that established the structure of the adduct.

The sample in DMSO-d yielded ^{1}H-NMR spectra with two major advantages over spectra acquired in ACN-d_3. First, the DMSO-d_6 ^{1}H-NMR spectrum showed a major rotamer relative to ACN-d_3, where equal rotameric populations were present. Because DMSO produced a major rotamer, the signal was enhanced, which enabled observation of weak correlations in 2D NMR spectra that would be absent in a 50:50 rotamer population. Also, exchangeable protons not observed in ACN spectra were observed in DMSO-d_6.

In the DMSO-d_6 sample, H8′ hydroxyl proton was observed. This was essential for establishing structural connectivity. In the ROESY spectrum, through-space correlation of H8′ to the methyl proton H9′ and to the new H7′ peak at 4.9 ppm were observed, establishing the proximal position of exchangeable proton H8′ to methylene H7′ and methyl H9′ (Figure 7.24). In the gHMBC spectrum (Figure 7.25), cross-peak connectivities were observed for H7′ and H8′ consistent with the attachment of the 7′ methylene to the α-Tocopherol aromatic ring. The gHMBC correlations observed between H7′ and C5′, C1′, C6′ established the new methylene carbon C7′ bound to the aromatic ring of α-Tocopherol. The qHMBC correlation of H7′ to a carbon at 136.3 ppm was consistent with an expected chemical shift for C5 on the thiazole ring of the adduct (Figure 7.22).

A mechanism of formation was proposed, which showed that vitamin E-TPGS can decompose to produce α-Tocopherol as a free-radical species that can rearrange to a reactive orthoquinone methide (o-QM) intermediate. The o-QM intermediate could then react with other formulation components, including the active pharmaceutical ingredient (API) yielding the adduct of faldaprevir and vitamin E (Figure 7.22) [45]. This finding demonstrated the potential chemical reactivity of vitamin E-TPGS as a formulation ingredient and the importance of isolating and identifying such compounds to ensure safety, efficacy, and shelf life of the API.

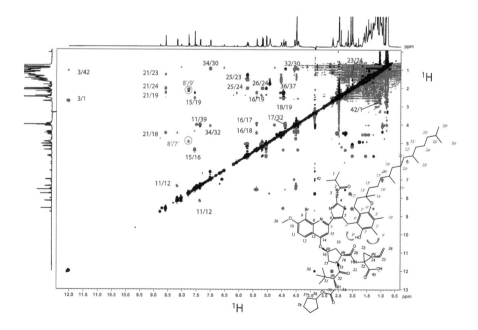

FIGURE 7.24 ROESY spectrum (600.2 MHz) of synthesized faldaprevir-(±)-α-Tocopherol adduct in dimethylsulfoxide-d_6 at 300K. Circled and numbered ROE cross peaks show through-space correlation of H8′ to the methyl proton H9′ and to the new H7′peak at 4.9 ppm, establishing the proximal position of exchangeable proton H8′ to methylene H7′ and methyl H9′.

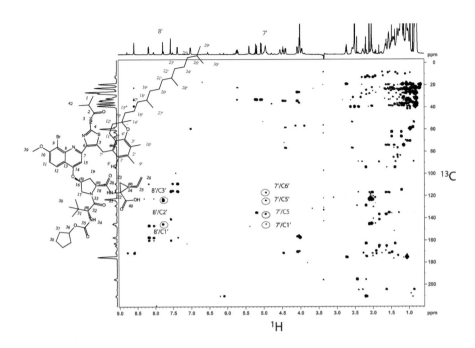

FIGURE 7.25 ¹³C qHMBC spectrum (600.2 MHz) of synthesized faldaprevir-(±)-α-Tocopherol adduct in dimethylsulfoxide-d_6 at 300K. Circled cross peak through-bond connectivities were observed for H7′ and H8′, consistent with the attachment to the α-Tocopherol aromatic ring. The gHMBC correlations between H7′ and C5′, C1′, C6′ established the new methylene carbon C7′ bound to the aromatic ring of α-Tocopherol. The correlation of H7′ to a carbon at 136.3 ppm is consistent with the expected chemical shift for C5 on the thiazole ring of the faldaprevir moiety.

7.6 TAUTOMER KINETICS

A paper by Zhou and Hill [49] reported the use of LC-NMR to study the keto-enol tautomerization of ethyl butyryl acetate **5** (Figure 7.26).

These authors described a new application where LC-NMR enabled the detailed kinetics of keto-enol tautomerization to be understood. This was the first report of LC-NMR being used to study the ketonization kinetics occurring in a thermodynamically favorable direction.

The enol solution was obtained by HPLC and the ketonization rate of the enol was monitored by NMR in a stopped-flow system. The ketonization reaction was examined in different acidic- and solvent-mediated conditions. The acid catalysis and solvent isotope effect showed that proton transfer is the rate-determining step in the reaction. The ketonization rate constants were found to correlate to the water percentage in an exponential relationship with D_2O/CH_3

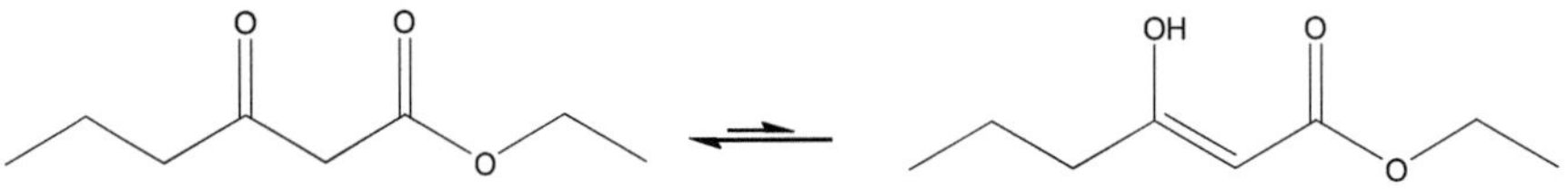

Compound **5**

FIGURE 7.26 Tautomer equilibrium favors keto over enol form of compound **5** [49].

CN and D_2O/CD_3OD systems, and little change in rate constants in the CD_3OD/CH_3CN system was observed.

Overall, the LC-NMR study showed that the ketonization reaction was enhanced by the presence of acid and aqueous solvent, which facilitated proton transfer. These studies provided important information in designing optimal experimental conditions for chemical synthesis.

7.7   UNSTABLE PRODUCTS

Carboxylate-containing drugs have been known to form acyl glucuronic acid conjugates in vivo. The acyl glucuronides are potentially reactive metabolites that have been shown to undergo hydrolysis (regeneration of parent compound) [50, 51], intramolecular rearrangement (acyl migration) [52, 53], and covalent adduct formation with low molecular weight nucleophiles (such as methanol) or with proteins [52, 54–56]. Hence, the identification and characterization of such compounds are of critical importance in pharmaceutical development.

Sidelmann et al. [57] have reported the use of LC-NMR in the "stopflow" mode to separate and structurally identify an equilibrated mixture of ester glucuronide isomers formed by intramolecular rearrangement reactions (acyl migration and mutarotation) of 2-fluorobenzoic acid β-1-glucuronide (1-O-(2 fluorobenzoyl)-D-glucopyranuronic acid) **6**. The equilibrated mixture of isomers was obtained by incubation of the synthetic 2-fluorobenzoic acid glucuronide in buffer solution (pH 7.4) at 25°C for 24 h. The β-anomer of the 1-O-acyl glucuronide and the 2-, 3-, and 4-positional glucuronide isomers (all three as both α- and β-anomers) (Figure 7.27) present in the equilibrium mixture were all characterized following separation in an isocratic chromatographic system containing phosphate buffer at pH 7.4 and 1% ACN in the mobile phase. ACN was chosen as the organic modifier, as methanol can obscure the acyl migration rates by methanolysis [58].

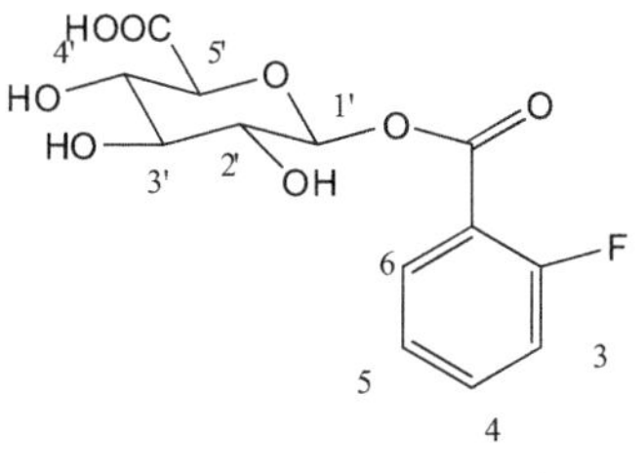

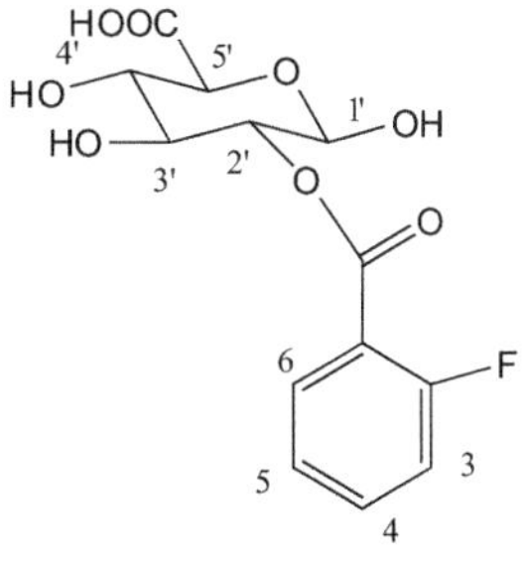

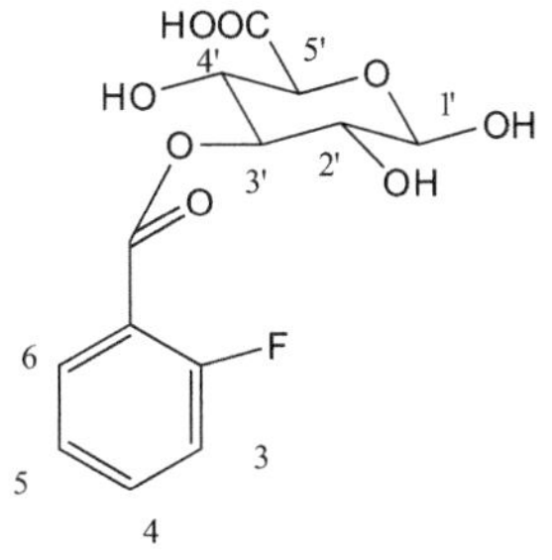

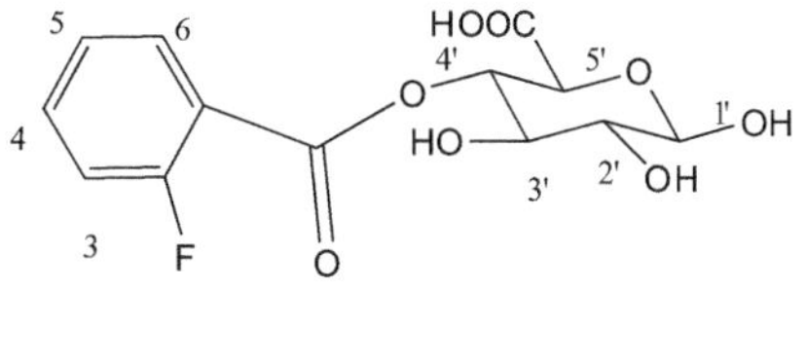

FIGURE 7.27   Acyl migration of 1-O-(2-fluorobenzoyl)-D-glucopyranuronic acid **6** to positions 2', 3', and 4' producing compounds **7**, **8**, and **9**, respectively [57].

Proton NMR spectra were obtained using a 100-μL flow probe. The investigators showed that the 4-*O*-acyl isomers eluted first from the chromatographic column, the α-anomer eluting before the β-anomer. The β-anomer of the 1-*O*-acyl isomer eluted next, followed by the 3-*O*-acyl isomers; this time the β-anomer before the α-anomer. Finally, the 2-*O*-acyl isomers were eluted with the α-anomer before the β-anomer as seen with the 4-*O*-acyl isomers. The ¹H-NMR spectra obtained from peaks corresponding to 4α, 3α, and 2α (compounds **7**, **8** and **9**, respectively) are shown in Figure 7.28 and compared with the ¹H-NMR spectrum obtained from the equilibrium mixture.

The LC-NMR investigations also elucidated the mutarotation of the positional glucuronide isomers. Because mutarotation rates are fast relative to the acyl migration rates, later eluting peaks undergo mutarotation while stopped on the chromatographic column. Even when the peaks were not stopped on the column, significant mutatrotation was found to occur due to column α- and β-isomerization. For NMR detection, this resulted in line broadening and had detrimental effects on the S/N and acquisition time. By increasing the ACN content to 5%, the retention times of the α- and β-anomers were reduced, enabling less time spent on the column. Capture of the compound in the flow cell allowed full structure assignment by LC-NMR.

These results show that this LC-NMR method was successfully employed to study acyl migration reactions of nonsteroidal anti-inflammatory drug glucuronides. Such studies are

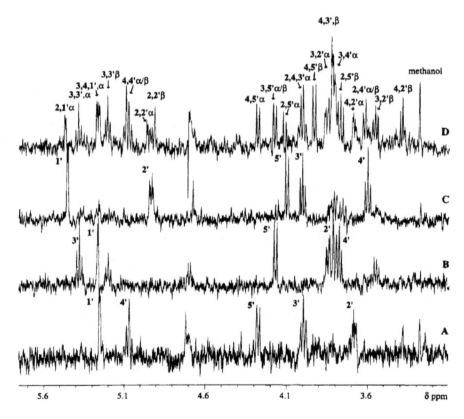

FIGURE 7.28 ¹H-NMR spectra obtained from the chromatographic peaks: (a) peak corresponding to the α-4-*O*-acyl isomer **7**, (b) peak corresponding to the α-3-*O*-acyl isomer **8**, and (c) peak corresponding to the α-2-*O*-acyl isomer **9** shown and compared with (d) the ¹H-NMR spectrum obtained from the equilibrium mixture. (Reprinted in part from Sidelmann et al. [57]. Copyright 1995, with permission from the American Chemical Society.)

important due to the possible relationship between acyl migration and toxicological properties. Because of the dynamic equilibrium of glucuronides in acyl migration transformations, the products of interest may be elusive to conventional isolation methods and structural characterization procedures. LC-NMR, however, was shown to have an advantage in situations that would involve extremely time-consuming purification of compounds that are in dynamic equilibrium and are constantly interconverting. In addition, LC-NMR has an advantage over LC-MS, which would not be able to detect positional isomerism for such cases.

Hence, LC-NMR was found to be of significant value in studying transient chemical isomerization reactions such as acyl migration. Application to the investigation of drug glucuronide reactivity and related protein binding and toxicological problems has been shown to be quite effective using this technology.

7.8 METABOLITES

The integration of SPE with LC-NMR to analyze metabolites in biological samples was first reported by Bruker scientists in 2004 [59]. The study involved the identification of known metabolites of paracetamol in human urine. The example provided a clear demonstration of the utility of SPE trapping over conventional LC-NMR. The use of SPE enabled the chromatography separation to be conducted without deuterated water, preventing deuterium exchange. Subsequent elution of the SPE cartridge with deuterated solvents not containing exchangeable protons (i.e., ACN-d_3) reduced dynamic range issues in detection and enabled NMR observation of exchangeable protons. The possibility of multiple trapping of the same peak on the same cartridge following multiple chromatographic injections vastly improved the amount of component in the NMR detection flow cell, producing significant gains in signal intensity.

A study reported by Schlotterbeck et al. described the use of LC-NMR to identify the bioactive metabolites of a lead mGlu5 receptor antagonist [60, 61]. Incubation of a thiazole-derived compound with the isolated CYP (cytochrome P450) isoforms responsible for the metabolism was followed by the collection of the biomass-free supernatant, concentration, and LC-SPE-NMR-MS analysis. The eluate was trapped on a Spark Prospect 2 SPE system, with 5% of the eluted volume directed to an ESI mass spectrometer. After drying and elution (using deuterated solvent) of the SPE cartridge, NMR spectra were collected on a 5-mm TCI cryoprobe. High-quality 1D and 2D spectra were acquired. The spectra and structure of one of the key metabolites of a lead metabotropic glutamate receptor 5 inhibitor are shown in Figure 7.29 [62].

This study allowed the full elucidation of the major metabolic pathways affecting the lead molecule, including the single CYP isoform responsible for the formation of each metabolite. The work led to the resynthesis of two major metabolites that were found to contribute to the efficacy of the parent compound in *in vivo* models. In addition, the information enabled the design of more metabolically stable analogs.

An important goal in metabonomics studies is to fully understand the impact of genetic modifications and toxicological interventions on the network of transcripts, proteins, and metabolites found within cells, tissues, or organisms. Because the aim of metabonomics is to detect low levels of small-molecule metabolites and xenobiotics from biofluids to organisms, MS has a sensitivity advantage over ^{1}H-NMR spectroscopy. This sensitivity disadvantage of NMR is being addressed by the development of magnets with increased field strength, cryogenically cooled probes [63], and microprobes [64]. An advantage of NMR spectroscopy, compared with other analytical tools, is that the technique is noninvasive, leading to possible medical applications designed to detect molecules *in vivo* (magnetic resonance imaging and magnetic resonance spectroscopy). In addition, high-resolution ^{1}H-NMR spectroscopy can be used in conjunction with statistical pattern recognition to globally profile metabolites, thereby reducing the need for the quantities of material that are required for full structure elucidation.

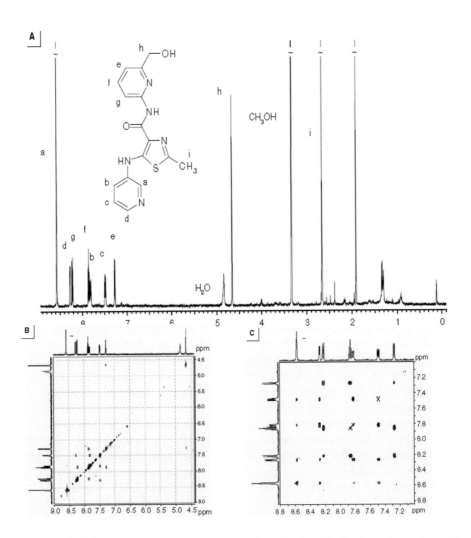

FIGURE 7.29 Nuclear magnetic resonance spectra acquired after liquid chromatography–solid phase extraction yielded assignment of a metabolite of a lead metabotropic glutamate receptor 5 inhibitor. (a) ^{1}H-NMR spectrum at 600 MHz, (b) nuclear Overhauser enhancement spectroscopy, (c) total correlation spectroscopy. (Reprinted in part from Schlotterbeck and Ceccarelli [62]. Copyright 2009, with permission from the American Chemical Society.)

NMR-based metabonomics has been found to be an ideal technique for screening human populations for common metabolic disorders. Expert systems have been built that predict both the occurrence and severity of coronary artery disease using blood plasma samples [65] (Figure 7.30). Development of such systems for a clinical application can result in significant financial savings over invasive angiography, currently the gold standard for diagnosis.

Although most metabolomics studies have been carried out using ^{1}H-NMR spectroscopy, there is a relatively small chemical shift range for the ^{1}H nucleus resulting in significant overlap and spectral complexity. An alternative, using ^{13}C NMR spectroscopy, where the resonances are spread over an ~200 ppm range seems attractive; however, ^{13}C natural abundance is only 1.1%, creating a severe sensitivity issue for direct observation. To address this issue, cryoprobes have been employed that allow natural abundance detection of metabolites

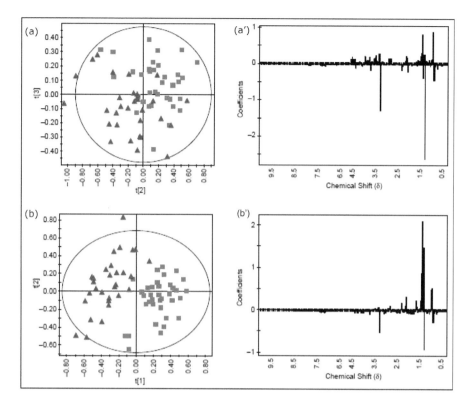

FIGURE 7.30 Comparison of patients with severe atherosclerosis (TVD) and patients with normal coronary arteries (NCA). (a) Multivariate classification scores plot showing the considerable separation achieved between NCA (▲) and TVD (■) samples. (a') Regression coefficients derived from the PLS-DA model, where each bar represents an NMR spectral region covering 0.04 ppm, showing how the ^{1}H-NMR profile of the TVD samples differed from the ^{1}H-NMR profile of the NCA serum samples. A positive value indicated there was a relatively greater concentration of metabolite (assigned using NMR chemical shift assignment tables) present in TVD samples, and a negative value indicated a relatively lower concentration. (b) Multivariate classification scores plot after application of the orthogonal signal correction data filter to remove uncorrelated variance components. Considerable improvement in separation was achieved (compared with a), (b') The regression coefficients for the multivariate classification method model using the OSC transformed dataset. (Reprinted by permission from Macmillian Publishers Ltd., Nature Medicine, Nature Publishing Group, 2003) [65].

as noted previously. The cryoprobe approach relies on cooling the NMR radiofrequency detector and preamplifier to 20 K. Because thermal noise is reduced by a factor equivalent to $T^{1/2}$, the thermal noise is reduced 4-fold, giving a 16-fold reduction in acquisition time for the same S/N using a conventional probe. Using cryoprobe technology, Keun and colleagues readily detected hepatic toxicity, employing ^{13}C NMR spectroscopy of urine, to detect metabolites via natural abundance ^{13}C nuclei [63].

NMR-based metabonomics shows great promise of correlating metabolic phenotypes with other -omic technologies, such as transcriptional analysis and proteomics during drug toxicity. Because DNA microarrays and global protein profiling techniques are technically challenging and relatively costly, it is often prudent to target the analysis to key time points. Metabonomics provides a mechanism for identifying key time points and metabolic events to be further investigated.

Work conducted by Griffin et al. [66] has examined orotic acid-induced fatty liver disease in rats using metabonomics, transcriptomics, and proteomics. In these studies, NMR spectroscopic changes in blood plasma and urine could be monitored alongside a three-tiered analysis of liver tissue, placing changes in the liver in context with the overall global metabolism of the animal. By providing a metabolic phenotype, different time points and strains of rats could be compared directly in the subsequent analyses, rather than trying to assess different strains as fast or slow responders to the insult. Transcriptional changes could be modeled in terms of a natural "metabolic time" rather than an artificial sampling time. Using this approach, comparison of the in-bred Kyoto strain and the out-bred Wistar strain of rat showed that Kyoto rats were susceptible to fatty liver accumulation, and metabonomic analysis identified that this strain of rat had an increase in cytosolic lipid triglyceride content over the outbred Wistars. The metabolic changes detected in liver tissue consisted of increases in lipid triglycerides and cholesterol esters, as well as choline-containing metabolites and their degradation products. These changes could be correlated with key transcriptional changes from a microarray study, identifying metabolic pathways that were perturbed by the orotic acid exposure. Using regression tools, the time profiles of both transcriptional and metabonomic data sets were modeled together, separating transcripts according to which pathways they were most correlated.

Joint metabonomic and transcriptomic approaches to drug toxicity have been used in drug assessment. Ringeissen and coworkers [67] investigated the action of peroxisome proliferator-activated receptor (PPAR) ligands, using an NMR-based metabonomic study of urine to define the key metabolite changes during peroxisome proliferation. The aim of the study was to identify cellular changes or pharmacologically related changes in the endogenous metabolism of rats treated with PPAR ligands that might correlate with the extent of peroxisome proliferation. These studies were of interest due to the compelling evidence suggesting peroxisome proliferation is correlated with hepatocarcinogenesis. However, the mechanism by which this class of chemicals induce liver tumors is not understood.

The investigators were able to correlate changes in N-methylnicotinamide (NMN) and N-methyl-4-pyridone-3-carboxamide (4PY) concentrations, with peroxisome proliferation as measured by electron microscopy. Using real-time polymer chain reaction (PCR) to confirm transcriptional changes of key enzymes involved in producing NMN and 4PY, the tryptophan-NAD+ pathway responsible for the observed changes was identified. In addition, a ^{1}H-NMR spectroscopy study for monitoring adverse drug reactions/pathologies was applied to identify urinary or plasma biomarkers that may be directly correlated with the pathology. Spectra were generated that contained data on many endogenous metabolites, and provided information on patterns of excretion of endogenous metabolites or single molecules. Hence, these investigators showed how metabonomics could be used to go from a complex multivariate problem involving systemic metabolism changes to identifying two biomarkers that could be measured by HPLC to monitor peroxisome proliferation.

Metabonomics has been used to determine the metabolic phenotype or "metabotype" of two different strains of mice, the AlpK:ApfCD (white) and C57BL107 (black) mouse [68]. From principal component analysis (PCA) of the urinary NMR spectra, it was possible to separate the two strains and to predict the strain of the mouse in 98% of cases using partial least squares-discriminant analysis (PLSDA) (Figure 7.31a). From comparison of the ^{1}H-NMR spectra, AlpK:ApfCD mice had higher elevated levels of 2-oxoglutarate, citrate, trimethylamine-N-oxide, and guanidinoacetic acid (Figure 7.31b), whilst C57BL107 mice had higher levels of taurine, creatinine, dimethylamine, and trimethylamine (Figure 7.31c). Perturbations in urinary metabolites were postulated to be the result of strain differences in enzyme activity, pathway flux, and metabolite excretion.

The ability of NMR and pattern recognition to identify genotype clearly has promising applications when studying genetic polymorphism and genetically modified animals that are

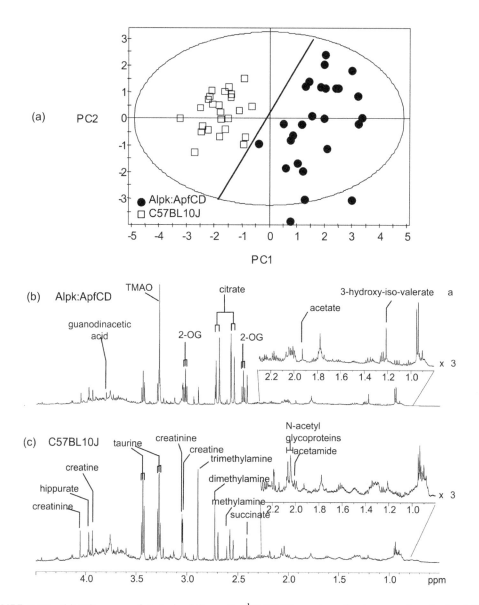

FIGURE 7.31 (a) PC scores plot derived from the ^{1}H-NMR spectra of urine samples obtained from Alpk:ApfCD and C57BL10J mice. (b) 500-MHz ^{1}H-NMR spectra of typical urine samples obtained from Alpk:ApfCD mouse, and (c) C57BL10J mouse. Visual inspection of the ^{1}H-NMR urine spectra revealed changes in the patterns associated with a variety of differences in the two strains. (Reprinted from *FEBS Lett.* 484(3), Gavaghan CL, Holmes E, Lenz E, Wilson ID, Nicholson JK. An NMR-based metabonomic approach to investigate the biochemical consequences of genetic strain differences: application to the C57BL10J and Alpk:ApfCD mouse. (Copyright (2000), with permission from Elsevier.) [68].

often used as models of human diseases [69, 70]. Metabonomic techniques have also been used to differentiate morphologically indistinguishable but phenotypically different species of earthworm, based upon the analysis of tissue extracts and celomic fluid, using ^{1}H-NMR spectroscopy and multivariate statistics [71].

The advantages of increasing the sensitivity and throughput of NMR-based metabolomics have been reported by Grimes and O'Connell [72]. In their paper, the application of microcoil NMR probe technology was compared with a standard 5-mm NMR probe in metabolomics studies of urine and serum. Sample concentration was evaluated to demonstrate the utility of the greatly improved mass sensitivity of microcoil probes. Concentrating the samples was found to improve the signal detection, but the benefits did not follow a linear increase. Although absolute quantitation was compromised, an analysis of relative concentrations was nonetheless possible.

To evaluate the potential for sample concentration with biofluids, a set of samples from five subjects were analyzed with zero, twofold, fourfold, and sixfold concentration. Spectral examples are shown for urine in Figure 7.32. Sample concentration greater than sixfold was not carried out, since it was expected that the salt concentration would be too high to obtain quality spectra.

The results from this study showed that concentration of urine and serum samples alters the absolute concentrations of the metabolite levels differently. The investigators reported that the absolute concentration of valine in a sixfold concentrated sample of urine was approximately sixfold greater than the unconcentrated sample, but for tyrosine, the concentration was reduced by 30%. These discrepancies in absolute concentration vs. sample concentration may be attributed to factors such as protein binding and precipitation of metabolites. Hence the absolute concentrations of metabolites can be affected by sample concentration, which raises considerations regarding the benefits of obtaining absolute concentration with the 5-mm probe vs. the mass sensitivity of the microcoil probe. Further investigation showed that concentration does not increase the noise of the data. To address this issue, the coefficient of variance of each of the metabolites at the concentration levels was calculated. No significant increase in the variability of the data at the higher concentration levels was detected, and hence it was concluded

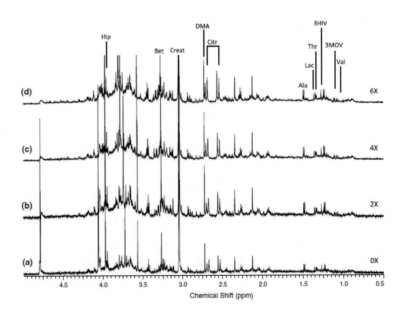

FIGURE 7.32   Urine spectra over a range of sample concentrations. The samples were 2×, 4×, and 6× concentrated. The top spectrum shows quantifiable metabolites annotated. Val, valine; 3MOV, 3-methyl-2-oxovalerate; 3HIV, 3-hydroxyisovalerate; Thr, threonine; Lac, lactate; Ala, alanine; Citr, citrate; DMA, dimethylamine; Creat, creatinine; Bet, betaine; Hip, hippurate. (With kind permission from Springer Science + Business Media. Grimes and O'Connell, Figure number 3.) [73].

that if concentration will enhance metabolite detection, the relative concentrations across samples can be compared with no anticipated increase in variance. Overall, the data demonstrated that the microcoil probe is a highly valuable technology for improving sensitivity in metabolomics studies, especially when sample volumes are limited. Increasing sample concentration was shown to be a viable option to increase the sensitivity and throughput in NMR studies of metabolites. However, care must be exercised, since the increase is not linear and changes with both the matrix and specific metabolites can occur.

7.9   cITP ISOLATES

Separation and analysis of nanomole quantities of heparin oligosaccharides was reported by Korir et al. in 2005 [73]. These investigators employed the use of online capillary isotachophoresis and NMR spectroscopy to isolate and structurally characterize these molecules. To boost the mass sensitivity of NMR detection, solenoidal microcoil probes were used with nanoliter to microliter detection volumes [74–78].

It should be noted that microcoil NMR probes have poorer concentration detection limits compared to probes with a traditional Helmholtz coil design. Therefore, capillary isotachophoresis (cITP) was used to concentrate charged analytes prior to NMR detection. This helped to improve the detection limits of the microcoil probes [79].

Using cITP, analytes were separated according to their electrophoretic mobilities by applying a high electric field (10–30 kV) across a capillary containing a discontinuous buffer system composed of a leading electrolyte (LE) and a trailing electrolyte (TE). Careful selection of LE and TE was needed to optimize the separation conditions. When conditions were optimal, effective concentration resulting in enhanced sensitivity of NMR detection could be achieved [80].

A major challenge of this study resulted due to the high negative charges of glycosaminoglycans (GAGs), which necessitated the anionic mode of cITP for the separation of these compounds. Anionic cITP NMR experiments require the development of suitable buffer systems and overcoming electroosmotic flow (EOF), which opposes the migration of anions and degrades the ability of cITP to separate and focus the analyte bands. Figure 7.33 shows the structures of the oligosaccharides **10**, **11,** and **12** used in the study.

Online cITP NMR spectra were obtained for the analysis of disaccharide, **12**, as shown in Figure 7.34.

Spectra a and b contain the resonances of the LE, while spectra c–g are a time profile of the focused compound **12** band. Spectrum g is a time average of the sample/TE interface, with the TE detected for this sample by the change in imidazole chemical shift between spectra g and h. The acetyl proton signal, which appears at 2.1 ppm in the analyte spectra c–g, distinguishes **12** from **11**. The authors also reported attempts to study the tetrasaccharide composed of the disaccharide **10**; however, the results were disappointing due to poor S/N of the tetrasaccharide spectra even with increased signal averaging.

Postacquisition coaddition of spectra of **11** is shown in Figure 7.35a with a similar post acquisition coaddition of the analyte **10** spectra shown in Figure 7.35b. These combined spectra provide enhanced S/N.

Comparison of the combined spectra in Figure 7.35 showed the spectra were distinctly different with the structural difference of a sulfate group in compound **10** instead of a hydrogen atom in compound **11**. Although the spectral features in the coadded NMR spectra are useful in distinguishing the disaccharides, they do not provide sufficient information for complete structure elucidation. Stopped-flow 2D NMR experiments and higher sample concentrations would be required to enable full structure characterization. Overall, it was encouraging that the investigators were able to begin using NMR to explore the relationship between biological function and the microstructures of GAGs. However, sensitivity still appears to be an acute limiting factor requiring further experimental design and development.

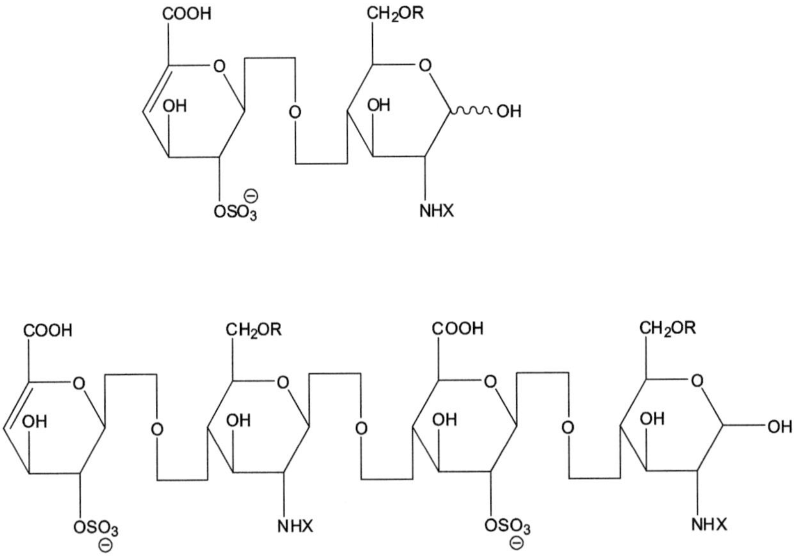

	Disaccharide	R	X
10	α-δUA-2S[1 → 4]-GlcNS-6S	SO_3^-	SO_3^-
11	α-δUA-2S[1 → 4]-GlcNS	H	SO_3^-
12	α-δUA-2S[1 → 4]-GlcNAc	H	$-COCH_3$

FIGURE 7.33 Structure of (a) the heparin disaccharides **10**, **11**, and **12** and (b) tetrasaccharide used in the study [73].

Because NMR has relatively poor sensitivity compared with other techniques such as MS, applicability in impurity analyses has historically been limited. This limitation was addressed through the online coupling of microcoil NMR with cITP, a separation method that can concentrate dilute components by two to three orders of magnitude. With this approach, ^{1}H-NMR spectra can be acquired for microgram (nanomole) quantities of trace impurities in a complex sample matrix [81].

cITP-NMR was used to isolate and detect 4-aminophenol (PAP) in an acetaminophen sample spiked at the 0.1% level, with no interference from the parent compound. Analysis of an acetaminophen thermal degradation sample revealed resonances of several degradation products in addition to PAP, confirming the effectiveness of online cITP-NMR for trace analyses of pharmaceutical formulations. Subsequent LC-MS/MS analysis provided complementary information for the structure elucidation of the unknown degradation products, which were dimers formed during the degradation process.

The results of the degradation of acetaminophen are shown in Figure 7.36. This study involved a forced degradation of the sample in D_2O. The forced degradation conditions were based on literature studies that document that the degradation of acetaminophen is accelerated at high temperatures and low pH [82]. The spectra in Figure 7.36 shows the cITP-NMR data obtained for thermally degraded acetaminophen in the tablet matrix. The aromatic resonances of the focused PAP begin to appear in spectrum A and become more intense in spectra B–F. In spectra G–K, a set of resonances are observed from unknown degradation products stacked behind the PAP. The changes in relative resonance intensity in spectra G–K suggest that these

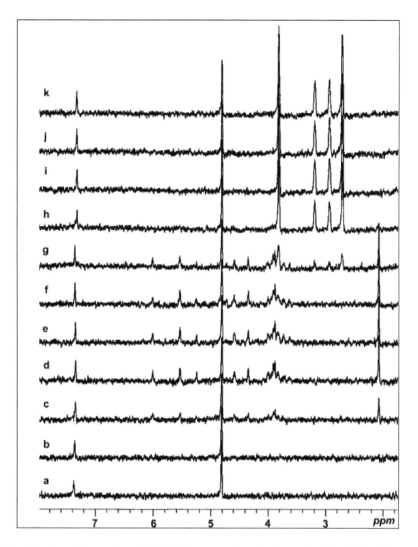

FIGURE 7.34 Results of the online cITP NMR spectra for the disaccharide α-δUA-2S[1→4]-GlcNAc, **12**. (Reprinted with permission from Korir et al. [73].)

spectra are produced by more than one component. By spectrum L, the focused analyte band has left the NMR microcoil, and only the HOD resonance is detected. In this experiment, cITP-NMR was able to cleanly concentrate, separate, and detect trace amounts of PAP in a real degradation sample and provide spectral evidence for the presence of additional degradation products. In these studies, the unknown compounds in spectra G–K cannot be fully characterized, since isolation of the pure components in sufficient quantities would be required. However, some structural properties can be inferred. Since these compounds are positively charged at the operating pD (poly dispersity), they have a lower electrophoretic mobility than PAP and, therefore, likely have a higher molecular weight. The compounds have aromatic resonances more complicated than PAP, suggesting dimerization as a possible mechanism for their formation, and at least one component retains an acetyl moiety, as indicated by the resonance at 2.61 ppm. Further characterization of these unknown components was achieved through isolation and analysis by MS.

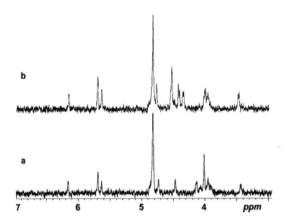

FIGURE 7.35 ¹H-NMR spectra of the disaccharides (a) compound **11** and (b) compound **10** obtained by postacquisition coaddition of the cITP NMR spectra for each analyte. (Reprinted with permission from Korir [73].)

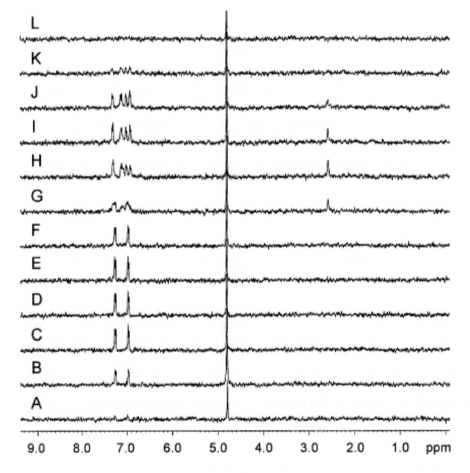

FIGURE 7.36 A portion of the online cITP-NMR spectra showing resonances of 4-aminophenol (PAP) (spectra B–F) and unexpected unknown components (spectra 2G–K) from the degradation sample in D_2O. Spectra A and L were measured for the LE and TE, respectively. (Reprinted with permission from Eldridge et al. [81]. Copyright (2007) American Chemical Society.)

Isolation of the focused analyte in off-line cITP experiments using nondeuterated water was carried out, and pooled isolates from 10 replicate separations were dissolved in 60 μL of 0.1% formic acid, and 5 μL of this sample was injected into the LC-MS. The LC-MS/MS results for the four unknown compounds, along with their proposed structures, were solved (Figure 7.37). The unknown products were expected to consist of two pairs of compounds

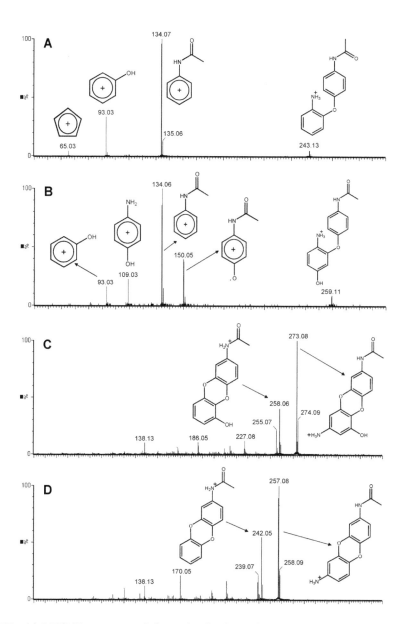

FIGURE 7.37 (a) MS/MS spectrum of the molecular ion *m/z* 243.125. The fragmentation pattern in (a) is complementary to that of (b), *m/z* 259.113. (c) MS/MS spectrum of molecular ion *m/z* 273.080, which is also similar to (d) *m/z* 257.077. The proposed structures for the unknown thermal degradation products are labeled on the far right next to the corresponding parent ion. The structures postulated for specific fragments are also shown for each mass spectrum. The MS data supported the LC-NMR results. (Reprinted with permission from Eldridge et al. [81]. Copyright (2007) American Chemical Society.)

with similar structures, m/z 243.125 and 259.113 and m/z 273.080 and 257.077, with a mass difference of 16, separating the members of each pair. The MS/MS fragmentation patterns for each pair of compounds were highly complementary. The fragment ions m/z 65.032, 93.029, 134.057, and 150.055 were acetaminophen fragments, suggesting that these products retain a part of the acetaminophen structure [82–84]. The MS/MS fragmentation patterns for peaks at m/z 273.080 and 257.077 were also complementary. Hence structures deduced by LC-MS for the four dimeric degradation products were consistent with the LC-MS/MS fragmentation patterns observed as well as the cITP NMR results.

7.10   NATURAL PRODUCTS

Plants are known to be a valuable resource for humans due to their chemical components that possess therapeutic properties. Of the 350,000 known plant species, only a small percentage has been investigated for their pharmacological potential [85]. This is the case in spite of the fact that over 40% of registered drugs are either natural products or derived from natural-product scaffolds [86]. Because there is a wealth of potential for the discovery of new natural products with biological activity, this area represents an important source for lead finding in drug discovery [87].

Many plants contain compounds that could have biological activity against key therapeutic targets, but they may also contain known therapeutic entities or known compounds that hold no interest as therapeutic targets or leads. Since isolation and characterization of every component in natural-product extracts can be tedious and time consuming, a process to profile plant extracts for new metabolites becomes important. Metabolite profiling of crude plant extracts can be challenging, since the components may possess very different structures; hence each needs to be independently isolated and characterized. Several rapid and powerful technologies have been applied to identify the components of such mixtures. These include LC-UV-DAD) [88, 89]; however, these technologies do not always provide conclusive or unambiguous results. In addition, structural identification can benefit from comparisons with spectra in commercial databases, which are not readily available for many technologies.

With respect to LC-NMR, however, this technology can yield a powerful means of rapidly scanning structural features of compounds as they elute off of a column. Continuous-flow LC-NMR experiments have been reported that put forth a strategy for efficient dereplication of natural-product crude extracts along with online identification of bioactive constituents based upon combinations of LC-NMR with LC-UV/DAD, LC-MS, and LC-MS MS [90].

Because plant extracts are usually obtained by maceration of dried plant material and extracted with either methylene chloride or methanol, these extracts may contain hundreds of compounds. Such compounds are usually separated using reverse-phase chromatography with gradients of MeCN:D_2O or MeOH:D_2O [91]. Sensitivity issues with respect to LC-NMR require on-column loading that delivers the highest amount of separated analyte in the lowest possible elution volume into the NMR flow cell. Crude plant extracts may require milligrams per injection. To address sample-loading issues, columns with increased length (e.g., 250 mm) or large inner diameter (8 mm) may be used. In addition, since the components of these crude extracts may have poor solubility, dissolution in DMSO may be required for injection onto the column.

When considering a continuous-flow experiment, chromatographic conditions need to be established that enable separation of large amounts of sample with satisfactory LC resolution. Such experiments often allow ^{1}H-NMR characterization of the main constituents; however, the minor components may be lost. To improve NMR detection, low flow (<0.1 mL/min) or stopped flow may be used. Both collection modes enable a higher number of transients to be recorded.

A natural-product isolation and characterization of plant extracts was reported by Zanolari *et al.* [92], where LC-NMR was used in the analysis of the alkaloid from *Erythoxylum vacciniifolium*, a Brazilian Erythroxylaceae, called "catuaba". These investigators used a combination of LC-UV-DAD and LC-APCI-MS to examine the extracts. Preliminary analysis showed similar UV spectra and MS^2/MS^3 showed similar low-molecular-weight fragments that suggested a common core structure in all of the components.

Continuous-flow LC-NMR and stopped-flow LC-NMR were subsequently performed at 1 mL/min on 3 mg of the extract. Based upon chemotaxonomic data on the genus *Erythoxylum* [93], the main components were proposed to be tropane alkaloids esterified with pyrrolic acid moieties. The main constituent was subsequently isolated for complete 2D NMR structural characterization. The structures of all other components were identified based upon comparisons with ^1H 1D NMR data, which showed that the other components were analogues of the parent catuabine D (Figure 7.38).

Examination of the data in Figure 7.33 showed that the continuous-flow LC-NMR ^1H spectrum recorded for compound **24** at 58 min was similar to the parent compound **22**. In both cases, a tropane moiety was observed that was deoxygenated at C-3 and C-6. The difference between the two compounds was in their ester substituents. Two methyl groups

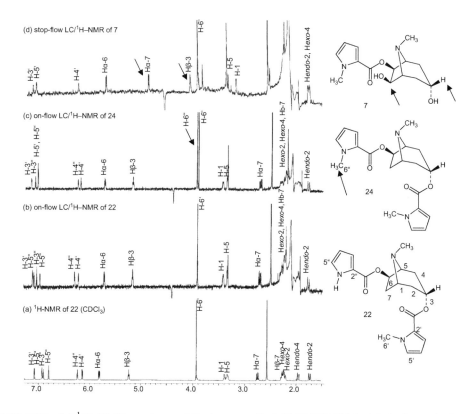

FIGURE 7.38   (a) ^1H–NMR spectrum of the tropane alkaloid **22** ($CDCl_3$) isolated after a single step of MPLC. (b) Continuous-flow $LC/^1H$–NMR spectrum of **22**. (c) Continuous-flow $LC/^1H$-NMR spectrum of **24**. (d) Stopped-flow $LC/^1H$-NMR of **7**. Comparison of the different spectra enabled the identification of the major constituents of the alkaloid extract of *Erythroxylum vacciniifolium* (Erythroxylaceae). (The main differences are indicated with arrows.) (Reprinted from Wolfender et al. [90]. Copyright 2005, with permission from John Wiley & Sons Inc.)

were observed for **24** at δ 3.93 ppm (3H, s, N-CH₃) and 3.94 ppm (3H, s, N-CH₃) in the continuous-flow LC-NMR ^{1}H spectrum, indicating two methylpyrrrole acid moieties. The work was subsequently corroborated by MS ([M+H]$^+$: 372.1954 $C_{20}H_{25}N_3O_4$). A similar type of analysis was carried out for compound **7**.

These same investigators also reported the implementation of LC/^{1}H-NMR and corresponding bioassays for the rapid identification of bioactive compounds. After extensive LC-UV-MS studies, it was concluded that the isolates primarily consisted of prenylated isoflavanones or isoflavones. Continuous-flow LC/^{1}H-NMR was then carried out on an 8-mm diameter C-18 column. Ten milligrams of material was injected and separation was achieved at 0.1 mL/min. The NMR number of scans was adjusted to 256 to improve signal. During the LC-NMR analysis, an LC microfractionation of peaks was performed every 10 min for antifungal bioautography assays against *C. cucumerinum*. The assay results showed distinction in the antifungal activities and the NMR results confirmed MS findings that the compounds were in fact prenylated isoflavanones or isoflavones. Two compounds **27** and **28** were examined in detail (Figure 7.39). Compound **28** exhibited the most potent antifungal activity. Extensive UV and MS evaluation of **28** along with extensive analysis of NMR coupling patterns and chemical shifts all supported the identification of **28** as isowighteone, a known prenylated isoflavone.

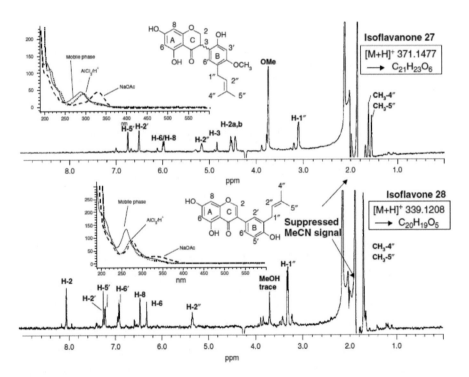

FIGURE 7.39 On-flow LC-NMR spectra of an isoflavanone **27** and the antifungal isoflavone **28**. Complementary UV-DAD and shifted UV-DAD spectra recorded online as well as exact molecular formula based on LC/APCI-Q-TOF-MS data are displayed. The LC-NMR continuous-flow experiments were conducted under the following conditions: HPLC was conducted using a C-18 column, Waters RCM 8 d 10 (100 d, 8 mm inner diameter (ID)) and MeCN:D₂O gradient (5: 95 to 100: 0; 19 h); 0.05% TFA; 0.1 mL/min. NMR spectra were acquired using: a 500-MHz spectrometer, 265 scans/increment; 60-μL flow cell (3 mm ID). (Reprinted from Wolfender et al. [90]. Copyright 2005, with permission from John Wiley & Sons Inc.)

Structural characterization of **27** showed a UV spectrum and ^{1}H-NMR signals at δ 4.39 ppm (H2a), δ 4.47 ppm (H-2b), and δ 4.80 ppm (H3) expected for an isoflavanone core. The protonated molecule at m/z 371.1477 [M=H]$^+$ suggests a molecular formula of $C_{21}H_{22}O_6$ with the consequential groups of 3 OH, 1 OMe, and 1 prenyl. The NMR 1D ^{1}H spectrum supports structure **27** with a singlet at methoxy δ 3.66 ppm (3H) and two broad aromatic singlets at δ 6.47 ppm and δ 6.70 ppm. However, the positions of the prenyl group and the methoxy group could not be unambiguously assigned on the basis of ^{1}H-NMR data only. The structure shown was based on comparison of calculated chemical shift values, but the full structure determination by NMR would require isolation of sufficient quantities of material or chemical synthesis.

Because of the low levels of metabolites, typically encountered in natural-product mixtures, full structure elucidation by NMR is a challenge. The solution lies in stop-flow methods if sufficient quantities of material are available or in employing SPE. An alternative approach is to employ HPLC microfractionation, drying, and reinjection of the concentrated LC-peak in deuterated solvent in microflow capillary LC-NMR probes. The work requires more manipulation of the sample, and deuterated solvent may be used, so solvent suppression is not needed, thereby facilitating comparison of chemical shifts with reported literature values. In addition, the concentrated sample provides higher sensitivity such that indirect carbon assignments may be obtained through HSQC or HMBC spectra enabling full structure characterization.

Lin et al. [94] have described the use of a microscale LC-MS-NMR platform. The platform consisted of two innovations in microscale analysis, nanoSplitter LC-MS and microdroplet NMR, and has been applied to the identification of unknown compounds found in natural products mixtures. The nanoSplitter provides the high sensitivity of nano electrospray MS while allowing the HPLC effluent from a large bore LC column to be collected and concentrated for NMR. In the approach described, 98% of the HPLC effluent is directed to a fraction collector for subsequent NMR and bioassay studies, while the remaining 2% is directed to the nanoSplitter for LC-MS analysis. A diagram of the system is given in Figure 7.40.

The system was tested with a series of experiments, wherein separation, fraction collection, pre-concentration, and microcoil NMR acquisition were performed using a mixture of commercial drugs. The systems LOD was determined at the 50-ng level with sample recovery around 98%. An example of the technology was described involving an extract of cyanobacterium *F. ambigua* showing antibacterial activity against *Myobacterium tuberculosis*. A highly active fraction from a silica gel SPE extraction eluted with 100% dichloromethane. Based on the UV chromatogram, peak fractions were collected. Each fraction underwent a 2-h ^{1}H-NMR acquisition. The four largest peaks were identified as known isonitrile-containing indole alkaloids (isonitriles of ambiguines A, E, and I and hapalindole H) by comparison with published data [95–100]. In addition, an unknown compound was found similar to the ambiguines, but not found in the literature or natural-product databases. This product was then flagged for subsequent scale-up and isolation for NMR characterization (see Figure 7.41). The scale up involved a 3-L growth for 32 days, which yielded 0.85 mg of product. Evaluation on a 900-MHz NMR spectrometer with a cryoprobe was used to establish the novel ambiguine K isonitrile. Hence the use of LC-MS-NMR enabled prioritization of samples for scale-up by facilitating the identification of known components. In addition, significantly lower limits of detection were possible by pooling LC runs. Since LC separation times are typically 1 h/fraction for multiple fractions, this is much less than analysis times required in NMR and, hence, this strategy becomes practical and time efficient. Overall, this process eliminated time and expense for large-scale purification of previous known leads and allowed resources and effort to be focused on the characterization of new active compounds. This process had the advantage of streamlining and reducing the natural-product discovery process.

LC-MS-μNMR Platform

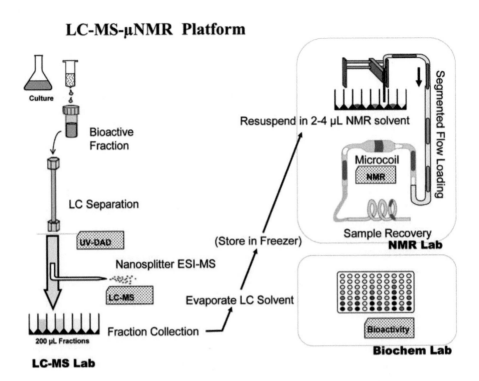

FIGURE 7.40 Schematic diagram of the LC-MS-NMR platform, as applied to natural-product discovery. The complex sample (bioactive fraction) is separated using high-resolution LC, with UV and MS data acquired online. 98% of the eluent is directed to a UV-guided fraction collection. Fractions are concentrated by drying and may be stored. For NMR, fractions are resuspended in a small volume (2–5 μL) of deuterated solvent and loaded into a microcoil NMR probe, with an observed volume of 1–2 μL, using microplate automation. Samples may be recovered after NMR analysis for further analysis and bioassay. (Reprinted with permission from Lin et al. [94]. Copyright (2008) American Chemical Society.)

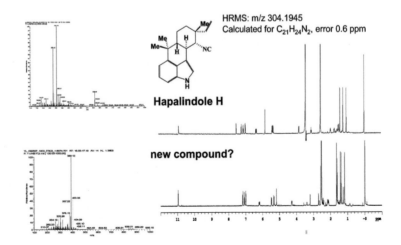

FIGURE 7.41 (Top) MS and NMR spectra of the 18.3-min LC peak identified from the literature as hapalindole H. (Bottom) MS and NMR spectra of the 17.3-min LC peak not found in the literature or natural-product databases. The unknown peak was prioritized for detailed NMR structure studies following scale-up purification. (Reprinted with permission from Lin et al. [94]. Copyright (2008) American Chemical Society.)

Gronquist et al. [101] have reported studies aimed at lowering the threshold of material required for structural analysis of small-molecule mixtures. These researchers employed the use of capillary-NMR probe technology to identify a large number of steroids from a severely mass-limited biological sample extracted from the firefly *Lucidota atra*. This challenge was of particular importance, since many natural products having potent biological activity are produced in relatively miniscule amounts. Hence many organisms producing potentially interesting metabolites cannot be studied using NMR spectroscopy because of the low sensitivity of the technology. The goal then became to expand the mass sensitivity of NMR spectroscopy and maintain high spectral resolution.

Because *L. atra* is relatively rare, large numbers of specimens cannot be collected. ^{1}H-NMR was carried out on a crude extract from five *L. atra* insects, which revealed the presence of steroidal pyrones, as evidenced by resonances consistent with the pyrone spin system as well as angular methyl groups in the steroidal skeleton. The preliminary NMR data revealed the fireflies contained 10 or more novel steroids [Figure 7.42]. However, the complexity of the NMR spectra for the mixtures of components did not permit complete characterization of the components.

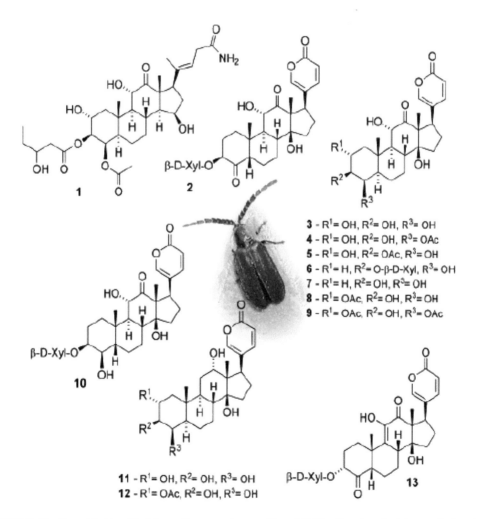

FIGURE 7.42 Steroids identified from *L. atra* using capillary NMR. (Reprinted with permission from Gronquist et al. [101]. Copyright (2005) American Chemical Society.)

To enable separation, HPLC was employed using the whole-body extracts of 50 specimens. Eleven fractions resulted where each fraction contained one to three steroidal pyrones
and related derivatives. The fractions that were isolated were in the 40–150-nM range, well
below the amount of material required for full structural characterization using 5-mm
probes, even with low volume (Shigemi tubes). Because cryoprobes can have radiation-
damping issues with respect to the large solvent peak, thereby introducing broad lineshape,
this can complicate the analysis of complex mixtures as well as introduce signal loss for
spectra such as HMBC, requiring long delays in the pulse sequence. The alternative solution
was to employ the CapNMR probe. This probe features a 5-μL flow cell with a 3-μL active
volume. The flow cell was connected to 2-m-long fused capillaries serving as inlet and
outlet ports. CapNMR offers a four- to fivefold higher S/N for HMBC and HMQC spectra
of a 17.2-μg sucrose sample compared to using a standard 5-mm inverse detection probe
with a Shegemi tube. It should be noted that while mass sensitivity is significantly higher
for the CapNMR probe, concentration sensitivity is actually only a fraction of that of
5-mm probes. By reducing volume from roughly 200 μL for a Shegemi tube to 5 μL in the
CapNMR probe, one increases concentration by a factor of about 40, while sensitivity only
increases by a factor of 5. However, because natural products are usually mass limited, the
CapNMR appeared to be a viable option.

Eleven HPLC fractions obtained from *L. atra* were dissolved in ~5 μL solvent, taking
care to minimize losses during concentration and injection. The resulting NMR spectra
allowed complete assignment of all major components in each fraction leading to the identification of 13 new steroids. This process was possible due to the excellent lineshape and
very low level of artifacts in the spectra which compared favorably with the much higher
artifact levels experienced when using cryoprobe systems. In fact, three compounds coeluted
during purification and were analyzed as a three component mixture without further purification. Direct comparison of CapNMR with a 5-mm probe and Shigemi tube yielded
roughly a threefold gain in S/N.

Twelve of the compounds characterized represented steroidal pyrones analogous to
those from other species of firefly. For one of the structures, however, instead of a pyrone
substituent, the molecule contained a pentenoic acid amide moiety corresponding to a ring
opened and reduced pyrone. Overall, the CapNMR probe enabled the discovery of new and
interesting chemistry by affording a relatively inexpensive means of increasing S/N.

Other natural-product studies have been reported where the investigators use semipreparative HPLC and capillary NMR to isolate and structurally characterize novel compounds. One study involved *Arctostaphylos Pumila Nutt.* (Ericaceae), a rare plant species,
from which two acylated caprylic alcohol glycosides were isolated [102]. The compounds
were initially extracted from dried plant material with EtOH:ETOAc (50:50) followed by
H_2O:MeOH (30:70) yielding >5 g of dry organic extracts. The organic extracts were then
subjected to flash-prep chromatography, and fractions were collected and 50 mg were frac-
tionated by preparative C18 HPLC (30–70 ACN in water), collecting 40 fractions. Fractions were tested for antibacterial properties against Gram-positive methicillin-resistant
Staphylococcus aureus. The active fraction was further purified using semipreparative
HPLC containing a C-18 column (8 × 250 mm inner diameter (ID), 5mm) fitted with
a DAD and an ELSD detector. The mobile phase consisted of an isocratic gradient of
75% ACN in water acidified with 0.01% formic acid. Flow rate was 3 mL/min, 0.33 min
per collection time at 40°C, and the samples were collected in minitubes. The separation
yielded two compounds that were subsequently identified as 2,6-diacetyl-3,4-diisobutyl-
1-*O*-octylglucopyranoside **1** and 2,6-diacetyl-3,4-dimethylbutyl-1-*O*-octylglucopyranoside
2 (Figure 7.43).

The molecular weights of the compounds were obtained from positive-mode high-
resolution electrospray ionization MS (HR-ESI-MS). Full NMR structure assignment and

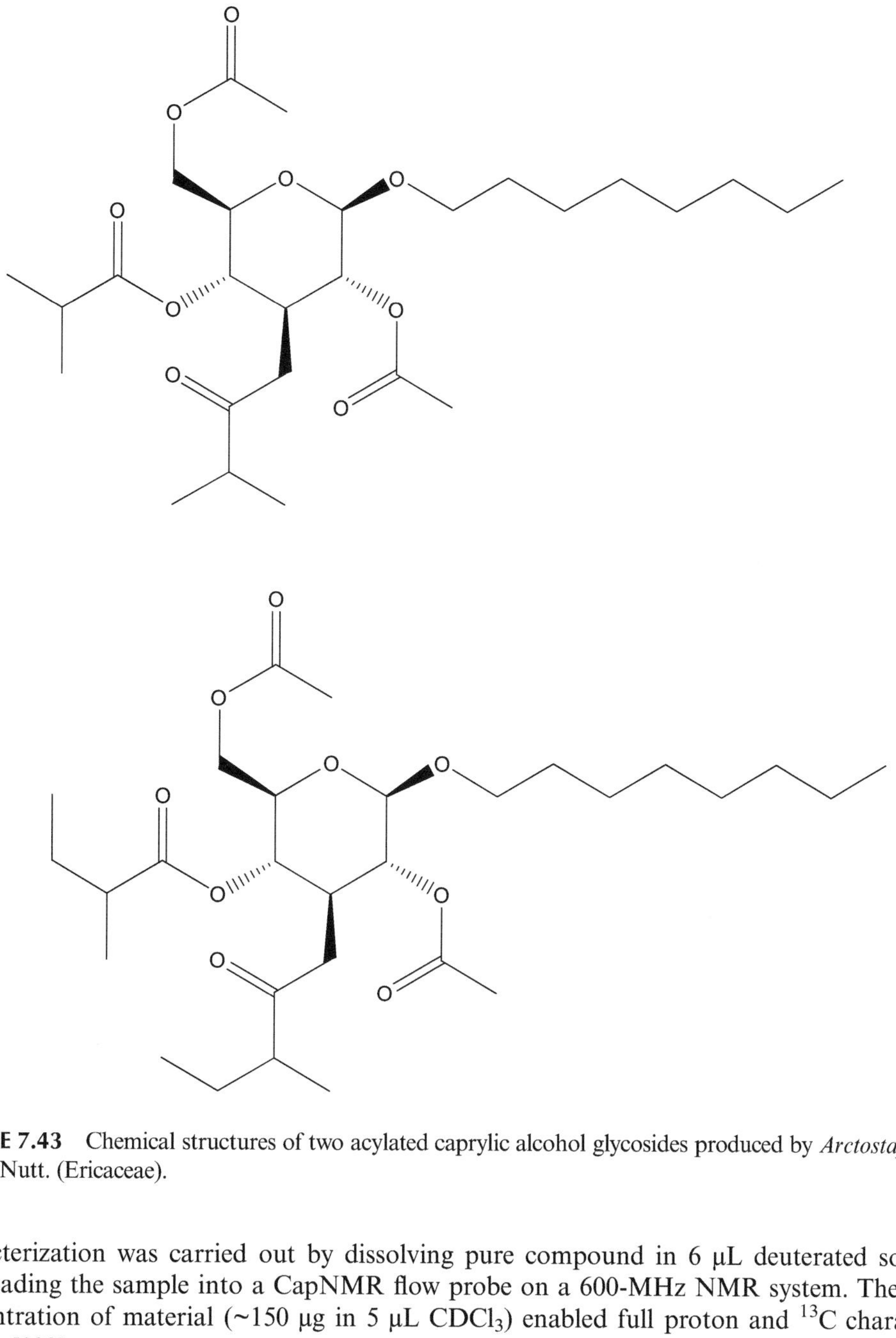

FIGURE 7.43 Chemical structures of two acylated caprylic alcohol glycosides produced by *Arctostaphylos pumila* Nutt. (Ericaceae).

characterization was carried out by dissolving pure compound in 6 μL deuterated solvent and loading the sample into a CapNMR flow probe on a 600-MHz NMR system. The high concentration of material (~150 μg in 5 μL CDCl$_3$) enabled full proton and ^{13}C characterization [102].

Another study involved developing high-throughput methods for the production, analysis, and characterization of libraries of natural products for high-throughput screening in the pharmaceutical industry. The library production integrated automated flash chromatography, SPE, filtration, and high-throughput parallel four-channel preparative HPLC to obtain libraries in 96- or 384-well plates. The libraries consisted of 36,000 fractions of approximately one to five compounds per well. Each library was screened for biological

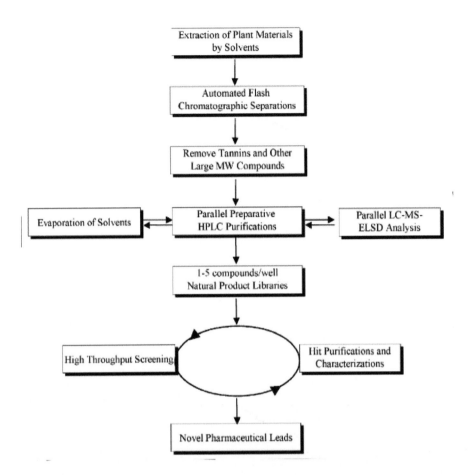

FIGURE 7.44 Schematic representation of high-throughput methods applied to the process of drug discovery from natural resources. (Reprinted with permission from, Eldridge et al. [103]. Copyright (2002) American Chemical Society.)

activity. Following biological screening, the active fractions were rapidly purified at the microgram level and structures obtained using NMR and MS [103].

Characterization and structure elucidation of individual compounds was done using LC-ELSD-MS data and ^{1}H and COSY NMR on as little as 5 µg of purified material using a 5-µL microcoil flow probe. Structures of novel compounds could be elucidated with about 50 µg of isolated material in a timely manner, whereas comparative identities could be obtained from samples containing only 5 µg of compound. The process employed in this study is illustrated in Figure 7.44.

A typical ^{1}H-NMR spectrum of 50 µg of paclitaxel in a 5-µL flow probe is shown in Figure 7.45.

7.11   PROTEINS/PEPTIDES

The rate-limiting step in biophysical characterization of membrane proteins by NMR is usually the availability of suitable amounts of quality protein. To address this problem, microcoil NMR technology was employed to screen microscale quantities of membrane

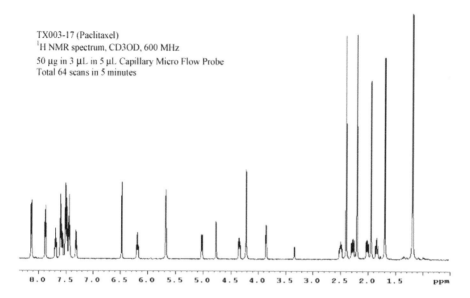

TX003-17 (Paclitaxel)
^{1}H NMR spectrum, CD3OD, 600 MHz
50 µg in 3 µL in 5 µL Capillary Micro Flow Probe
Total 64 scans in 5 minutes

FIGURE 7.45 ^{1}H-NMR spectrum of 50 µg of paclitaxel in 3 µL CD3OD acquired using the 5-µL microcoil flow probe on the 600-MHz NMR spectrometer. (Reprinted with permission from, Eldridge et al. [103]. Copyright (2002) American Chemical Society.)

proteins to determine proper folding for samples destined for structural studies. Micoscale NMR was used to screen a series of zwitterionic phosphocholine detergents for their ability to reconstitute membrane proteins, using the previously well-characterized OmpX, an eight-stranded β-barrel structure. Fold screening was achieved on microgram amounts of uniformly ^{2}H,^{15}N-labeled OmpX [104].

The NMR data were obtained at 25°C on a 700-MHz NMR spectrometer equipped with a 1-mm TXI microprobe. The 1-mm NMR capillaries were filled with 7 µL of the solution containing the mixed OmpX-detergent micelles. The 2D [^{15}N,^{1}H]-TROSY correlation experiments were recorded as described previously [105].

OmpX was homogeneously reconstituted in new detergents such as 1,2-dihexanoyl-*sn*-glycero-3-phosphocholine (DHPC). However, there was no precipitation during the microscale reconstitution in 138-Fos or in 179-Fos, whereas more than 50% of the protein precipitated during each cycle of reconstitution in DHPC. These results showed that reconstitution in 138-Fos or in 179-Fos can efficiently be achieved, whereas with DHPC, several recovery cycles were needed in order to obtain an NMR sample. The 2D [^{15}N,^{1}H]-TROSY correlation spectra of the mixed micelles with OmpX/115-Fos, OmpX/TPC, OmpX/34-Fos, and OmpX/185-Fos (Figure 7.46 d-g) showed a cluster of broad lines in the center of the spectrum and additional, resolved cross peaks of unequal shapes and intensities, indicating that these protein samples were not homogeneously folded and suggest formation of nonspecific soluble aggregates.

Overall, the combined use of preliminary screening of a library of candidate detergents for OmpX refolding using SDS gel electrophoresis and 2D [^{15}N,^{1}H]-TROSY correlation NMR spectroscopy resulted in the identification of two new detergents, 138-Fos and 179-Fos, that afford reconstitution of OmpX in its native form more efficiently than DHPC. These two single-chain phosphocholine detergents are the structural mimics of the lysopholipids that have been shown to retain the folded conformation of several R-helical membrane proteins in solution NMR studies [99]. The microcoil NMR technology was key

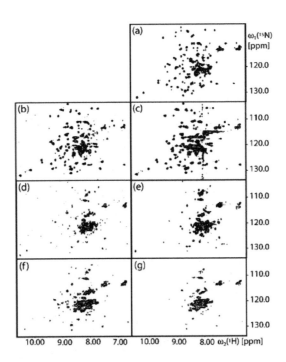

FIGURE 7.46   2D [^{15}N,^{1}H]-TROSY correlation NMR spectra of uniformly[^{2}H,^{15}N]-labeled OmpX reconstituted in mixed micelles with different detergents: (a) DHPC; (b) 138-Fos; and (c) 179-Fos, where the vertical band of peaks near 8.0 ppm represents t_1-noise from the signal of the amide proton of 179-Fos; (d) 115-Fos; (e) TPC; (f) 34-Fos; (g) 185-Fos. The spectra were collected with the following parameters: data size 50 × 1024 complex points; 300 scans per increment. Acquisition time was 9 h per experiment. (Reprinted with permission from Zhang et al. [104]. Copyright (2008) American Chemical Society.)

to this success, reducing the cost of producing the protein as well as the novel amphiphiles, which are often initially made in small quantities [106–112].

Protein screening by NMR is also used in structural genomics centers to identify folded proteins that are promising targets for three-dimensional (3D) structure determination by X-ray crystallography or NMR spectroscopy. Both 1D ^{1}H-NMR spectra and 2D [^{1}H,^{15}N]-correlation spectroscopy (COSY) typically require milligram quantities of unlabeled or isotope-labeled protein, respectively. To enable analytical analysis of protein structure integrity, miniaturization of a structural genomics pipeline with NMR screening for folded globular proteins was employed. This was accomplished using a high-density micro-fermentation device and a microcoil NMR probe. The proteins were microexpressed in unlabeled or isotope-labeled media, purified, and then subjected to 1D ^{1}H-NMR and/or 2D [^{1}H,^{15}N]-COSY screening. To demonstrate that the miniaturization is functioning effectively, nine mouse homologue protein targets were processed and the results were compared with those from a "macro-scale" structural genomics high-throughput pipeline. The results from the two pipelines were comparable, illustrating that the data were not compromised in the miniaturized approach.

In this study, the NMR sample was prepared using 10 μL of protein in 10-mM Tris buffer, 100-mM NaCl at pH 7.8, and supplemented with about 1 μL of D$_2$O. About 8 μL of protein was injected into a flow cell of the microcoil NMR probe. Preparation of the flow cell required cleaning with 30 μL of 10-mM aqueous Tris buffer at pH 7.8 and 100 mM NaCl to ensure removal of residual protein from previous measurements. Verification of cleaning was obtained by recording a 1D ^{1}H-NMR spectrum of the cleaning solvent. These flow cells periodically require cleaning with 100 μL of a 1% Mico-90 detergent solution followed by extensive washing with water and standard buffer. All NMR measurements were carried out using a quadruple-tuned HCN z-gradient microcoil probe (CapNMR, MRM/Protasis Inc.) with a total cell volume of 5 μL and an active volume of 1.5 μL. All data were acquired at 298 K on a 600-MHz spectrometer using a three-channel setup. Water suppression in the 2D [^{1}H,^{15}N]-HSQC measurements was achieved using a watergate sequence and relaxation delays of 1.0–1.2 s. Excellent signal was achieved as shown in Figure 7.47.

Overall, the miniaturization process reduced the protein requirement from milligram amounts to the microgram range. Further progress to improve low-loss protein purification for small sample volumes and improvements in NMR techniques (e.g., higher-field magnets, improved microcoil probes, improved cryogenic probes) to improve sensitivity are needed. These studies have shown promise in the miniaturization of a structural genomics pipeline using microexpression and microcoil NMR [113].

The Northeast Structural Genomics Consortium (NESG) has implemented a "automated pipeline" for target selection, construct optimization, protein sample production, and efficient microscale screening of protein NMR targets using a microcryoprobe. This probe requires small amounts of protein, typically 10–200 μg of sample in 8–35 μL volume. Extensive automation has been implemented by combining database tools and use of a microcryoprobe with enhanced mass sensitivity. Initial screening was conducted using a room-temperature 600-MHz 1-mm probe, but it was later switched to a 600-MHz 1.7-mm mirocryoprobe. This probe provides a mass sensitivity (S/N per microgram of solute) that is one order of magnitude higher than conventional 5-mm probes, reducing the required volume from 300 to 30 μL. The 1.7-mm micro NMR cryoprobe on a 600-MHz spectrometer was used seamlessly for target screening. Acquisition of data for backbone and side-chain chemical shift assignments and structure determination of [U-^{13}C,^{15}N] proteins up to

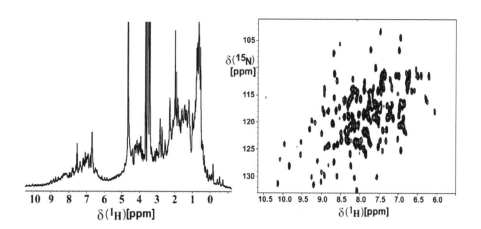

FIGURE 7.47 1D ^{1}H-NMR (left) and 2D [^{1}H,^{15}N]-correlation NMR spectra (right) of the Nostoc sp. hypothetical protein All2754. Both spectra have been recorded with a 2-mM protein solution in 10-mM Tris–HCl buffer at pH 7.8, 100-mM NaCl and 90%/10% H$_2$O/D$_2$O on a Bruker DRX 600 MHz spectrometer (Bruker, Billarica, USA) using an HCN z-gradient microcoil probe (MRM/Protasis, Savoy, USA), T = 298 K. 1D ^{1}H-NMR: measurement time 5 min. 2D [1H,15N]-COSY: soft water flipback, STATES-TPPI, 102,464 points, relaxation delay: 1.1 s, measurement time: 3 h. (Reprinted from Peti et al. [113]. With kind permission from Springer Science and Business Media.)

20 kDa is also possible. Reports from the NESG have described the general overall process of initial NMR sample characterization emphasizing the role of 1D and simple 2D NMR screening for selecting optimal constructs, solvent conditions, and oligomerization states. This process enables validation of protein targets for incorporation of isotopic labels (^{15}N, ^{13}C, ^{2}H) for NMR structure determination [114].

1D ^{1}H-NMR spectra with solvent presaturation was obatined for a block of protein samples. Sixty samples per 24 h were screened. Illustrative 1D proton spectra are shown in Figure 7.48.

The criteria for scoring are as follows: (a) S/N, (b) upfield-shifted methyl protons indicating a folded core formed by aromatic and methyl stacking, and (c) dispersion of the amide protons. Each sample contained 50-μM DSS internal standard. Most proteins may be easily classified by 1D proton NMR with optional follow-up by 2D ^{1}H-^{15}N HSQC analysis.

With microtubes, it was also noted that inspection of the sample for precipitation is often ineffective in establishing the integrity of the protein (Figure 7.49).

Rossi et al. have shown that some precipitation in the capillary tube may still contain soluble protein in sufficient amounts for structure determination by NMR. Overall, the researchers have shown that the microtechnologies were essential for the success of the NESG consortium. Reduction in scale of sample size provided a useful template for structural biology programs, enabling rapid exploration of samples and sample conditions for construct and/or buffer optimization.

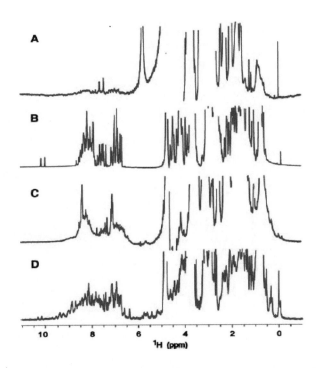

FIGURE 7.48 1D ^{1}H-NMR spectra with H_2O presaturation of representative NESG targets obtained with a 1.7-mm micro NMR cryoprobe at 20°C with corresponding NESG target IDs. (a) HR3159A spectrum scores as "poor" on account of broad poorly dispersed resonances. (b) LmR69A spectrum scores "unfolded" due to sharp and poorly dispersed peaks in all regions. (c) EwR71A spectrum scores as "promising" with the presence of upfield-shifted methyl peaks but crowding of the amide region (7–9 ppm), and relatively broad peaks. (d) NsR431C spectrum scores "good" with sharp uniform intensity and upfield-shifted methyls. (Reprinted from Rossi et al. [114]. With kind permission from Springer Science + Business Media.)

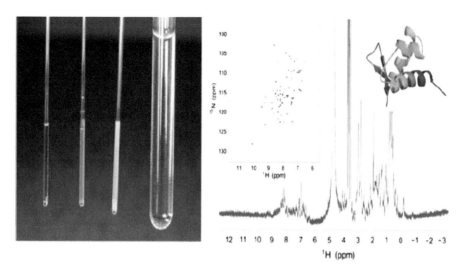

FIGURE 7.49 From the left, 1-mm microtubes showing no or increasing degrees of precipitation of target protein in different buffers. Signal is detected in the center tube, but not in the clear (left) or the heavily precipitated (right) microtubes. The best spectra from *S. typhimurium* (NESG target StR82) were recorded at 20°C in 450-mM NaCl at pH 6.5. The insets show the 2D ^{1}H-^{15}N HSQC and the ribbon diagram of the structure solved using optimal conditions (PDB ID:2JT1). (Reprinted from Rossi et al. [114]. With kind permission from Springer Science + Business Media.)

The protein kinase ZAP-70 is involved in T-cell activation and interacts with tyrosine-phosphorylated peptide sequences known as immunoreceptor tyrosine activation motifs (ITAMs). Hentschel and coworkers have reported on the study of regulatory phosphorylation sites in the tryptic fragment containing amino acids 485–496 (ALGADDSYYTAR). Four possible peptides with phosphorylation at none, one, or both of the Y-492 and Y-493 tyrosines were specifically synthesized and analyzed by ^{1}H/^{13}C-NMR at 600 MHz using a capillary HPLC-NMR microprobe. Unambiguous discrimination of the peptides was possible via the effects of chemical shifts of phosphorylation on the aromatic tyrosine protons. With the microprobe and the detection volume of 1.5 μL, structure elucidation was carried out with the very small amounts of peptides. Approximately 15 μg (0.5-mg peptide in 50-μL D$_2$O) of the analyte were used, corresponding to about 2 mg in 5-mm tubes. Capillary HPLC–NMR spectra were recorded in the stopped-flow mode on less than 400 ng of each peptide, using 1D and 2D NMR spectroscopy (^{1}H,^{1}H-COSY-90, ^{1}H/^{13}C-HSQC, and ^{1}H/^{13}C-HMBC).

Because of the low amounts of compound and respective relative concentrations, direct observation of ^{13}C was not feasible. Carbon assignments were obtained through indirect detection techniques. Figure 7.50 shows the ^{1}H–^{13}C HSQC 2D spectrum of a nonphosphorylated peptide and provides H–C correlations via 1/*JCH* for the assignment of the ^{13}C signals via the known ^{1}H assignments.

It is important to note a construction feature of the microprobe that affects ^{13}C measurements. The copper detection coil is surrounded by a fluid that is similar to the susceptibility of the copper in order to decrease field inhomogeneities. This fluid, FC-43, shows resonance signals between 80 and 115 ppm; hence, this range of the ^{13}C chemical shifts is not available for direct observation. Because the peptides under investigation did not have nuclei resonating in that region, the carbon assignments were not adversely affected [115].

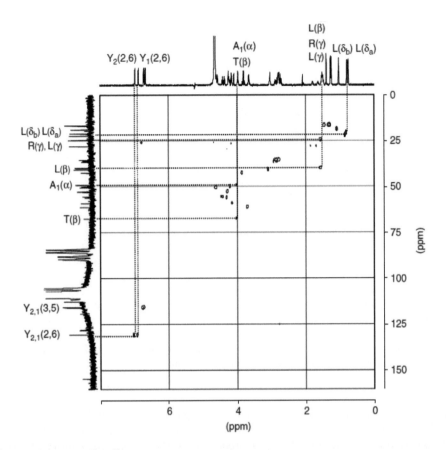

FIGURE 7.50 A 600-MHz ^{1}H–^{13}C HSQC 2D spectrum obtained from 15 μg of the nonphosphorylated peptide. The corresponding 1D ^{1}H and ^{13}C spectra are also shown on the horizontal and vertical axes, respectively. (Reprinted from Hentschel et al. [115]. Copyright 2005, with permission from John Wiley & Sons Inc.)

REFERENCES

1. Felton, L. 2006. A review of "pharmaceutical stress testing: predicting drug degradation". In *In Drug Development and Industrial Pharmacy*, ed. S.W. Baertschi, 505, Boca Raton, FL: Taylor & Francis.
2. ICH. 2003. Topic Q1A (R2): stability testing of new drug substances and products. *Proceedings of the International Conference on Harmonization* Published in the Federal Register, 21 November 2003 68(225):65717–65718. http://ich.org/fileadmin/Public_Web_Site/ICH_Products/Guidelines/Quality/QiA_R2/Step4/Q1A_R2_Guideline.pdf.
3. Breton, D., Buret, D., Mendes-Oustric, A.C., Chaimbault, P., Lafosse, M., and Clair, P. 2006. LC-UV and LC-MS evaluation of stress degradation behaviour of avizafone. *J. Pharm. Biomed. Anal.* 41:1274–1279.
4. Bakshi, M., and Singh, S. 2004. HPLC and LC-MS studies on stress degradation behavior of tinidazole and development of a validated specific stability-indicating HPLC assay method. *J. Pharm. Biomed. Anal.* 34:11–18.
5. Xu, X., Bartlett, M.G., and Stewart, J.T. 2001. Determination of degradation products of sumatriptan succinate using LC-MS and LC-MS-MS. *J. Pharm. Biomed. Anal.* 26:367–377.
6. Pan, C., Liu, F., Ji, Q., Wang, W., Drinkwater, D., and Vivilecchia, R. 2006. The use of LC/MS, GC/ MS, and LC/NMR hyphenated techniques to identify a drug degradation product in pharmaceutical development. *J. Pharm. Biomed. Anal.* 40:581–590.

7. Peng, S.X. 2000. Hyphenated HPLC-NMR and its applications in drug discovery. *Biomed. Chromatogr.* 14:430–441.

8. Fukutsu, N., Kawasaki, T., Saito, K., and Nakazawa, H. 2006. Application of high-performance liquid chromatography hyphenated techniques for identification of degradation products of cefpodoxime proxetil. *J. Chromatogr. A.* 1129:153–159.

9. Wu, L., Hong, T.Y., and Vogt, F.G. 2007. Structural analysis of photo-degradation in thiazole-containing compounds by LC-MS/MS and NMR. *J. Pharm. Biomed. Anal.* 44:763–772.

10. Hongu, M., Hosaka, T., Kashiwagi, T., Kono, R., and Kobayashi, H. 2002. Preparation of substituted imidazoles/oxazoles/thiazoles as large conductance calcium-activated K channel openers. *PCT Int. Appl.* WO2002(083111):A2 1–2301.

11. Gonnella, N.C., Mullis, J., Pennino, S., Campbell, S., Norwood, D., and Granger, A. 2012. Application of LC-SPE-NMR and micro cryo-probe technology in solving challenging structure elucidation problems. Paper presented at the Horváth Symposium, Hartford, CT.

12. Forsyth, D.A., and Sebag, A.B. 1997. Computed ^{13}C NMR chemical shifts via empirically scaled giao shieldings and molecular mechanics geometries: conformation and configuration from ^{13}C shifts. *J. Am. Chem. Soc.* 119:9483–9494.

13. Wann, M.-H. 2005. Application of LC-NMR in pharmaceutical analysis. *Sep. Sci. Technol.* 6:569–679.

14. Provera, S., Guercio, G., Turco, L., Curcuruto, O., Alvaro, G., Rossi, T., and Marchioro, C. 2010. Application of LC–NMR to the identification of bulk drug impurities in NK1 antagonist GW597599 (vestipitant). *Mag. Res. Chem.* 48:523–530.

15. Novak, P., Cindrić, M., Tepeš, P., Dragojević, S., Ilijas, M., and Mihaljević, K. 2005. Identification of impurities in acarbose by using an integrated liquid chromatography-nuclear magnetic resonance and liquid chromatography-mass spectrometry approach. *J. Sep. Sci.* 28:1442–1447.

16. Ziegelbauer, K., Babczinski, P., and Schonfeld, W. 1998. Molecular mode of action of the antifungal β -amino acid BAY 10-8888. *Antimicrob. Agents Chemother.* 42:2197–2205.

17. Mittendorf, J., Kunisch, F., Matzke, M., Militzer, H.C., Schmidt, A., and Schonfeld, W. 2003. Novel antifungal beta-amino acids: synthesis and activity against Candida albicans. *Bioorg. Med. Chem. Lett.* 13:433–436.

18. Sorbera, L.A., Castaner, J., and Bozzo, J. 2002. PLD-118: antifungal isoleucyl-tRNA synthetase inhibitor. *Drugs Fut.* 27:1049–1055.

19. Petraitiene, R., Petraitis, V., Kelaher, A.M., Sarafandi, A.A., Mickiene, D., Groll, A.H., Sein, T., Bacher, J., and Walsh, T.J. 2005. Efficacy, plasma pharmacokinetics, and safety of icofungipen, an inhibitor of Candida isoleucyl-tRNA synthetase, in treatment of experimental disseminated candidiasis in persistently neutropenic rabbits. *Antimicrob. Agents Chemother.* 49:2084–2092.

20. Parnham, M.J., Bogaards, J.J.P., Schrander, F., Schut, M.W., Oreskovic, K., and Mildner, B. 2005. The novel antifungal agent PLD-118 is neither metabolized by liver microsomes nor inhibits cytochrome P450 in vitro. *Biopharm. Drug Dispos.* 26:27–33.

21. Hasenoehrl, A., Galic, T., Ergovic, G., Marsic, N., Skerlev, M., Mittendorf, J., Geschke, U., Schmidt, A., and Schonfeld, W. 2006. In vitro activity and in vivo efficacy of icofungipen (PLD-118), a novel oral antifungal agent, against the pathogenic yeast Candida albicans. *Antimicrob. Agents Chemother.* 50:3011–3018.

22. Novak, P., Tepeš, P., Ilijaš, M., Fistrić, I., Bratoš, I., Avdagić, A., Hameršak, Z., Marković, V.G., and Dumić, M. 2009. LC–NMR and LC–MS identification of an impurity in a novel antifungal drug icofungipen. *J. Pharm. Biomed. Anal.* 50(1):68–72.

23. Rinaldi, F. 2007. The application of LC-SPE-cryoflow NMR spectroscopy in pharmaceutical development. *Amer. Drug Discov.* 3(1):33–37.

24. Corcoran, O., and Spraul, M. 2003. LC-NMR-MS in drug discovery. *Drug Discov. Today.* 8:624–631.

25. Corcoran, O., Wilkinson, P.S., Godejohann, M., Braumann, U., Hofmann, M., and Spraul, M. 2002. Advanced sensitivity for flow NMR spectroscopy; LC-SPE-NMR and capillary scale LC-NMR. *Am. Lab.: Chromatogr. Perspectives.* 34:18–21.

26. Spraul, M., Freund, A.S., Nast, R.E., Withers, R.S., Maas, W.E., and Corcoran, O. 2003. Advancing NMR sensitivity for LC-NMR-MS using a cryoflow probe: application to the analysis of acetaminophen metabolites in urine. *Anal. Chem.* 75:1536–1541.

27. Exarchou, V., Godejohann, M., Van Beek, T.A., Gerothanassis, I.P., and Vervoort, J. 2003. LC-UV-solid-phase extraction-NMR-MS combined with a cryogenic flow probe and its application to the identification of compounds present in Greek oregano. *Anal. Chem.* 75:6288–6294.

28. Nyberg, N.T., Baumann, H., and Kenne, L. 2001. Application of solid-phase extraction coupled to an NMR flow-probe in the analysis of HPLC fractions. *Magn. Reson. Chem.* 39:236–240.

29. Nyberg, N.T., Baumann, H., and Kenne, L. 2003. Solid-phase extraction NMR studies of chromatographic fractions of saponins from Quillaja saponaria. *Anal. Chem.* 75:268–274.

30. Xu, F., and Alexander, A.J. 2005. The design of an on-line semi-preparative LC-SPE-NMR system for trace analysis. *Mag. Res. Chem.* 43:776–782.

31. Sharman, G.J., and Jones, I.C. 2003. Critical investigation of coupled liquid chromatography-NMR spectroscopy in pharmaceutical impurity identification. *Magn. Reson. Chem.* 41:448–454.

32. Lindon, J.C., Nicholson, J.K., and Wilson, I.D. 2000. Directly coupled HPLC-NMR and HPLC-NMR-MS in pharmaceutical research and development. *J. Chromatogr. B.* 748:233–258.

33. Alsante, K.M., Boutros, P., Couturier, M.A., Friedmann, R.C., Harwood, J.W., Horan, G.J., Jensen, A.J., Liu, Q., Lohr, L.L., Morris, R., Raggon, J.W., Reid, G.L., Santafianos, D.P., Sharp, T.R., Tucker, J.L., and Wilcox, G.E. 2004. Pharmaceutical impurity identification: a case study using a multidisciplinary approach. *J. Pharm. Sci.* 93(9):2296–2309.

34. Camel, V. 2003. Solid phase extraction of trace elements. *Spectrochim. Acta, Part B.* 58:1177–1233.

35. Snyder, L.R., Kirkland, J.J., and Glajch, J.L. 1997. *Practical HPLC Method Development*, 2nd edition. New York: John Wiley & Sons, Inc.

36. Scott, R.P.W., and Kucera, P. 1976. Some aspects of preparative-scale liquid chromatography. *J. Chromatogr.* 119:467–482.

37. Cortes, H.J. 1989. *Multidimensional Chromatography: Techniques and Applications.* New York: Marcel Dekker.

38. Mondello, L., Lewis, A.C., and Bartle, K.D. 2001. *Multidimensional Chromatography.* Chichester: John Wiley & Sons, Inc.

39. Ramsteiner, K.A. 1988. Systematic approach to column switching. *J. Chromatogr.* 456:3–20.

40. Ackermann, B.L., Murphy, A.T., and Berna, M.J. 2002. The resurgence of column switching techniques to facilitate rapid LC/MS/MS based bioanalysis in drug discovery. *Am. Pharm. Rev.* 5:54–63.

41. Veuthey, J.-L., Souverain, S., and Rudaz, S. 2004. Column-switching procedures for the fast analysis of drugs in biological samples. *Ther. Drug Monit.* 26(2):161–166.

42. Alexander, A.J., Xu, F., and Bernard, C. 2006. The design of a multi-dimensional LC-SPE-NMR system (LC(2)-SPE-NMR) for complex mixture analysis. *Magn. Reson. Chem.* 44:1–6.

43. Camel, V. 2003. Solid-phase extraction. *Compr. Anal. Chem.* 41:393–457.

44. Colin, F.P., Ajith, D.G., and Revathy, S. 2000. Contributions of theory to method development in solid-phase extraction. *J. Chromatogr. A.* 885:17–39.

45. Gonnella, N.C., Rodriguez, S.A., Pennino, S., and Norwood, D. 2016. Structure elucidation of a complex faldaprevir-(±)-α-tocopherol addition product reveals reactivity of common excipient vitamin E-TPGS. *J. Heterocycl. Chem.* 53(6):1878–1891.

46. Llinas-Brunet, M., Bailey, M.D., Goudreau, N., Bhardwaj, P.K., Bordeleau, J., Bos, M., Bousquet, Y., Cordingley, M.G., Duan, J., Forgione, P., Garneau, M., Ghiro, E., Gorys, V., Goulet, S., Halmos, T., Kawai, S.H., Naud, J., Poupart, M.-A., and White, P.W. 2010. Discovery of a potent and selective noncovalent linear inhibitor of the hepatitis C virus NS3 protease (BI 201335). *J. Med. Chem* 53:6466–6476.

47. Lamarre, D., Anderson, P.C., Bailey, M., Beaulieu, P., Bolger, G., Bonneau, P., Bçs, M., Cameron, D.R., Cartier, M., Cordingley, M.G., Faucher, A.-M., Goudreau, N., Kawai, S.H., Kukolj, G., Lagace, L., LaPlante, S.R., Narjes, H., Poupart, M.-A., Rancourt, J., Sentjens, R.E., St. George, R., Simoneau, B., Steinmann, G., Thibeault, D., Tsantrizos, Y.S., Weldon, S.M., Yong, C.-L., and Llinàs-Brunet, M. 2003. An NS3 protease inhibitor with antiviral effects in humans infected with hepatitis C virus. *Nature.* 426:186–189.

48. Zhang, Z., Tan, S., and Feng, S.-S. 2012. Vitamin E TPGS as a molecular biomaterial for drug delivery. *Biomaterials.* 33:4889–4906.

49. Zhou, C.C., and Hill, D.R. 2007. The keto-enol tautomerization of ethyl butylryl acetate studied by LC-NMR. *Magn. Reson. Chem.* 45:128–132.

50. Spahn-Langguth, H., and Benet, L.Z. 1992. Acyl glucuronides revisited: is the glucuronidation process a toxification as well as a detoxification mechanism?. *Drug Metab. Rev.* 24:5–47.

51. Smith, P.C., Song, W.Q., and Rodriguez, R.J. 1992. Covalent binding of etodolac acyl glucuronide to albumin in vitro. *Drug Metab. Dispos.* 20:962–965.

52. Bradow, G., Kan, L.S., and Fenselau, C. 1989. Studies of intramolecular rearrangements of acyl-linked glucuronides using salicyclic acid, flufenamic acid, and (S)- and (R)-benoxaprofrn and confirmation of isomerization in acyl-linked. DELTA. 9-11-carboxytetrahydrocannabinol glucuronide. *Chem. Res. Toxicol.* 2:316–324.

53. King, A.R., and Dickinson, R.G. 1991. Studies on the reactivity of acyl glucuronides. I. phenolic glucuronidation of isomers of diflunisal acyl glucuronide in the rat. *Biochem. Phannacol.* 42:2289–2299.

54. Smith, P.C., McDonagh, A.F., and Benet, L.Z. 1986. Irreversible binding of zomepirac to plasma protein in vitro and in vivo. *J. Clin. Invest.* 77:934–939.

55. Smith, P.C., Benet, L.Z., and McDonagh, A.F. 1990. Covalent binding of zomepirac glucuronide to proteins: evidence for a Schiff base mechanism. *Drug Metab. Dispos.* 18:639–644.

56. Weiss, J.S., Guatam, A., Lauff, J.J., Sundberg, M.W., Jatlow, P., Boyer, J.L., and Saligson, D. 1983. The clinical importance of a protein-bound fraction of serum bilirubin in patients with hyperbilirubinemia. *N. Engl. J. Med.* 309:147–150.

57. Sidelmann, U.G., Gavaghan, C., Carless, H.A.J., Farrant, R.D., Lindon, J.C., Wilson, I.D., and Nicholson, J.K. 1995. Identification of the positional isomers of 2-fluorobenzoic acid 1-O-acyl glucuronide by directly coupled HPLC-NMR. *Anal. Chem.* 67:3401–3404.

58. Salmon, M., Fenselau, C., Cukier, J., and Odell, G.B. 1974. Rapid transesterification of bilirubin glucuronides in methanol. *Life Sci.* 14:2069–2078.

59. Godejohann, M., Tseng, L.H., Braumann, U., Fuchser, J., and Spraul, M. 2004. Characterization of a paracetamol metabolite using on-line LC-SPE-NMR-MS and a cryogenic NMR probe. *J. Chromatogr. A.* 1058:191–196.

60. Schlotterbeck, G., Ross, A., Hochstrasser, R., Senn, H., Kühn, T., Marek, D., and Schett, O. 2002. High-resolution capillary tube NMR. A miniaturized 5-microL high-sensitivity TXI probe for mass-limited samples, off-line LC NMR, and HT NMR. *Anal. Chem.* 74(17):4464–4471.

61. Ceccarelli, S.M., Schlotterbeck, G., Boissin, P., Binder, M., Buettelmann, B., Hanlon, S., Jaeschke, G., Kolczewski, S., Kupfer, E., Peters, J.-U., Porter, R.H.P., Prinssen, E.P., Rueher, M., Ruf, I., Spooren, W., Stämpfli, A., and Vieira, E. 2008. Metabolite identification via LC-SPE-NMR-MS of the in vitro biooxidation products of a lead mGlu5 allosteric antagonist and impact on the improvement of metabolic stability in the series. *Chem. Med. Chem.* 3:136–144.

62. Schlotterbeck, G., and Ceccarelli, S.M. 2009. LC-SPE-NMR-MS: a total analysis system for bioanalysis. *Bioanalysis.* 1(3):549–559.

63. Keun, H.C., Beckonert, O., Griffin, J.L., Richter, C., Moskau, D., Lindon, J.C., and Nicholson, J.K. 2002. Crogenic probe 13C NMR spectroscopy of urine for metabonomic studies. *Anal. Chem.* 74:4588–4593.

64. Griffin, J.L., Nicholls, A.W., Keun, H.C., Mortishire-Smith, R.J., Nicholson, J.K., and Kuehn, T. 2002. Metabolic profiling of rodent biological fluids via 1H NMR spectroscopy using a 1mm Microlitre probe. *Analyst.* 127:582–584.

65. Brindle, J.T., Antii, H., Holmes, E., Tranter, G., Nicholson, J.K., Bethell, H.W.L., Clarke, S., Schofield, P.M., McKilligin, E., Mosedale, D.E., and Grainger, D.J. 2002. Rapid and non-invasive diagnosis of the presence and severity of coronary heart disease using ¹H-NMR-based metabonomics. *Nat. Med.* 8:1439–1444.

66. Griffin, J.L., Bonney, S.A., Mann, C., Hebbachi, A.M., Gibbons, G.F., Nicholson, J.K., Shoulders, C.C., and Scott, J. 2004. An integrated reverse functional genomic and metabolic approach to understanding orotic acid-induced fatty liver. *Physiol. Genomics.* 17(2):140–149.

67. Ringeissen, S., Connor, S.C., Brown, H.R., Sweatman, B.C., Hodson, M., Kenny, S.P., Haworth, R.I., McGill, P., Price, M.A., Aylott, M.C., Nunez, D.J., Haselden, J.N., and Waterfield, C.J. 2003. Potential urinary and plasma biomarkers of peroxisome proliferation in the rat: identification of N-methylnicotinamide and N-methyl-4-pyridone-3-carboxamide by 1H nuclear magnetic resonance and high performance liquid chromatography. *Biomarkers.* 8:240–271.

68. Gavaghan, C.L., Holmes, E., Lenz, E., Wilson, I.D., and Nicholson, J.K. 2000. An NMR-based metabonomic approach to investigate the biochemical consequences of genetic strain differences: application to the C57BL10J and alpk: apfCDmouse. *FEBS Lett.* 484(3):169–174.

69. Griffin, J.L. 2004. Metabolic profiles to define the genome: can we hear the silent phenotypes. *Phil. Trans. Biol. Sci. R. Soc. Lond. B.* 2:1471–2970.

70. Griffin, J.L., Williams, H.J., Sang, E., Clarke, K., Rae, C., and Nicholson, J.K. 2001. Metabolic profiling of genetic disorders: a multitissue ¹H nuclear magnetic resonance spectroscopic and pattern recognition study into dystrophic tissue. *Anal. Biochem.* 293(1):16–21.

71. Bundy, J.G., Spurgeon, D.J., Svendsen, C., Hankard, P.K., Osborn, D., Lindon, J.C., and Nicholson, J.K. 2002. Earthworm species of the genus Eisenia can be phenotypically differentiated by metabolic profiling. *FEBS Lett.* 521(1–3):115–120.

72. Grimes, J.H., and O'Connell, T.M. 2011. The application of micro-coil NMR probe technology to metabolomics of urine and serum. *J. Biomol. NMR.* 49:297–305.

73. Korir, A.K., Almeida, V.K., Malkin, D.S., and Larive, C.K. 2005. Separation and analysis of nanomole quantities of heparin oligosaccharides using on-line capillary isotachophoresis coupled with NMR detection. *Anal. Chem.* 77:5998–6003.

74. Olson, D.L., Peck, T.L., Webb, A., Magin, R.L., and Sweedler, J.V. 1995. High-resolution micro-coil 1H-NMR for mass-limited, nanoliter-volume samples. *Science.* 270:1967–1970.

75. Lacey, M.E., Subramanian, R., Olson, D.L., Webb, A.G., and Sweedler, J.V. 1999. High-resolution nmr spectroscopy of sample volumes from 1 nL to 10 μL. *Chem. Rev.* 99:3133–3152.

76. Olson, D.L., Lacey, M.E., and Sweedler, J.V. 1998. The nanoliter niche. NMR detection for trace analysis and capillary separations. *Anal. Chem.* 70:257A-264A.

77. Webb, A.G. 1997. Radiofrequency microcoils in magnetic resonance. *Prog. Nucl. Magn. Reson. Spectrosc.* 31:1–42.

78. Wolters, A.M., Jayawickrama, D.A., and Sweedler, J.V. 2002. Microscale NMR. *Curr. Opin. Chem. Biol.* 6:711–716.

79. Kautz, R.A., Lacey, M.E., Wolters, A.M., Foret, F., Webb, A.G., Karger, B.L., and Sweedler, J. V. 2001. Sample concentration and separation for nanoliter-volume NMR spectroscopy using capillary isotachophoresis. *J. Am. Chem. Soc.* 123:3159–3160.

80. Wolters, A.M., Jayawickrama, D.A., Larive, C.K., and Sweedler, J.V. 2002. Capillary isotachophoresis/NMR: extensionto trace impurity analysis and improved instrumental coupling. *Anal. Chem.* 74:2306–2313.

81. Eldridge, S.L., Almeida, V.K., Korir, A.K., and Larive, C.K. 2007. Separation and analysis of trace degradants in a pharmaceutical formulation using on-line capillary isotachophoresis-NMR. *Anal. Chem.* 79:8446–8453.

82. Gilpin, R.K., and Zhou, W. 2004. Studies of the thermal degradation of acetaminophen using a conventional HPLC approach and electrospray ionization-mass spectrometry. *J. Chromatogr. Sci.* 42:15–20.

83. Fairbrother, J.E. 1974. Acetaminophen. In *Analytical Profiles of Drug Substances*, ed. K. Florey, 1–109, New York: Academic Press.

84. El-Obeid, H.A., and Al-Badr, A.A. 1985. Acetaminophen. In *Analytical Profiles of DrugSubstances*, ed. K. Florey, 551–596, Orlando, FL: Academic Press.

85. Hosttettmann, K., Potterat, O., and Wolfender, J.-L. 1998. The potential of higher plants as a source of new drugs. *Chimia.* 52:10–17.

86. Cragg, G.M., Newmann, D.J., and Snader, K.M. 1997. Natural products in drug discovery and development. *J. Nat. Prod.* 60:52–60.

87. Henkel, T., Brunne, R.M., Muller, H., and Reichel, F. 1999. Statistical investigation into the structural complementarity of natural products and synthetic compounds. *Angew. Chem. Int. Ed.* 38:643–647.

88. Huber, L., and George, S.A. 1993. *Diode Array Detection in HPLC.* New York: Marcel Dekker.

89. Niessen, W.M. 1999. State-of-the-art in liquid chromatography-mass spectrometry. *J. Chromatogr. A.* 856:179–197.

90. Wolfender, J.L., Queiroz, E.F., and Hotettmann, K. 2005. Phytochemistry in the microgram domain- a LC-NMR perspective. *Mag. Res. Chem.* 43:697–709.

91. Wolfender, J.-L., Ndjoko, K., and Hostettmann, K. 2001. The potential of LC-NMR in phytochemical analysis. *Phytochem. Anal.* 12:2–22.

92. Zanolari, B., Wolfender, J.-L., Guilet, D., Marston, A., Queiroz, E.F., Paulo, M.Q., and Hostettmann, K. 2003. On-line identification of tropane alkaloids from *Erythroxylum catuaba* ("catuaba") by LC-MSn and LC-NMR. *J. Chromatogr. A.* 1020:75–89.

93. Brachet, A., Munoz, O., Gupta, M., Veuthey, J.L., and Christen, P. 1997. Alkaloids of Erythroxylum Lucidum Steam-Bark. *Phytochemistry.* 46:1439–1442.

94. Lin, Y., Schiavo, S., Orjala, J., Vouros, P., and Kautz, R. 2008. Microscale LC-MS-NMR platform applies to the identification of active cyanobacterial metabolites. *Anal. Chem.* 80:8045–8054.

95. Raveh, A., and Carmelli, S. 2007. Antimicrobial ambiguines from the cyanobacterium Fischerella sp. collected in Israel. *J. Nat. Prod.* 70:196–201.

96. Smitka, T.A.B.R., Doolin, L., Jones, N.D., Deeter, J.B., Yoshida, W.Y., Prinsep, M.R., Moore, R. E., and Patterson, G.M.L. 1992. Ambiguine isonitriles, fungicidal hapalindole-type alkaloids from three genera of blue-green algae belonging to the Stigonemataceae. *J. Org. Chem.* 57:857–861.

97. Klein, D.D.D., Braekman, J.C., Hoffmann, L., and Demoulin, V. 1995. New hapalindoles from the cyanophyte hapalosiphon laingii. *J. Nat. Prod.* 58:1781–1785.

98. Moore, R.E.C.C., and Patterson, G.M.L. 1984. Hapalindoles:newalkaloids from the blue-green alga hapalosiphon fontinalis. *J. Am. Chem. Soc.* 106:6456–6457.

99. Park, A.M.R.E., Patterson, G.M.L., and Fischerindole, L. 1992. A new isonitrile from the terrestrial blue-green alga fischerella muscicola. *Tetrahedron Lett.* 33:3257–3260.

100. Stratmann, K.M., Bonjouklian, R.E., Deeter, J.B., Patterson, G.M.L., Shaffer, S., Smith, C.D., and Smitka, T.A. 1994. Welwitindolinones, unusual alkaloids from the blue-green algae hapalosiphon welwitschii and westiella intricata. relationship to fischerindoles and hapalinodoles. *J. Am. Chem. Soc.* 116:9935–9942.

101. Gronquist, M., Meinwald, J., Eisner, T., and Schroeder, F.V. 2005. Exploring uncharted terrain in nature's structure space using capillary NMR spectroscopy: 13 steroids from 50 fireflies. *J. Am. Chem. Soc.* 127:10810–10811.

102. Hu, J.-F., Yoo, H.-D., Williams, C.T., Garo, E., Cremin, P.A., Zheng, L., Vervoort, H.C., Lee, C. M., Hart, S.M., Goering, M.G., O'Neil-Johnson, M., and Eldridge, G.R. 2005. Miniaturization of the structure elucidation of novel natural products-two trace antibacterial acylated caprylic alcohol glycosides from arctostaphylos pumila. *Planta Med.* 71:176–180.

103. Eldridge, G.R., Vervoort, H.C., Lee, C.M., Cremin, P.A., Williams, C.T., Hart, S.M., Goering, M. G., O'Neil-Johnson, M., and Zheng, L. 2002. High throughput method for the production and analysis of large natural product libraries for drug discovery. *Anal. Chem.* 74:3963–3971.

104. Zhang, Q., Horst, R., Geralt, M., Ma, X., Hong, W.-X., Finn, M.G., Stevens, R.C., and Wüthrich, K. 2008. Microscale NMR screening of new detergents for membrane protein structural biology. *J. Am. Chem. Soc.* 130:7357–7363.

105. Pervushin, K., Riek, R., Wider, G., and Wüthrich, K. 1997. Attenuated T_2 relaxation by mutual cancellation of dipole–dipole coupling and chemical shift anisotropy indicates an avenue to NMR structures of very large biological macromolecules insolution. *Proc. Natl. Acad. Sci. USA.* 94:12366–12371.

106. Krueger-Koplin, R.D., Sorgen, P.L., Krueger-Koplin, S.T., Rivera-Torres, I.O., Cahill, S.M., Hicks, D.B., Grinius, L., Krulwich, T.A., and Girvin, M.E. 2004. An evaluation of detergents for NMR structural studies of membrane proteins. *J. Biomol. NMR.* 28:43–57.

107. Zhang, Q., Ma, X., Ward, A., Hong, W.X., Jaakola, V.P., Stevens, R.C., Finn, M.G., and Chang, G. 2007. Designing facial amphiphiles for the stabilization of integral membrane proteins. *Angew. Chem. Int. Ed.* 46:7023–7025.

108. McGregor, C.L., Chen, L., Pomroy, N.C., Hwang, P., Go, S., Chakrabartty, A., and Prive, G.G. 2003. Lipopeptide detergents designed for the structural study of membrane proteins. *Nat. Biotechnol.* 21:171–176.

109. McQuade, D.T., Quinn, M.A., Yu, S.M., Polans, A.S., Krebs, M.P., and Gellman, S.H. 2000. Rigid amphiphiles for membrane protein manipulation. *Angew. Chem. Int. Ed. Engl.* 39:758–761.

110. Popot, J.L. 2003. Amphipols: polymeric surfactants for membrane biology research. *Cell. Mol. Life Sci.* 60:1559–1574.

111. Schafmeister, C.E., Miercke, L.J., and Stroud, R.M. 1993. Structure at 2.5 Å of a designed peptide that maintains solubility of membrane proteins. *Science.* 262:734–738.

112. Zhao, X., Nagai, Y., Reeves, P.J., Kiley, P., Khorana, H.G., and Zhang, S. 2006. Designer short peptide surfactants stabilize G protein-coupled receptor bovine rhodopsin. *Proc. Natl. Acad. Sci. USA.* 103:17707–17712.

113. Peti, W., Page, R., Moy, K., O'Neil-Johnson, M., Wilson, I.A., Stevens, R.C., and Wüthrich, K. 2005. Towards miniaturization of a structural genomics pipeline using micro-expression and microcoil NMR. *J. Struct. Funct. Genomics.* 6:259–267.

114. Rossi, P., Swapna, G.V.T., Huang, Y.J., Aramini, J.M., Anklin, C., Conover, K., Hamilton, K., Xiao, R., Acton, T.B., Ertekin, A., Everett, J.K., and Montelione, G.T. 2010. A microscale protein NMR sample screening pipeline. *J. Biomol. NMR.* 46:11–22.

115. Hentschel, P., Krucker, M., Grynbaum, M.D., Putzbach, K., Bischoff, R., and Albert, K. 2005. Determination of regulatory phosphorylation sites in nanogram amounts of a synthetic fragment of ZAP-70 using microprobe NMR and on-line coupled capillary HPLC–NMR. *Magn. Reson. Chem.* 43:747–754.

8 Other Specialized Flow NMR

8.1 NMR AND PARALLEL DETECTION

It is generally acknowledged that NMR spectroscopy is a highly valuable technology for structural analysis. Spectral measurements are usually made on only a single sample at a time. Different types of NMR probes, however, have been developed that employ parallel, non-interacting sample coils or cells. Such probes (termed multiplex or parallel probes) were built to allow for simultaneous detection of multiple samples. This approach has the potential to significantly increase the throughput of NMR measurements, which is conventionally subject to serial data acquisition. The multiplex probe can interface with a normal NMR spectrometer, with minor modifications to the hardware.

Parallel detection methods in NMR represent a high-throughput approach for the simultaneous detection of more than one NMR sample at a time. The concept is very attractive for situations where multiple measurements are required due to large sample volume. Parallel methods can be applied in a large number of analytical techniques to improve sensitivity, precision, and throughput. In NMR, Fourier transform methods employing imaging detectors allow multiplex sampling of entire spectral regions. The same principle can be applied to the sampling of a collection of tubes or flow cells. This approach for data collection holds promise in areas such as combinatorial chemistry [1–3] and high-throughput screening [4], which rely on the rapid analysis of large numbers of samples that can benefit greatly from the development of such methods. While parallel detection methods, as applied to multiple samples, have been implemented in a number of different analytical procedures [5, 6], applications in NMR face a unique set of challenges.

Initial attempts to use multiplex sampling employed the use of a single coil; however, because of a poor filling factor, the sensitivity was compromised. In addition, there was no way to separate the spectrum of signals from multiple samples [7]. To address the sensitivity issue, individual solenoid coils were wrapped around capillary tubes containing the different samples. The signal arising from each coil could then be recorded using multiple methods [8]. A detection scheme was introduced that was based on the application of pulse field gradients. These field gradients could be applied to separate signals from samples based upon their spatially dependent frequency offset, a property that is unique for each sample. The application of field gradients allows for the signals from multiple samples to be deconvoluted using a single receiver only. This gradient approach simplified probe construction and improved the ability to generate relatively high-resolution ^{1}H-NMR spectra.

One of the first multiplex probes built with optimized signal-to-noise (S/N) for individual spectra [8] was a four-channel probe that enabled the simultaneous acquisition of NMR data. This multiplex probe consisted of four non-interacting sample coils that were each capable of detecting NMR signals at the same resonant frequency with good sensitivity and resolution. Hence, the simultaneous detection of spectra from four individual samples could be acquired. With this probe, the signals could be differentiated using field gradients and analyzed with software yielding either a table of peaks assigned to each sample or individual spectra. The construction of this probe (described below) clearly demonstrated the potential of this configuration for increasing the throughput of NMR detection, an important issue in applications that require screening of large molecular libraries.

8.2 DATA PROCESSING AND DECONVOLUTION OF PARALLEL DATA

Although simultaneous data collection of multiple samples has the potential to vastly increase throughput, a major complication resides in the deconvolution or separation of spectra from the individual samples. Multiple approaches were reported for deconvolution of spectra obtained from multiple samples using a probe with z-gradients, a single or multiple coils, and a single receiver. One approach involved peak picking each spectrum and correlation of the results with calculated frequency shifts expected for each sample. These frequency shifts are based upon the strength of the applied gradient and the position of the coils relative to the center z-gradient. This method, however, is tedious. Hence, a subtraction data processing method was subsequently developed [8]. The subtraction method used z-gradient values in the range of 0 to 48 mG/m. Application of a gradient (e.g., 48 mG/m) shifts the frequencies by an amount equal to the calculated shift of a sample in a coil at position z relative to the center of the gradient coil. Spectra were then obtained using a 0 gradient and the two spectra (with and without gradient applied) were subtracted. Negative peaks were removed through appropriate setting of the threshold and the resultant spectrum was subtracted from the 0-gradient spectrum to give the spectrum of the sample of interest. While subtractions were relatively free of artifacts, this method does not produce quantative integrations plus the resonance lines were broadened, which further complicates analysis especially for complex samples. Furthermore, the increased line broadening results in reduced S/N [8].

The third method for separating the spectral mixture was to use a process called reference deconvolution [9]. This approach employs a mathetical process that enables elimination of deleterious contributions to peak shape and linewidth caused by the response of the instrument. Reference deconvolution may be performed by multiplying the free induction decay (FID) by the complex ratio (CR) of an ideal FID and a reference FID. The reference FID (r-FID) is constructed by taking the experimental spectrum and zeroing all signals with the exception of the selected reference signal and taking its inverse Fourier transform. The ideal FID (i-FID) is generated by placing a single point (delta function) at the peak of the reference signal and zeroing the rest of the spectrum, followed by inverse Fourier transformation.

$$CR = i - FID/r - FID \qquad (8.1)$$

The corrected experimental FID is calculated by multiplying the experimental FID by CR, followed by Fourier transformation that yields the corrected FID. Reference deconvolution may then be incorporated into the subtraction procedure previously described. Deconvolution of both spectra (with 0 gradient and full gradient) to the same linewidth before proceeding with the subtraction algorithm gave the best results. A noted complication, however, was that due to the non-uniform line-broadening across the sample coils, the spectra needed to be deconvolved using a standard from each coil and the reference peaks from each coil must be well separated from the rest of the spectrum. Because of these issues, the use of an external standard is considered to be a more robust approach.

In 2006, a technique entitled Spectral Unraveling by Space-selective Hadamard Spectroscopy (SUSHY) was reported [10]. This technique enabled recording and deconvolution of NMR spectra of multiple samples loaded in multiple sample tubes in a conventional 5 mm solution NMR probe equipped with a tri-axis pulsed field gradient coil. The individual spectrum from each sample can be extracted by adding and subtracting data that are simultaneously obtained from all the tubes based on the principles of spatially resolved Hadamard spectroscopy. The SUSHY method is easily incorporated in multi-dimensional, multi-tube NMR experiments.

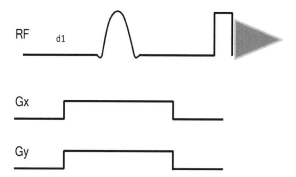

FIGURE 8.1   The 1D SUSHY pulse sequence: The first shaped pulse is a hyperbolic-secant inversion pulse applied in the presence of transverse gradients whose effective gradient direction separates the tubes. The rectangular 90° pulse is applied in the absence of gradient pulses and excites all the spins. The FID is observed in the absence of any gradients.

Considering a dual sample system, a simple 1D SUSHY pulse sequence will use a hyperbolic-secant inversion pulse applied in the presence of transverse gradients whose effective gradient direction (e.g., an orthogonal orientation) separates the two tubes. The rectangular read pulse is a 90° pulse, applied in the absence of pulsed-field gradient pulses, excites all the spins and the FID is observed with no applied gradients (Figure 8.1). In the presence of the gradient pulses, the shaped-pulse is calibrated to invert the spins, either from both sample tubes or just one sample tube. The non-selective 90° observe pulse in the absence of gradient pulses reads the state of the magnetization of all the spins in both the tubes. Addition and subtraction of the spatially phase-encoded spectra produce NMR data from the individual tubes

This method reduces the total experimental time by up to a factor of four in a four-tube system. SUSHY retains the sensitivity advantage of FTNMR as all the spins in all the tubes are excited for detection, and suitable data addition and subtraction given by a Hadamard scheme yield NMR spectra from each individual sample. This method combines the principles of imaging and spectroscopy for spatial selection and uses standard spectrometer hardware and probe configuration to obtain the NMR data.

8.3   PARALLEL PROBE CONSTRUCTION

8.3.1   Multiplex Four-Coil Probe

The initial construction of the NMR multiplex flow probes used excitation and detection coils with solenoid geometry. The construction of such probes was based on previously reported probe design studies [11]. The first four-coil parallel flow probe that was constructed used coils from polyurethane-coated high-purity (>99%) 36-gauge copper wire that was wrapped around fused silica capillaries. The length of the entire assembly was 20 mm with a 1.6 mm o.d., and 0.8 mm i.d. The cylindrical casing served as both the coil form and the sample holder. The inductance of each coil was roughly 20 nH. The coils were attached to the capillary tubes using a cyanoacrylate adhesive. Each of the rf coils had a length of about 0.7 mm, four turns, and an inner diameter of 1.6 mm. Sample tubes were mounted in a coil holder made of PVC that held the capillary tubes and the center-to-center intercoil spacing was set at 3.2 mm. The entire coil array was placed in a removable PVC chamber containing Fluorinert FC-43 (Syn Quest Laboratories, Alachua, FL), a susceptibility matching fluid that has been shown to improve magnetic field homogeneity by minimizing field distortions induced by copper NMR coils [12]. Care had to be taken in the construction of the PVC chamber to prevent leakage of the Fluorinert fluid. In addition with parallel construction, the coil leads had to be cut to the same length so that the resistance and inductance of each of the coils were similar.

A single resonant circuit was constructed using non-magnetic tunable capacitors to tune and match the circuit. The variable capacitors were located directly beneath the sample region. The tuning range for all four coils in parallel was approximately 2 MHz centered around 300 MHz with a resonant Q of 60. The coil housing was mounted on a home-built 39 mm–diameter probe body and it used a semi-rigid copper coaxial line to connect the resonant circuit at the top of the probe to a BNC connector at the base. Teflon tubes (2.0 mm OD) were used to enable flow into and out of the probe. The tubes were connected to the capillaries using polyolefin heat shrink and sealed with Torr-Seal (Varian Associates, Palo Alto, CA). A picture of this four-coil multiplex probe is shown in Figure 8.2.

The first spectra using this probe were collected on a spectrometer operating at 300 MHz for ^{1}H. Initial samples contained about 500 mM of analyte in D_2O. The probe was centered in the magnet by loading four H_2O/D_2O samples and adjusting the linear gradient (z1 shim) to separate the peaks from the individual coils. The center was chosen as the point at which the top and bottom coils were shifted in frequency by an equal and opposite amount and initial shimming to a resolution of 2–4 Hz typically required 2–3 h, clearly a drawback with respect to high throughput.

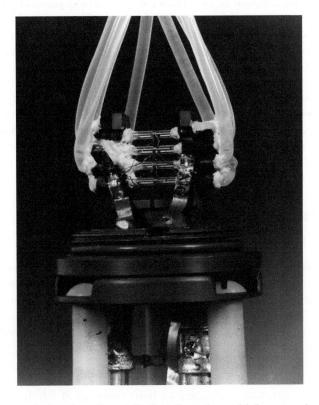

FIGURE 8.2 Picture of the first four-coil flow-through multiplex sample NMR probehead. A description is found in the text. (Reprinted from *Analytica Chimica Acta* 397, MacNamara, E., Hou, T., Fisher, G., Williams, S., and Raftery, D., Multiplex sample NMR: an approach to high-throughput NMR using a parallel coil probe 9–16, 1999 with permission from Elsevier.)

Nonetheless, this pioneering work demonstrated the possibility of introducing significant parallelism into the excitation and detection of multiple NMR samples. A parallel probe with simplified hardware requirements (reaching resolution of ~3 Hz without much difficulty) could be produced. Additional progress needed to be made in the area of improved resolution since much of the current limitations in linewidth for the parallel probe resulted from small imperfections in the coil construction as well as the small aspect ratio of the current coil configuration [13,14]. This was apparent from early experiments that used static samples sealed with parafilm. In such cases, the linewidths were not reproducible because of small changes in coil geometry and position upon the loading and sealing of new samples in the capillaries. The flow-through probe, however, reduced this problem but did not eliminate it and attention to the construction of the coils was needed to significantly improve resolution.

The mass sensitivity of the parallel probe was quite impressive when compared with that of a standard commercial broad-band switchable 5 mm high-resolution liquid probe, primarily due to the reduced coil diameter and was similar to previous microcoil probes. Comparison with a commercial 5 mm probe showed the conventional probe had a mass sensitivity, Sm of 48, which was calculated from the S/N of 78 measured for 0.1% ethylbenzene in $CDCl_3$. In comparison, the four-coil probe had an Sm of 4200 for the t-butanol methyl proton peak after a single scan. The observed volume of each of the samples in the four-coil probe was approximately 500 nL and the concentrations of 500 mM resulted in 0.25 mmol of sample.

The limit of detection (LOD) for this four-coil parallel probe was found to be about 710 pmol. When compared with the LOD (52 pmol) of a microcoil probe reported by Olson and co-workers [12] and considering differences in linewidths and accumulated transients between the results, the LODs and sensitivity were considered comparable. A small degradation of sensitivity in the spatially resolved sub-spectra was reported but this was attributed to the increased linewidth in the gradient-acquired spectrum. The features that would need to be improved to increase mass sensitivity would include increasing the filling factor or going to smaller coil diameters.

While it was determined that the parallel coil configuration did not adversely affect the sensitivity, increasing the parallelism of this method >4 has other challenges that need to be addressed. One important consideration when scaling up the number of coils would be the number of coils that will fit into the homogeneous region of the magnet. For a wide bore magnet, this region extends over 20 mm giving such magnets a spatial advantage in coil capacity accommodation.

Based on their initial work, MacNamara and coworkers concluded that the parallel coils do not adversely affect the sensitivity. They also predicted that a 10-coil probe design would be possible. Such a probe would provide a 10-fold improvement in throughput, which in conjunction with flow-through samples, would represent a significant advance in high-throughput screening using NMR.

As more coils are added, however, the probability of overlap between resonances increases and while improvements in the resolution would improve, spectral quality issues would need to be addressed with respect to the number of coils and the complexity of deconvolution. Applications such as SAR by NMR were cited where one is interested in identifying only the molecules that interact strongly with large proteins. These molecules typically have significantly shorter relaxation times than other, non-interacting molecules [15]. Hence, spectral editing methods such as saturation transfer difference spectroscopy (STD) [16] may be used to discriminate against the non-interacting molecules. In such cases, the use of parallel sample coils could possibly advance the throughput. The multiplex sample probe could also be of interest for applications such as quality control process monitoring. For example, cases where the primary goal is to monitor deviations of the NMR spectrum from a known, standard spectrum (such as identification of a potential problem in a batch process or synthetic route) could significantly benefit from use of a Multiplex sample probe.

8.3.2 MULTIPLEX DUAL PROBE

About the same time that the four-coil probe was introduced, another probe design that used a dual-coil probe with two independent circuits to tune and match the sample coils to the same resonant frequency was reported by the same group [17]. Just as with the four-coil probe, this dual probe was designed to interface to a normal NMR spectrometer, with minor modifications to the hardware. An advantageous property of the dual-coil configuration was that the coils and resonant circuits were constructed to prevent interference or coupling with each other. The signal from each coil could be measured using simultaneous 90° pulses on both resonant circuits. The signals were then detected using separate NMR receivers; hence, an instrument modification had to be made that involved incorporation of a home-built second receiver.

The data were acquired at 7.4 T with the probe tuned to 75.44 MHz for ^{13}C. The ^{13}C FIDs for two samples, methanol and carbon tetrachloride, both 99% enriched with ^{13}C, were acquired simultaneously using a single transmitter pulse and separate NMR receivers. S/N measurements were comparable to those observed using single-coil probes. All experiments were performed using a standard 90° pulse without proton decoupling. No evidence of cross-talk was found in the spectra even after considerable signal averaging. This probe also demonstrated the feasibility of significant parallelism in NMR as well as its potential in high-throughput analysis.

Although the achievement of very high spectral resolution was not the primary aim of this work, the resolution that was obtained had proved to be a limitation. The experiments showed that the spectral resolution in the two coils was primarily limited by magnetic susceptibility mismatches between the coils, the glass sample holders, and the samples. Initial attempts to use larger fill factors resulted in larger linewidths. Remedies to this situation were proposed that included the use of smaller samples and coils or the incorporation of a susceptibility matching fluid around the sample region and coils [12]. A limitation to using a smaller coil with this configuration would be that the LOD for non-^{13}C-enriched samples is compromised as the sample sizes become smaller. Use of zero magnetic susceptibility magnet wire appeared to be a more viable alternative [17].

Because multiple receivers were incorporated in this dual-coil probe design, this condition was viewed as a complicating design factor for multiple parallel NMR detection schemes. The dual probe consisted of a 4-turn, 4 mm solenoid coil wrapped from 20-gauge-insulated magnet wire with fixed tuning capacitors incorporated. The inductors were attached to a 30 mm long glass tube measuring 4 mm o.d., and 2 mm i.d. using epoxy cement. Samples were introduced in sealed glass sample tubes (2 mm o.d. and 6 mm long). The coils were electrically isolated from one another by a horizontal piece of copper-plated circuit board that was grounded to the probe body. This prevented cross-talk between the two circuits. Incorporation of fixed and variable capacitors was achieved by placing them in a single aluminum box located just below the NMR magnet to allow for easy tuning and matching of the circuits. The regions containing the variable capacitors were electronically isolated from one another by an aluminum plate. This design yielded two coils that were independently tunable minus deleterious coupling due to mutual inductance when they are tuned and matched to the same resonant frequency.

In this work reported by Fisher et al., two coils were electrically isolated from one another by using a copper ground plane, with the impedance-matching network similarly isolated [17]. By using two independent duplexer/preamplifier stages and two receivers, simultaneous acquisition of two ^{13}C spectra was possible. Although multinuclear and multiplex probe design had been achieved, the system still had not achieved high-throughput capability.

8.4   MICROPROBES

Following the initial studies on parallel detection in early 1999, Li et al. reported on the construction of high-sensitivity multiple solenoidal microprobes [18]. Solenoidal micro-coils were employed for their high sensitivity, ease of electrical decoupling, and efficient use of the homogeneous region of the magnet. The design used two radio-frequency switches with high isolation between channels that enabled the acquisition of even multi-dimensional data sets.

Two designs for incorporating multiple solenoidal microcoils into a single-probe head were described by Li and coworkers. By combining radio-frequency switches and low-noise amplifiers, multiple NMR spectra were acquired in the same time required for a single spectrum from a conventional probe consisting of one coil. This method did not compromise sensitivity with regard to the individual microcoils, and throughput was increased linearly with the number of coils. For this probe only one receiver was needed, and data acquisition parameters were optimized for each sample.

The design as reported by Li et al. consisted of a four-coil system for proton NMR at 250 MHz using a wide-bore magnet. This probe had an observe volume of 28 nL for each microcoil. Signal-cross contamination was ~0.2% between individual coils. Simultaneous one- and two-dimensional spectra were obtained from samples of fructose, galactose, adenosine tri-phosphate, and chloroquine; each sample containing about 7 nmol of compound. A more compact two-coil configuration was also reported for operation at 500 MHz, with observe volumes of 5 and 31 nL for the two separate coils. Performance of this probe was demonstrated with one- and two-dimensional spectra acquired from samples of 1-butanol (55 nmol) and ethylbenzene (250 nmol) [18]. The data showed minimal loss in sensitivity (<5%) compared to spectra obtained from a single coil, resulting in an increase in throughput effectively equal to the number of coils used.

Overall, the introduction of solenoidal microcoils showed that it was possible to acquire multidimensional high-resolution NMR spectra from multiple samples in the same time that it takes to acquire a single spectrum. This process combined the intrinsic high sensitivity of micro-coils with high-throughput capacity, thus expanding the capability of modern high-field NMR spectrometers. As expected it was proposed that the configuration could be extended to include more than four coils with either multiple receivers or time-domain multiplexing into a single receiver, thereby further increasing the throughput advantages offered by this approach. The major challenges of finite space within the magnet bore and the electrical decoupling of a large number of coils would still need to be addressed. This is because placing the coils close together in the most homogeneous part of the magnet could lead to degradation of spectral resolution due to perturbations in the magnetic field from the proximity of multiple conductor elements. In such cases, possible improvements in spectral resolution might require improved development of magnetic susceptibility matching fluids [19].

Following the major accomplishments with multi-coil probes in 1999, other multiple-coil probe construction was carried out leading to the expansion of coils beyond the four-coil configuration and incorporation of other probe features such as making probe tunable to nuclei other than ^{1}H and ^{13}C [20]. Likewise, the design and construction of versatile dual-volume heteronuclear double-resonance microcoil NMR probes further improved NMR detection of mass-limited samples.

8.5   ADVANCEMENTS IN MULTIPLE-COIL PROBE DESIGN

With the advent of multiple-coil probes, numerous strategies to improve upon the multiple-coil probe design became apparent. Beginning with the multiple-frequency microcoil NMR idea, other variations or improvements were proposed. These included:

1) Reducing observe volumes for mass-limited samples or increasing the observe volume for concentration limited samples.
2) Adding different frequencies between the multiple coils for a variety of experiments.
3) Inclusion of a lock frequency for maintaining the line shapes for longer acquisitions.
4) Addition of sample transfer lines for both analysis of flow samples and easier hyphenation of NMR with other analytical tools.
5) Addition of two ^{1}H channel receivers that are configured to connect to each of the ^{1}H channels of the probe. This would allow faster analysis of flow samples.

In 2009, the design and construction of a double-resonance 300 MHz dual-volume microcoil NMR probe with thermally etched 440 nL detection volumes and fused silica transfer lines for high-throughput stopped-flow or continuous-flow sample analysis was reported [21]. The probe consisted of two orthogonal solenoidal detection coils. Shielded inductors enabled construction of a probe with negligible radiofrequency cross-talk. The dual probe had an upper coil tuned to ^{1}H–^{2}H resonant frequencies and a lower coil tuned to ^{1}H–^{13}C. A single channel receiver was employed in the system design. The system was able to perform 1D and 2D experiments with an active deuterium lock. The spectra that were obtained had acceptable line widths with good mass sensitivity for both ^{1}H and ^{13}C NMR detection. ^{13}C-directly detected 2D HETCOR and INADEQUATE spectra of 5% v/v ^{13}C labeled acetic acid were obtained in minutes of data acquisition. Hence, many of the challenges previously facing probe design were being solved.

In the construction of this probe, an oval sample cell was designed to improve the fill factor using a quick-thermal-etching technique. Because the detection coils could be tuned to multiple resonances, it was possible to obtain homonuclear and heteronuclear 1D or 2D NMR experiments for structural analysis. The 2D lock channel was incorporated into the resonant circuit of the upper coil to improve the line shape for longer 2D acquisitions on either sample coil. This configuration was suited for a variety of experiments using one or both of the ^{1}H/^{13}C frequencies, and a combination of experiments could be run with or without the lock. The upper and lower coils were tuned and matched using a double-resonance RF design which features LC trap and pass elements tuned for use at 7 T. A semi-rigid coaxial cable with a 5 cm length and an outer diameter of 0.181 in was used to connect the coils to the rest of the resonant circuit. With this transmission line in place, the final coil/lead inductance was measured to be approximately 100 nH. Fixed-value capacitors, shielded 5 mm tunable inductors, and fixed and tunable capacitors that provide tuning and matching were used.

The single-coil double-resonance circuit tuned and matched to ^{1}H and ^{2}H frequencies at 300 and 46.05 MHz, respectively, or ^{1}H/^{13}C at 300 and 75.44 MHz, respectively, was modeled after the previously reported NMR resonant circuit designs [22–24]. This probe design incorporated a high-frequency trap, low-band-pass filter, and a variable length transmission line [25, 26] that was used to tune the circuit and to connect the sample coil leads to the other circuit components.

Tunable non-magnetic-shielded inductors provided flexibility in tuning the trap and pass components to their respective frequencies, and minimized the rf cross-talk of the components. Shielded inductors proved to be an advantageous approach while putting together all the tuning, matching, trapping, and passing elements for the four resonances in a single-probe circuit board of 38 mm circumference. A grounded copper sheet was placed between the two circuits to minimize the radiofrequency cross-talk. The combination of the grounded shield and the shielded inductors was important in keeping the proton resonances from interacting. The detection cells and their transmission lines needed to be surrounded by a delrin container holding FC-43 Fluorinert to reduce the line-broadening effects that can arise due to susceptibility mismatching of the probe materials [12].

Measurements of the electrical performance of the probe showed good radio-frequency design and construction. The power isolation of any two intraports, ^1H to ^2H (upper coil) and ^1H to ^{13}C (lower coil), was measured to be less than 0.75% (~21 dB). The power isolation of any two inter-ports between the upper and lower coil was found to be less than 0.004% (~44 dB). Proper intra-port isolation within each coil ensures maximum power transfer to the sample coil from one port with minimal power leakage to the other port, while the proper inter-port isolation between upper and lower coils minimizes the radiofrequency cross-talk.

The probe coils were shimmed with a sample consisting of a 1:9 H_2O/D_2O mixture with the optimized shim center placed at the lower coil. Each coil required a separate set of shim values for optimal performance in line width. Linewidths were measured to be 0.8 Hz (lower coil) and 1.1 Hz (upper coil). These linewidths, however, depended on the position of the coil in the center of the magnet. For example, when the upper coil was centered in the magnet, the upper coil could be shimmed to 0.8 Hz but the lower coil could only be shimmed to about 2 Hz line width without any distortion in the line shape. The lock in this system was shown to have good stability.

A 40 mM sucrose sample was used to test the ^1H limit of detection (LOD). Spectra were obtained using solvent suppression and 30 min of data acquisition. The amount of material in both coils was approximately 330 pmol. This picomole detection limit is similar to LODs reported for previous microcoil probe efforts [27]. Using a sample of 5% v/v $^{13}CH_3{}^{13}COOH$ in D_2O, the LOD for the ^{13}CO peak at 176 ppm for this sample located in the lower coil was 3.5 nmol with 40 min data acquisition using a 1D ^{13}C experiment. The ^1H decoupling efficiency of the lower coil was tested by acquiring a standard WALTZ decoupled ^{13}C experiment, which required a decoupling power of only 17 dB (12.6 mW) to achieve effective decoupling.

A photograph of the dual probe and an example of a ^{13}C directly detected 2D HETCOR spectrum of the labeled acetic acid sample obtained from the lower coil are shown in Figure 8.3. Both coupled and decoupled HETCOR spectra could be obtained with good sensitivity

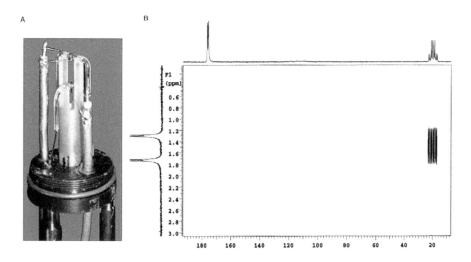

FIGURE 8.3 (A) Image of a probe head showing the cell and transmission line tuning element configuration. (B) HETCOR (coupled) spectrum of 5% doubly ^{13}C labeled acetic acid acquired <5 min (32 increments and 8 scans per increment). Correlation between the methyl carbon and its attached protons appears as a quartet of doublets. (Reprinted from *Journal of Magnetic Resonance* 197, Kc, R., Henry, I.D., Park, G.H.J. and Raftery, D. Design and construction of a versatile dual volume heteronuclear double resonance microcoil NMR probe 186–192, 2009 with permission from Elsevier.)

and resolution in less than 5 min using 32 increments and 8 scans per increment. The HETCOR plot shown below correlates the carbon multiplet at 20 ppm to the proton multiplet at 1.5 ppm.

A phase-sensitive 2D INADEQUATE spectrum with ^{1}H decoupling was also reported using only 64 increments and 36 scans per increment. The reported spectral data were processed with sinebell apodization. A long pre-delay of 5 s had to be used to prevent overheating of both the probe and the sample during the long experimental period. Couplings between the two labeled ^{13}C nuclei (J_{cc}) on the methyl and carbonyl carbons were measured to be the same value of 56.6 Hz at both peak positions in the F2 dimension. The sensitivity, resolution, flow design, and variety of one-dimensional and two-dimensional experiments as well as homonuclear and heteronuclear NMR experiments that can be performed by the dual-coil probe set the stage for a variety of separation-based applications. Clearly, multidimensional chromatographic-NMR studies can be employed with this dual-coil probe either in stopped-flow or continuous-flow mode. Possible uses for such a probe could include studies of metabolic eluates transferred from a chromatographic column to the top coil of the probe for 1D ^{1}H characterization while an interesting/unknown peak could be sent to the lower coil for multinuclear 2D NMR structural analysis. The probe also offers itself as a system for analyzing collected chromatographic bands stored in collection loops. The system is well positioned to interface with an automatic sample handler that would be able to transfer the bands to a designated coil in the dual-volume probe.

8.6   BIOLOGICAL SCREENING AND FLOW NMR

An interesting application of a dual-flow NMR system has been employed in a biological screening process that utilizes NMR to determine the binding affinities of weakly binding low molecular weight compounds (fragments) for a particular biological target. The technology was applied to a process, known as fragment-based drug discovery (FBDD), which has become a widely accepted tool that is complementary to a biological or mass spectrometry–based high-throughput screening (HTS) in developing small-molecule inhibitors of pharmaceutical targets.

Unlike conventional screening methods that search for molecules with high binding affinity for a particular target, the FBDD method is sensitive to weak target binders. This fragment approach is similar to the first NMR fragment screen which was based upon the premise that small molecular scaffolds with weak binding affinity for a target may be synthetically modified using either structure-based approaches or via combinatorial or parallel synthesis to ultimately generate a potent biological inhibitor. In this area, NMR plays a particularly important role because of its ability to identify weakly binding ligands [28].

Originally described as SAR by NMR [28], screening for simple, low affinity ligands as starting points for drug discovery has been recognized as a viable drug discovery approach. The "fragment-based approach" (FBA) has been successfully employed to generate small molecule inhibitors of pharmaceutical targets that were not found using conventional high-throughput screening (HTS) approaches. Some of the successes from the FBA include developing new chemical scaffolds for targets that yielded limited or no chemical candidates from HTS [29], identifying true hits by analysis of compounds from other screening campaigns to remove false positives [30] and identifying small molecule inhibitors of protein–protein interactions [31]. The FBA is particularly attractive from the standpoint of being able to design ligands that possess drug-like properties as described in Lipinski's "rule of five" [32]. Therefore, FBA holds the promise to deliver compounds with higher solubility, bioavailability, and lower toxicity, thus increasing the likelihood that a drug candidate will successfully complete a clinical trial in the drug-development process.

Because NMR-based ligand screening has gained considerable recognition for its high potential in the early stages of drug discovery, different types of NMR-screening methods were developed. To this end, a method was developed that makes efficient use of a single sample of a scarce target, or one with poor or limited solubility, to screen against an entire compound library. The comparative method is based on immobilizing the target for the screening procedure. To achieve this screening process, a dual-cell, flow-injection probe with a single receiver coil was constructed. The flow-injection probe was interfaced with a single high-performance pump and sample-handling system to enable the automated analysis of large numbers of compound mixes for binding to the target. The probe used in this process contains an 8 mm $^1H/^2H$ dual-tuned coil, triple-axis gradients, and was easily shimmed. NMR spectra of comparable quality to a standard 5 mm high-resolution probe were reported with this probe. The lineshape in the presence of a solid support shows identical quality relative to that obtained in glass NMR tubes in a 5 mm probe. Using this dual-flow probe, ligand binding in a complex mixture could be readily detected in ~30 min and was readily applied to drug-discovery efforts.

The ligand-screening method, called TINS (target immobilized NMR screening), reduces the amount of target required for the FBA to drug discovery [33]. Binding is detected by comparing 1D ^1H NMR spectra of compound mixtures in the presence of a target immobilized on a solid support to a control sample. The screening process employs a probe that was developed and designed to be used with the TINS method but with the performance features of probes typically used in high-resolution NMR. The probe was built for a 500 MHz spectrometer as an 8 mm selective probe with one RF coil. The coil is double tuned for proton (^1H) observation and decoupling with a deuterium (^2H) lock channel. The probe contains a saddle coil geometry to take advantage of the vertical magnet for inserting and ejecting the sample and was built with a Q factor of 247 on the ^1H channel. Because the RF circuitry was selected to achieve RF equivalency to a high-resolution NMR probe, special measures were taken in selecting low-susceptibility and/or susceptibility compensated materials for the coil and adjacent components. The shim properties of the probe were reported to be similar to those of a 500 MHz high-resolution probe. The probe also contained triple axis, XYZ gradients.

Probe performance was initially assessed using a suite of tests with standard NMR samples in an 8 mm sample tube. The lineshape test revealed an overall performance of the 8 mm probe comparable to a high-resolution 5 mm probe and the sensitivity was considerably better. In addition, water suppression, which is a major factor in the acquisition of NMR data of biological samples in H_2O/D_2O mixtures, gave excellent results.

The fluid-flow path consisted of a single pump that was compatible with the available controlling software (Bruker Biospin, HyStar, version3.1). An HPLC pump was used, operating at a moderate flow rate of up to 2 mL/min with a back pressure of 60–90 bar. A flow splitter was placed downstream of the HPLC pump. In order to minimize potential differences in flow over the two cells of the sample holder, PEEK tubing with an inner diameter of 90 μm was used. At a flow rate of 0.5 mL/min, this tubing was found to generate back pressure of about 59 bar while the sample holder filled with resin generated less than 1 bar. With this configuration, even if one of the cells is poorly packed resulting in a twofold increase in back pressure, there is less than 2% difference in flow over the cells. Considering the width of the peak of mixtures flowing through the system, this difference can be easily accommodated, thereby ensuring that each cell is optimally equilibrated for every experiment. Two sample loops of 400 μL followed after the flow splitter. The loops were controlled by a two-way valve that alternately places the loops in a serial or parallel flow configuration. When configured for serial flow, the loops are overfilled with 0.95 mL of the compound mixture using a standard liquid handler (Gilson 215 autosampler). Automated valve switching returns the loops to parallel flow for sample injection. Timing of the delivery of the sample to the cells was calibrated by following the amplitude of the water signal in each cell using a1D imaging-type experiment.

Upon delivery of a sample to the flow cell, the flow is stopped and the NMR data are acquired. The sample is then washed with 900 µL of buffer per cell (which was 4.5 times the cell volume) to ensure the chamber is purged of the previous sample components. With this application, care must be taken to keep the air out of the system since this could lead to various problems with the resin in the cells, including cracking or increased susceptibility mismatch leading to line broadening. Following the wash cycle, the flow cells are then ready to repeat the sample delivery cycle. The entire process for this system was manufactured by Bruker Biospin, hence, is under control of HyStar software which can interface with automation software to automate the complete process of sample injection and NMR acquisition.

The key element of the system is the dual-cell sample holder. A number of factors had to be considered when designing the sample holder. First the geometry of the cells must not generate broad resonances while at the same time maximizing the volume of each sample. Next, the geometry of the cells had to show minimal signal cross-talk using only standard triple-axis gradient hardware (50 G/cm). Finally, it was necessary to be able to open the sample holder in order to fill and empty the cells with the solid support for the immobilized target. Because the sample holder needed to be opened to insert solid support, glass was not a practical option. Instead a machined cylinder of KelF (8 mm in diameter) was used. Two cylindrical holes of 3.2 mm were bored out to form the individual cells that hold the solid support. Each cell contained a volume of approximately 190 µL of which 135 µL was active volume within the coil when placed inside the probe. Smooth fluid flow was achieved by tapering tthe cells toward the end that was formed by a porous frit designed to retain beads of at least 50 µm in diameter. The sample holder used standard threads for PEEK fittings and tubing.

To enable the cells to be filled with the solid support, a reservoir was designed that mounts onto the sample holder in place of the cap. A 50% slurry of the solid support may be transferred via pipette into the sample holder–reservoir combination. After settling, the reservoir can be attached to an HPLC pump and the resin in each cell simultaneously packed under pressure using the flow splitter [34]. A diagram of the system is shown in Figure 8.4.

The binding integrity of the target in the TINS method was validated by the detection of a variety of ligands for protein and nucleic acid targets (K_D from 60 to 5000 µM). The ligand binding potency of a test biological target was found to be undiminished even after 2000 different compounds had been applied. These studies showed the potential to apply the assay for screening typical fragment libraries. TINS can be used in competition mode, allowing rapid characterization of the ligand binding site. A very important application of the TINS technology is that it provides a means of screening of targets that are difficult to produce or that are insoluble, such as membrane proteins.

Because a fragment campaign can only be successful if true binders are discovered, TINS was evaluated against established methods. Three commonly used methods for hit discovery in FBDD, high concentration screening (HCS), solution ligand observed nuclear magnetic resonance (NMR), and surface plasmon resonance (SPR), were used. For the NMR screening, the commonly used saturation transfer difference (STD) NMR spectroscopy and the proprietary target immobilized NMR screening (TINS) were compared. Using a target typical of FBDD campaigns, it was shown that HCS and TINS were the most sensitive to weak interactions. It was also found that a good correlation between TINS and STD for tighter binding ligands existed, but the ability of STD to detect ligands with affinity weaker than 1 mM K_D was poor. Similarly, it was found that SPR detection is most suited to ligands that bind with K_D better than 1 mM. In addition, the good correlation between SPR and potency in a bioassay made this a good method for hit validation and characterization studies [33].

The initial test system for TINS used the well-characterized binding of spermidine to the phosphate backbone of DNA [35]. A 5′ biotinylated oligonucleotide was synthesized and immobilized on streptavidin Sepharose. Two samples of streptavidin Sepharose in standard

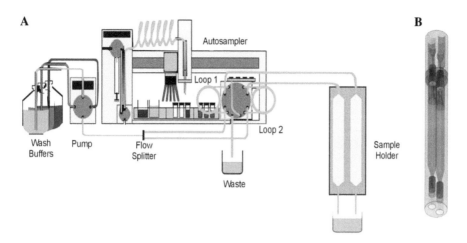

FIGURE 8.4 Liquid path for the TINS ligand screening apparatus. (A) The automation system includes an HPLC pump that splits the flow in two. An autosampler loads the compound cocktail to be tested into two loops that alternate between serial and parallel flow. Each mix of test compounds is injected in serial into the two loops. The valve is then switched and the contents of each loop are injected in parallel into the dual-cell sample holder, which is placed in the NMR probe for the duration of the screening process. (B) An illustration of the dual-cell sample holder is shown. (Reprinted from *J. Magn. Reson.* 182, Marquardsen, T., Hofmann, M., Hollander, J. G., Loch, C. M., Kiihne, S. R., Engelke, F., and Siegal, G. Development of a dual cell, flow-injection sample holder, and NMR probe for comparative ligand-binding studies 55–65, 2006 with permission from Elsevier.)

NMR tubes were prepared, one of which contained the 5′ biotinylated oligo nucleotide. Control and experimental resins were equilibrated with the mix of compounds, and 1D ^{1}H spectra of each were recorded. Subtraction of the spectrum recorded in the presence of the oligonucleotide from the control spectrum yielded the TINS spectrum (Figure 8.5). This spectrum contained only resonances of spermidine, known to bind the phosphate backbone of DNA. The affinity of the purely electrostatic interaction is highly dependent on salt concentration and pH [35]. Under the conditions used, the spermidine binding was expected to be weak (KD ~ 5 mM). The presence of multiple binding sites (about nine per DNA molecule) ensures that the binding equilibrium was shifted toward the complex. The spermidine DNA interaction demonstrates that weak binding of one compound in a mixture to an immobilized target can readily be detected using simple, static NMR methods [36].

TINS ligand screening was also assessed against the ability to immobilize functional Escherichia coli membrane proteins, DsbB, and OmpA [37] and the stability of such an immobilized preparation. Although biophysical methods had already been successfully applied to a number of soluble protein targets [28], there still remained obstacles when screening technologies such as SAR by NMR were applied to membrane proteins. The reasons for this failure are twofold. First, insufficient quantities of the target membrane proteins can be mass produced. Second, ther are problems related to the solubilization of the protein and appropriate media to enable solubilization without compromising function. While standard biophysical methods for screening proteins require tens or even hundreds of milligrams of purified, functional protein, most membrane proteins are difficult to produce in these quantities, and once produced, it is difficult to retain functional stability. Although advances in membrane protein production have enabled the production of low milligram quantities of a variety of membrane proteins [38–40], the proteins that can be produced in sufficient quantity need to be solubilized in a surfactant that maintains their functional state.

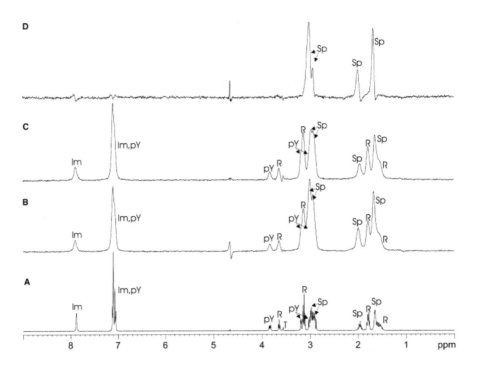

FIGURE 8.5   (A) 1D ^{1}H spectra of a mixture of small compounds in the presence of an immobilized oligonucleotide. ^{1}H spectrum of a mixture of 1 mM each of the following: imidazole (Im); phosphotyrosine (pY); arginine (R); and spermidine (Sp) in 25 mM Tris-d$_{11}$ (pH 7.5) and 140 mM NaCl dissolved in D$_2$O. (B) The same mixture equilibrated with streptavidin Sepharose. (C) The same mixture equilibrated with streptavidin Sepharose to which a 5′ biotinylated, double-stranded oligonucleotide was bound. All ^{1}H NMR spectra were recorded with identical parameters (9.4 T, 256 transients, total time ~4.5 min). (D) The difference spectrum ([B] and [C]) showing only peaks from spermidine, a known binder of the DNA backbone. (Reprinted from Vanwetswinkel, S., Heetebrij, R.J., van Duynhoven, J., Hollander, J.G., Filippov, D.V., Hajduk, P.J., and Siegal, G. TINS, Target Immobilized NMR Screening: An Efficient and Sensitive Method for Ligand Discovery, 12 *Chemistry & Biology* 207–216, 2005 with permission from Elsevier.)

This process can be difficult to achieve. Attempts to screen ligands against membrane proteins also resulted in nonspecific partitioning of fragments into the surfactant leading to high levels of false positives.

Because the use of detergent micelles to solubilize membrane proteins has only met with limited success in retaining the native function of the protein and since the use of micelles can interfere with biophysical assays, an alternate solution needed to be devised. A solution to this dilemma was to employ nondetergent media to functionally solubilize membrane proteins. The nanodisc (ND) was developed as a surfactant-free approach to solubilize membrane proteins. NDs consist of a lipid bilayer that is surrounded by an amphiphilic α-helical membrane scaffold protein (MSP). A variety of proteins have been functionally solubilized in NDs [41–43], which are much better mimics of the native state of the protein.

To test the TINs technology with a membrane protein, a pharmaceutical target was selected that could be functionally solubilized in detergent micelles and NDs. The inner membrane protein of E. coli disulphide bond forming protein B (DsbB) and its homologs in other gram-negative bacteria was chosen because of its oxidoreductase functionality that is essential for protein disulfide bond formation in the periplasm [44]. The periplasmic DsbA functions as the catalyst for protein disulfide bond formation. DsbA is reoxidized by DsbB with concomitant reduction of bound ubiquinone or menaquinone. Because bacterial virulence factors are secreted proteins that require disulfide bonds for proper folding and function, the DsbA/DsbB system was identified a potential antimicrobial drug target [45–47].

Another reason the DsbB membrane protein was an ideal candidate to test the TINS methodology was because it could be biosynthesized in large quantities. In addition, the solubilization of this protein in detergent micelles does not inhibit the enzymatic activity. This activity can be readily checked through screening assays. This target has also been extensively studied and reports on the biochemical data that describes the enzymatic activity of the wild-type as well as numerous relevant mutants are known [44,48–50]. This membrane protein has also had 3D structures generated for wild-type DsbB bound to its redox partner DsbA [51] and structures of a mutant representing an enzymatic intermediate are available [52].

In TINS, binding of a fragment was described using a T/R (target/reference) ratio, defined as the average ratio of the amplitude of peaks in the presence of the target, which in this study was DsbB, to that in the presence of the reference, OmpA. Binding studies were conducted that showed this membrane protein's functionality and stability remained relatively constant throughout the screen that was completed in ~6 days.

The fragment library was screened for binding to DsbB at 500 mM each. A total of 182 mixtures were used. A spatially selective Hadamard NMR experiment [10] was used to simultaneously acquire a 1D ^{1}H spectrum of compounds in the presence of DsbB/DPC or OmpA/DPC. Interpretation of this data had the advantage that deconvolution was not necessary to discern protein binding. This was because fragments could be directly identified by comparing peaks from TINS spectra with reference spectra of the individual fragment [37]. The screen resulted in 93 hits for DsbB, defined as fragments which had a T/R ratio less than 0.3, as shown by an example of a mix containing two hits in Figure 8.6.

To avoid the possibility of false positives in the biochemical assay, 13 fragments showing strong inhibition in the single concentration point assay were selected for further analysis. These 13 fragments were assayed for potency (IC_{50}) by dose-response experiments (Figure 8.7). Dose-response experiments were carried out with increasing fragment concentrations, ranging from 0.0001 to 10 mM, while both DsbA and UQ1 (a known binding ligand) were kept in excess. UQ1 was used to monitor the activity integrity of the protein over the course of the experiment. Three of the 13 fragments were found to be false leads and showed signs of protein precipitation at higher compound concentration and/or steeper than expected Hill coefficients. The remaining 10 fragments titrated over 2 log orders and exhibited a Hill coefficient close to unity.

These criteria were used as an indication that each of the 10 binding ligands were reversible inhibitors with a 1:1 stoichiometry. The eight most potent compounds exhibited IC_{50} values between 7 and 170 mM (weak binders) and consisted of a variety of different synthetic templates. The calculated binding efficiency index [53] indicated that these fragments were excellent starting points for hit to lead programs. Hence, these results were able to clearly establish the feasibility of using a FBA for finding starting compounds for subsequent development of candidates targeting membrane proteins.

A second study involved use of the micelle solubilized β1-adrenergic StaR (β1AR) and was based on direct covalent attachment using Schiff base chemistry. About 80% of the *n*-dodecyl-β-D-maltopyranoside (DDM)-solubilized β1AR was immobilized and subsequently, unreacted aldehydes on the resin were blocked using deuterated Tris buffer. The immobilized protein was assayed for functionality by binding of [^{3}H]-dihydroalprenolol, a well-characterized, high-affinity ligand that binds with 1:1 stoichiometry. Measurement of [^{3}H]-dihydroalprenolol binding after immobilization showed that ~100% of the immobilized receptor was functional. After 4 days, ~50% of the immobilized receptor remained functional for high-affinity ligand binding if stored at temperatures up to 10°C.

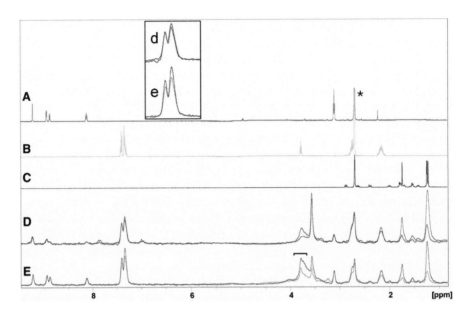

FIGURE 8.6 Detection of ligand binding to immobilized DsbB using TINS. The 1D ^{1}H NMR spectrum of three different fragments in solution (A–C) is shown for reference. The ^{1}H NMR spectrum of a mix of the three fragments in the presence of DsbB/DPC (red spectrum) or OmpA/DPC (blue spectrum) that have been immobilized on the Sepharose support is shown in (D). The spectra of the same mix recorded in the presence of DsbB/ND (green) or/ND (magenta) is shown in (E). The asterisk indicates the resonance from residual ^{1}H DMSO and the bracket shows residual sugar ^{1}H resonances from the Sepharose media. The residual H$_2$O resonance at 4.7 ppm has been filtered out. (Reprinted from *Chem. Biol.* 17, Fruh, V., Zhou, Y., Chen, D., Loch, C., Ab, E., Grinkova, Y. N., Verheij, H., Sligar, S.G., Bushweller, J. H., and Siegal, G. Application of fragment-based drug discovery to membrane proteins: Identification of ligands of the integral membrane enzyme DsbB. 881–891, 2010. with permission from Elsevier.)

For assays such as TINS, in which the target is reused, the need to monitor protein viability is imperative. Therefore, it becomes necessary to have a known ligand that can be readily removed without denaturing the protein as a test compound. Accordingly, the affinity of both dopamine and dobutamine were investigated as potential test inhibitors for β1AR, since these two compounds are on the biosynthetic route to epinephrine. Binding assays for both dopamine and dobutamine showed that they inhibited dihydroalprenolol binding with IC$_{50}$s of 60mM and 0.60 mM, respectively. In order to determine the feasibility of detecting weak ligand binding to immobilized β1AR, dopamine binding was examined using the TINS assay. The TINS method [36] uses a reference protein to cancel out nonspecific binding, thereby reducing the hit rate of false positives. The immobilized proteins, target and reference, were packed into separate cells of a dual-cell sample holder [34] and placed inside the magnet. OmpA was determined to be an appropriate reference for micelle-solubilized membrane proteins and therefore was used.

For this study with β1AR, the binding assay was conducted by injecting mixtures of compounds into both cells simultaneously. The flow was stopped at a predetermined point when the sample entered the flow cell. A spatially selective one-dimensional ^{1}H Hadamard experiment [10] was used to acquire an NMR spectrum of the compounds in solution. Since the NMR relaxation of a spin is approximately 1000 times faster in the solid state than in solution, strong binding of a ligand to the immobilized protein should result in the complete disappearance of resonances from the bound molecule in the NMR spectrum. Therefore, various binding

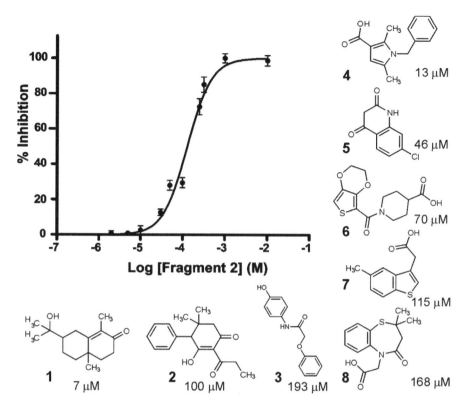

FIGURE 8.7 Potency determination of selected hits from the TINS screen. An example of an inhibition curve used to determine the IC$_{50}$ for compound 2. The curve represents the mean of three independent experiments performed in triplicate. Eight of most potent compounds are shown with the respective IC$_{50}$ values. (Reprinted from *Chem. Biol.* 17, Fruh, V., Zhou, Y., Chen, D., Loch, C., Ab, E., Grinkova, Y. N., Verheij, H., Sligar, S.G., Bushweller, J. H., and Siegal, G Application of fragment-based drug discovery to membrane proteins: Identification of ligands of the integral membrane enzyme DsbB. 881–891, 2010 with permission from Elsevier.)

strengths could be detected as a reduction in the amplitude of the signals from the compounds bound to the target relative to those in the presence of the reference.

Specific binding of dopamine was assayed to immobilized β1AR using immobilized OmpA as the reference protein. The amplitude of all of the NMR signals from dopamine was significantly reduced in the presence of β1AR, while the signals derived from soluble DDM was the same in both samples. This data showed that TINS is capable of detecting weak but specific binding typical of fragments. It was estimated that approximately 10 compounds at a time may be screened with TINS. Under these conditions, a throughput of about 500 compounds per day could be realized. With the rate determined, it was estimated that a library of 5000 compounds could be screened in about 10 days.

8.7 NMR AND MICROREACTORS

Microreactors enable the use of minimal amounts of reagents under precisely controlled conditions; hence, they are well suited to rapidly screen reaction conditions and kinetics studies [54]. A report by Kakuta et al. [55] described the coupling of a micromixer and a solenoidal NMR microcoil, having a detection volume of 800 nL, to study methanol-induced conformational changes of ubiquitin on a timescale of seconds. Ciobanu and coworkers [56] used a multiple-microcoil NMR setup to study the kinetics of D-xylose-borate reaction by NMR spectroscopy. Two reagents were mixed in a Y-shaped mixer, then passed through a capillary around which three identical solenoidal microcoils (each of which had a detection volume of approximately 31 nL) were wrapped. The reaction times were calculated based upon the combination of the flow rates and the distance between the mixer and each NMR microcoil. Bart et al. [57] employed a glass microreactor coupled to a stripline NMR chip with a detection volume of 600 nL for the real-time monitoring of the acetylation of benzyl alcohol with acetyl chloride in the presence of N,N-diisopropylethylamine. The ^{1}H-NMR spectra recorded using continuous flow allowed for the detection of reaction intermediates, which were almost unobservable when the reaction was performed in a 5 mm NMR tube.

Microwave-assisted continuous-flow organic synthesis was also reported where the flow system was hyphenated with nanoliter-volume NMR spectroscopy. This combination appeared to be a promising tool for the rapid optimization of reaction conditions [58]. In this experiment, a fused-silica capillary was wrapped around a WeflonTM bar, yielding a reaction volume of 1.6 µL, and coupled to a microfluidic NMR chip equipped with a planar transceiver microcoil (with a detection volume of 6 nL). A Diels–Alder cycloaddition was chosen as a model reaction, and the conversion was calculated at three different temperatures. Because the volume of the NMR chip was much lower than the microwave reaction volume, the latter could be divided into several zones. Under flow conditions, each of the zones was exposed to the microwave irradiation for different times. By choosing the appropriate flow rate and the NMR acquisition parameters, the necessary data points for determining the conversion were obtained in a single constant-flow experiment.

A fully integrated microreactor chip/microcoil NMR setup was reported by Wensink et al. [59] for a kinetic study of the imine formation from aniline and benzaldehyde by ^{1}H-NMR spectroscopy at 1.4 T. A planar microcoil was integrated on top of a microfluidic chip containing two inlets, a mixing zone, a detection zone, and one outlet. The total channel volume between the mixing point and the coil was 0.57 µL, and the detection volume was 56 nL. The residence time in the detection volume ranged from 0.9 s to 30 min, depending on the flow rate.

The combination of chip-based microfluidics with NMR was initially reported by Trumbull et al. [60], who integrated a planar NMR coil on a capillary electrophoresis (CE) chip. These authors showed ^{1}H spectra of a 30 nL water sample acquired at 250 MHz with a frequency-domain signal-to-noise (S/N) ratio of 23.5 per scan, and a minimal linewidth of 1.39 Hz. Comparable results have been reported showing a S/N of 117 per scan but with a large linewidth of 30 Hz for 30 nL water in a micromachined glass chip with integrated microcoil, acquired at 300 MHz [59]. The main factor that limits the performance of micromachined NMR probes in terms of sensitivity and spectral resolution was identified to be probe-induced static magnetic field inhomogeneity, and routes for improvement were proposed in the areas of probe design and materials, magnetic field shimming, and signal processing [61].

8.8 SIGNAL-TO-NOISE RATIO (MICROCOILS)

For an optimal signal-to-noise (S/N) ratio, the planar microcoil has to be properly designed. The parameters that define the coil are the number (N), width (w), height (h), and separation (s) of the coil windings. In addition, the the inner diameter (i.d.) has to be optimized (see Figure 8.8).

Different effects contribute to and detract from the optimization. For example, increasing the number of coil windings will increase the signal level but will also increase the coil resistance, and hence the noise level will increase. It is difficult to calculate which parameters yield the maximum S/N, because the coil resistance depends non-linearly on the number of coil

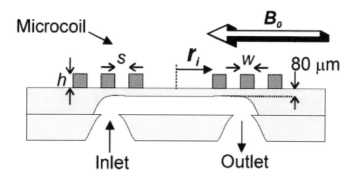

FIGURE 8.8 Schematic cross-section of the microcoil with three windings on top of a microfluidic chip. (Wensink, H., Benito-Lopez, F., Hermes, D.C., Verboom, W., Gardeniers, H.J. G.E., Reinhoudt, D.N., and van den Berg, A, Measuring reaction kinetics in a lab-on-a-chip by microcoil NMR, 5 *Lab Chip* 280–284, 2005 reproduced with permission of the Royal Society of Chemistry.)

windings, due to the frequency-dependent skin effect and a proximity effect from the neighboring coil windings.

It has been reported that the optimum coil geometry may be determined by using finite element simulations [62]. The design rules that govern this work may be described as follows:

(1) The windings and the separations between the windings should be as small as possible; this puts demands on the tolerances of the microfabrication process.
(2) The coil height has only a small influence on the resistance and should be about 20 to 50 mm.
(3) The inner coil radius should match the average radius of the sample volume to obtain a high filling factor.
(4) Although there is no general rule to determine the optimal number of windings, it should be determined by simulations for specific coil shapes.
(5) A simple model of a planar microcoil can be derived such that S/N scales with the inverse of the distance between NMR detection coil and sample,and that distance should be as small as possible.

Spectral resolution has a high dependence on the homogeneity of the magnetic fields over the interrogated NMR sample volume. It is known that materials introduced in a homogeneous field will distort the magnetic field. Exceptions occur when the magnetic permeability of the material matches the magnetic field. Materials such as glass and copper will disturb the homogeneity of a magnetic field that can lead to a decrease in the spectral resolution. This is a typical problem for NMR detection on a chip, where the volume of glass surrounding the sample is large on a microchip relative to a conventional 5 mm NMR probe. These field distortions are exacerbated by abrupt changes in material geometry near the NMR detection area. To avoid this problem, it was recommended that electrical and fluidic connections be placed away from the detection area. Although the fluidic channel that contains the sample will also disturb the magnetic field, for reasons of sensitivity this needs to be close to the detection coil.

Two approaches to maintain a homogeneous field inside the sample volume of interest have been proposed. The first approach is based upon simulations that indicate the field inside a spherical or cylindrical volume will be very homogeneous [61,63]. A second approach involves placing a straight channel in line with the magnetic field. Field disturbances created by the ends of the channel will not affect the central region of the sample chamber where the detection coil is placed. The straight channel approach has the advantage that it is easier to fabricate on the microscale than perfectly spherical or cylindrical channels.

An example of the successful use of this technology involved monitoring a reaction in which an imine (Schiff base) formation took place (see Figure 8.9). This reaction had been performed previously using a microfluidic chip; however, in the previous work the reaction was analyzed by mass spectrometry [64] as well as Raman microscopy [65].

For the NMR study, the chip inlets were connected to two 100 µL syringes containing 4.95 M benzaldehyde and 0.475 M tetramethylsilane (TMS; NMR reference) in deuterated nitromethane (CD_3NO_2) and 4.95 M aniline, 0.523 M TMS in CD_3NO_2, respectively. Imine spectra in Figure 8.10 show the ^1H NMR spectra of the individual solutions measured in the chip.

Because of the high concentrations used for the reactant materials, a substantial amount of water was formed during the reaction. Since water is immiscible with the high organic content in the sample, it migrated to the channel walls, particularly at the corners. Therefore, a high flow rate flush was applied between the scans to prevent the dominance of water resonance in the ^1H spectra. After flushing, the flow rate was returned to a lower value to achieve the desired residence time. The measurement procedure required that the two reactants were loaded in the chip with a chosen flow rate. The flow rate determines the residence time in the channel section between the mixing point and the detection area, and this residence time is taken as the reaction time. A NMR scan was taken with an aquisition time of 0.2 s, and 2000 data points followed by a delay between scans. The delay time had to be scaled with the flow rate to ensure new sample in the detection volume, with a maximum delay time of 10 s. A total of 32 NMR acquisitions were averaged for each reaction measurement.

Analysis of the reaction kinetics focused on two rather broad peaks by conventional NMR standards. These resonances were the aldehyde peak of benzaldehyde at 9.9 ppm and the imine peak of the product at 8.4 ppm. The results showed an increase of the imine peak and decrease of the aldehyde signal with increasing residence time. The peak area of the two peaks of interest was calculated to follow the course of the reaction. The conversion of the reaction was plotted as the ratio of each peak area to the sum of both peak areas, as a function of the residence time. The results showed a fit with a second-order rate equation, a rate constant of 6.6×10^{-2} $M^{-1}min^{-1}$, and a correlation coefficient of 0.93. The reaction shown in Figure 8.9 was also

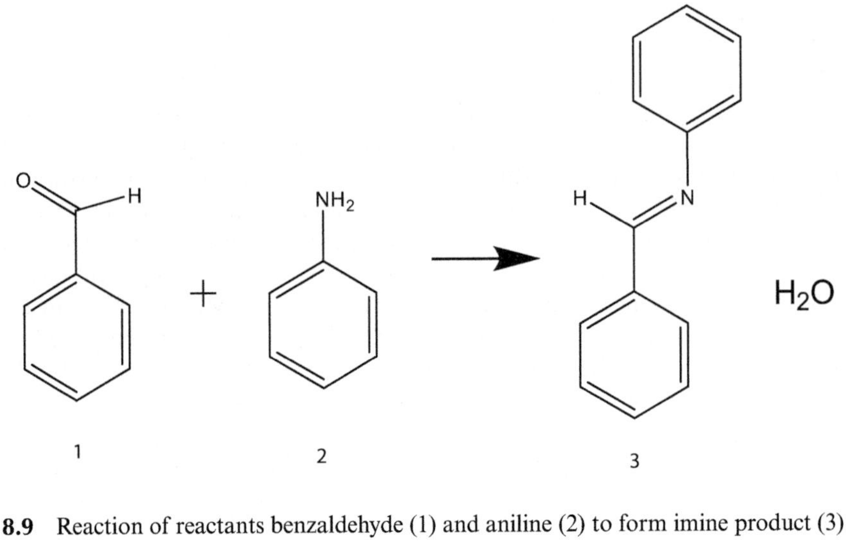

FIGURE 8.9   Reaction of reactants benzaldehyde (1) and aniline (2) to form imine product (3) [59].

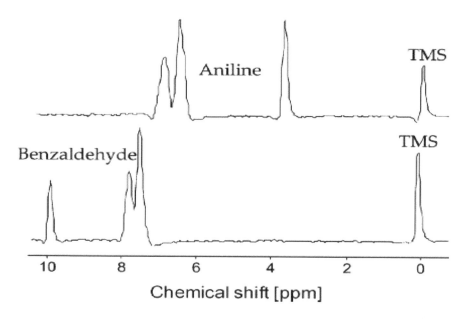

FIGURE 8.10 ^{1}H NMR spectra of the individual solutions analine (top) and benzaldehyde (bottom) measured in the chip. Spectral resolution is completely determined by magnetic field imhomogeneity in the chip. (Wensink, H., Benito-Lopez, F., Hermes, D.C., Verboom, W., Gardeniers, H.J.G.E., Reinhoudt, D.N., and van den Berg, A, Measuring reaction kinetics in a lab-on-a-chip by microcoil NMR, 5 *Lab Chip* 280–284, 2005 reproduced with permission of the Royal Society of Chemistry.)

carried out in a conventional 5 mm probe using a 400 MHz NMR spectrometer and the same concentrations as used in the chip. The reaction was followed by taking a single scan every 5 s, and the experiment was repeated three times. For the tube-based study, the second-order rate constant was found to be $(3.35 \pm 0.05) \times 10^{-2}$ M^{-1} min^{-1}. This value is consistent with the values measured by Raman spectroscopy for imine formation (in chloroform) [66]. When compared with the reaction rate constant obtained in the microchip, there is a two-fold increase over the larger-scale system. This result was attributed to an improved mixing performance in the chip.

The degree of mixing may be described as the penetration depth for diffusion (PDD) of a species from one liquid to another. For a simple one-dimensional diffusion model without chemical reaction, the PDD is equal to $\sqrt{2Dt}$, where **t** the time of contact of the two liquids and **D** the diffusivity of the species of interest. The diffusivity of the reactants benzaldehyde and aniline is 1.5×10^{-9} m^2 s^{-1} and 1.96×10^{-9} m^2 s^{-1}, respectively [67]. Since the time needed for complete mixing in a 160 μm wide channel would be estimated at about 2 s, then the mixing in the microcell can be considered to be nearly instantaneous in the time range in which the reaction kinetics were measured. The reaction time range was comparable to that reported Kakuta et al. [55], who used a mixing time of 1.4 s. Hence, this work demonstrated that monitoring a simple chemical reaction on the microscale was possible. However, corrections need to be made for comparisons with reaction rates from larger vessels. Improvements of this chip would include adding extra functionality such as mixers and heaters.

It is also possible that 2D homonuclear pulse sequences, like COSY and NOESY, may be used with the system. In fact, a basic 2D COSY experiment with ^{13}C-labeled acetic acid on a microfluidic chip with an integrated Helmholtz microcoil (i.e., a dual microcoil) has been demonstrated [63].

Additional functionality that can aid in the measurement of reaction kinetics could be the implementation of two or more coils at different positions along a microchannel. A configuration reported by Ciobanu et al. involved wrapping three solenoidal microcoils at different locations around a flow through capillary [56]. These researchers studied xylose–borate reaction kinetics as their model system. The reactants in this method were kept in separate containers and then were rapidly mixed together using a Y-mixer. After mixing, the solution was pumped through the microcoils while NMR data were being acquired. The mixture passed through a capillary around which were wound multiple, physically distinct NMR detector coils. The distance between the mixer and each individual NMR coil, together with the flow rate used, determined the postreaction-time point being measured in each of the coils. Signal averaging could be performed for as long as necessary to obtain an adequate S/N ratio, because the time of the data measurement and the reaction time are decoupled through the continual mixing and flow achieved with this experimental method. However, if longer total data-acquisition time is needed then more reactants would be needed to feed the flow stream. The use of multiple small-volume microcoils, however, minimizes the amount of reactants.

The design of the NMR probe used in this study consisted of three identical 370 μm diameter solenoidal RF microcoils, constructed of 50 μm copper wire wrapped around polyimide tubing. Each coil was about 1 mm long and had seventeen turns. The sample flowed through a 200 μm internal diameter capillary, which resulted in a detection volume of approximately 31 nL. The three-coil-probe assembly was immersed in FC-43 susceptibility fluid to minimize susceptibility-mismatch distortions. Linewidths at half height of 1–2 Hz were obtained for HDO (2% H_2O in D_2O) for each single coil. Simultaneous data acquisition performed by using optimized shimming for all three coils yielded linewidths of approximately 4–6 Hz.

Experiments were performed in which an initial borate concentration was higher than the d-xylose concentration, to enable examination of only 1:1 complexes of borate and xylose. As the reaction progresses, a decrease in the height of the α and β anomeric protons and the appearance of a new peak at 5.55 ppm becomes apparant. NMR data corresponding to successive reaction times were recorded independently and simultaneously with all three RF coils. When the data from all three coils were examined, it was found that the ratio of peak heights (anomeric proton of product/anomeric proton of starting material) relative to time was identical for spectra obtained from all three coils, thus validating the use of distinct NMR coils at multiple locations along the capillary/reaction-time coordinate.

8.9 MICROPROBE DESIGN (MICROCOIL/MICROSLOT/MICROSTRIP)

While the performance of the microchip for NMR spectroscopy was adequate for the relatively simple imine-formation reaction kinetic study, where low molecular weight species at high concentrations were studied, to work at more practical concentrations and with more complex molecules will require vastly improved spectral resolution. Since in the present design the spectral resolution is completely determined by magnetic field inhomogeneity in the chip, improving the design of the chip is a primary goal.

The planar designs for microprobes need to be evaluated with respect to their orientation in the magnetic field. The capillaries in microstrip or microslot probes can be oriented parallel to B_0 because of their planar design. Figure 8.11 shows the magnetic fields created by the NMR magnet, B_0, the microstrip, B_1, and the current running through the separation capillary during a separation, B_2, for both a solenoidal microcoil (Figure 8.11a) and a microslot (Figure 8.11b) [68].

Since the direction of B_2 is perpendicular to B_0 for the microslot, the separation current should not contribute to the inhomogeneity of B_0. When compared with solenoidal coils, better quality spectra would be expected. Experiments coupling CE and cITP to microstrip or microslot NMR probes should be able to experimentally verify this hypothesis.

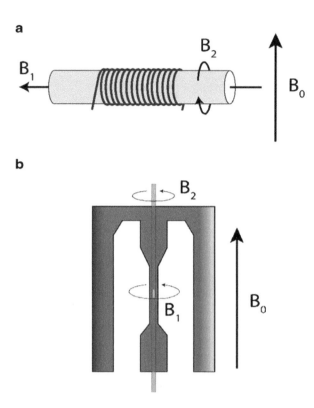

FIGURE 8.11 Illustrations of (a) a solenoidal microcoil and (b) microslot. The direction of the magnetic fields created by the coil or slot (B_1), the current running through the separation capillary during a separation (B_2), and the magnet (B_0) are shown. (Reprinted from Jones, C.J. and Larive, C.K, Could smaller really be better? Current and future trends in high-resolution microcoil NMR spectroscopy, *Anal Bioanal Chem*. 402, 2012 With kind permission from Springer Science and Business Media.)

The planar design of microstrip and microslot probes makes them amenable to coupling to lab-on-a-chip applications. Kentgens and coworkers have demonstrated that microfluidics can be readily bonded to microstrip and microslot chips without compromising the electrical properties of the chip [69].

Because lab-on-a-chip devices have been gaining increased importance in commercial development for a wide range of chemical and biochemical studies, the incorporation of this technology into NMR systems could represent a major breakthrough in the tools available for chemical and structural elucidation. A promising feature of the microstrip and microslot probe designs is their ease with which they may be built. In fact, these microstrip and microslot probes can be produced using well-established lithographic methods making them amenable to facile production in a commercial design process. Relative to solenoidal microcoils, which can undergo automated fabrication, such coils often still are hand wound by skilled probe engineers. A fabrication difference between the microstrip probe and the microslot probe is that the microstrip probe may be constructed using only a single-layer lithographic process, and hence offers a full range of scalability. For the microslot probe however, scalability is limited by the laser wavelength used to create the slot in the microstrip.

As noted previously, the microprobes hold great promise in reaction monitoring, even allowing multiple reactions to be monitored simultaneously with a single NMR spectrometer.

A major advantage of the microprobes is their high mass sensitivity. When combined with their microsize, reaction monitoring can be performed using small amounts of compounds.

Even though microcoils are attractive in terms of sensitivity and excitation bandwidths, they have not been enthusiastically embraced as a general tool in mainstream NMR research. One main reason is due to the compromises that must be made in spectral resolution.

In NMR probe design, high sensitivity demands a high filling factor, so the sample needs to be in close contact with the rf coil. In general, the magnetic susceptibility of the metal wire will be different from that of the coil form support (or air for a freestanding coil) and the sample holder. This condition will introduce local variations of the static magnetic field in the sample, which causes a reduction in the spectral resolution. Because of the small dimensions of a microcoil, resultant field gradients will vary over very short distances producing a field profile that cannot be easily shimmed by macroscopic shim coils that are relatively distant from the sample. Optimization of field uniformity in the sample may be achieved by immersing the microcoil in a medium that has the same susceptibility as a copper metal [12]. Perfluorocarbons are particularly suited for this purpose as these liquids have the same magnetic susceptibility as copper. These perfluorocarbons introduce no proton background signal and are commonly used in commercial microcoil probes for capillary NMR. The successful susceptibility matching of these solenoidal coils has enhanced the resolution of these probes so that they could be effectively applied to a variety of high-throughput applications. An alternative to using perfluorocarbons would be use of zero-susceptibility wire. This material is also effective in preventing deterioration of the spectral resolution that results from the close proximity of the coil to the sample. Zero-susceptibility wire is made from a coaxial structure of metals which have opposite signs in susceptibility such as copper and aluminum [13]. Susceptibility-matched coils have found their way into a number of homebuilt and commercial probes showing that high resolution close to that in conventional 5mm probes is possible using solenoid microcoils [70].

In contrast to helical coils, the microchip possesses properties that make the susceptibility problem easier to address. The first aspect is that the axis of the microchip can be oriented parallel to the static field B_0. Likewise, the magnetization of the copper strip is homogeneous when oriented parallel to the external field. Calculation of the local field variations for a plane just above a stripline in the microchip showed that the susceptibility broadening could be limited to below 0.1 ppm [71]. Furthermore, it was shown that etching only needed to occur on a narrow strip section which means a minimal disruption of a long uniform strip. A resolution of 1 Hz was achieved for protons at 600 MHz with this configuration.

The development of flat microcoils has mainly been driven by the possibility to integrate them into microfluidic devices with the aim to add NMR as a key analysis technique in the "lab on a chip" approach. The growing interest in small-volume chemistry performed in dedicated chip sets requires that analytical technologies such as NMR be able to handle the corresponding volumes. Although Trumbull et al. [60] were the first to address this issue by making a single turn planar coil, they succeeded in getting well-resolved (1.4–3 Hz) spectra for approximately 400 nL of water and ethanol. However, sensitivity was found to be relatively poor. Massin and coworkers [61] succeeded in increasing the sensitivity by an order of magnitude using two- and three-turn helical coils; however, spectral resolution was compromised with this system. Wensink et al. [59] were able to improve both sensitivity and resolution with a flat helical coil, operating at very low field of 1.4 T, above a straight microfluidic channel. Later it was determined by Kontgens and coworkers that the stripline geometry [71, 72] would not only be able to improve on sensitivity but more importantly resolution, since susceptibility broadening can be avoided to a large extent due to the symmetry of the design. A prototype probe was built to demonstrate the sensitivity of the stripline design detecting 12 nL ethanol in a single scan. The LOD was good, however, the resolution was poor since the probe was constructed for solids applications.

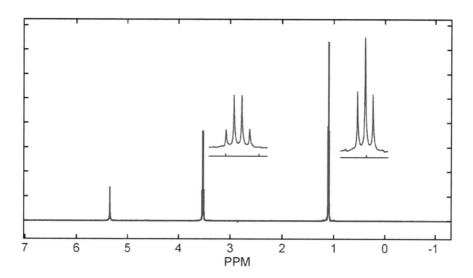

FIGURE 8.12 Single scan ^{1}H spectrum of a 600 nL ethanol sample analytical grade obtained in a microfluidic stripline probe. The signal-to-noise ratio was ~5050 and the resolution better than 1 Hz. The expansion the proton J-multiplets are baseline resolved. (Reprinted with permission from Kentgens, A.P.M, Bart, J., van Bentum, P.J.M., Brinkmann, A., van Eck, E.R.H., Gardeniers, J.G.E., Janssen, J.W. G., Knijn, P., Vasa, S. and Verkuijlen, M.H.W. High-resolution liquid- and solid-state nuclear magnetic resonance of nanoliter sample volumes using microcoil detectors. J. Chem. Phys. 128:052202-1-052202-17, 2008, Americal Institute of Physics.)

A fully integrated stripline design that achieves tremendous resolution was later constructed. Figure 8.12 shows a 600 MHz proton spectrum of 600 nL ethanol that was obtained under flowing conditions in a single scan using a microchip probe. Resolution of 1 Hz was achieved which allows resolving the proton *J*-multiplets nearly down to the baseline.

The chip was made using silicon where surface charges induce some losses resulting in a lowering of the Q of the resonator. Because of this, the sensitivity experienced some reduction relative to theoretical prediction. Further optimizations of the production process are underway. The related microslot design introduced by Maguire et al. also shows favorable numbers for sensitivity and resolution. However, because the rf field was not confined, the rf homogeneity was poorer [73]. Considering the successful implementation of the microfluidic chip with the stripline detector, a probehead that allows the placement of a microfluidic reactor close to the NMR chip has been built. With this technology in hand, kinetic studies of a series of on chip reactions, such as Diels–Alder reactions, alcohol protections, and oxidation reactions, can be realized.

It has been further projected that the scalable design of microstrip and microslot "coils" combined with their easy coupling to microfluidics and ease of fabrication position this technology for expansion into NMR microarrays. Development of such probes would allow multiple analytes to be detected simultaneously.

8.10 FUTURE DIRECTIONS

8.10.1 PORTABLE NMR AND SENSITIVITY

Portable NMR spectrometers offer promise for experiments in which chemical and structural information is needed, but where a full-sized NMR magnet is either unavailable or impractical

[74–76]. With such systems, a small scale of microcoils becomes critical in the manufacture of the device. While portable NMR is not a widely accepted tool among NMR spectroscopists, a system is now commercially available from Picospin. Studies have been reported by Diekmann and coworkers where they successfully coupled CE to a portable NMR for the separation and detection of fluorinated compounds [77]. Although this study employed a full-sized CE to enable separations, it is expected that incorporation of USB-powered high-voltage power supplies would enable the CE instrumentation to also be reduced to a movable scale. Incorporating microstrip or microslot probes into portable NMR systems would likewise contribute to the compact and mobile nature of such systems. While the development of a portable NMR system is attractive from numerous standpoints, it must be recognized that issues with respect to sensitivity and resolution would result from using a lower field magnet in order to achieve true portability. However, the system can provide an inexpensive, rapid, and fully portable chemical analysis tool (with and without separation capabilities) that could potentially meet the needs of laboratories where optimal resolution and sensitivity are not critical factors.

8.10.2 DNP AND NMR SENSITIVITY

An emerging approach to improving NMR sensitivity involves hyperpolarization, of which the most common method is dynamic nuclear polarization or DNP. The way DNP works may be described as follows:

To observe an NMR signal from spin ½ nuclei, there must be a difference in the number of spins in the two states as defined by the Boltzmann distribution:

$$N_\alpha/N_\beta = e^{\Delta E/k_B\,T} \tag{8.2}$$

where N_α and N_β represent the number of nuclei in the α and β spin states, respectively; ΔE is the energy difference between the two spin states; k_B is the Boltzmann constant; and T is the temperature. In NMR, the ΔE term is generally small compared to other spectroscopic methods, limiting the sensitivity of the NMR measurement (see Chapter 2). DNP attacks the basic NMR sensitivity problem in that it selectively increases the population of one nuclear spin state over the other. This is accomplished by first polarizing the electron spins of a radical either through microwaves or optical pumping. This polarization is then transferred from the electrons of the radical to the nuclei of the analyte, and followed by a fast NMR acquisition before the nuclei have time to relax. Although DNP can achieve signal enhancements of up to 44,400 for ^{13}C and 23,500 for ^{15}N compared to spectra acquired at thermal equilibrium in a 9.4 T magnetic field, spectra were confined to one-dimensional experiments due to the short lifetimes of the hyperpolarized state [78,79]. However, with the advent of fast one-scan NMR experiments, the acquisition of 2D-NMR spectra coupled with DNP is now feasible [80].

Since DNP technologies are in early stages of development, it may be expected that the coupling of DNP with microcoil NMR will be realized to address the NMR sensitivity issue. Further improvements to detect a wider range of heteronuclei are an equally important part of expanding the use of microcoil NMR offering great promise in the further expansion of this technology.

REFERENCES

1. Lam, K.S., Lebl, M., and Krchnak, V. 1997. The "one-bead-one-compound" combinatorial library method. *Chem. Rev.* 97:411–448.
2. Keifer, P.A. 2003. Flow NMR applications in combinatorial chemistry. *Curr. Opin. Chem Biol.* 7:388–394.

3. Ellman, J., Stoddard, B., and Wells, J. 1997. Combinatorial thinking in chemistry and biology. *Proc. Natl. Acad. Sci. USA.* 94(7):2779–2782.

4. Kalelkar, S., Dow, E.R., Grimes, J., Clapham, M., and Hu, H. 2002. Automated analysis of proton NMR spectra from combinatorial rapid parallel synthesis using self-organizing maps. *J. Comb. Chem.* 4:622–629.

5. Ueno, K., and Yeung, E.S. 1994. Simultaneous monitoring of DNA fragments separated by electrophoresis in a multiplexed array of 100 capillaries. *Anal. Chem.* 66:1424–1431.

6. Asano, K., Goeringer, D., and McLuckey, S. 1995. Parallel monitoring for multiple targeted compounds by ion trap mass spectrometry. *Anal. Chem.* 67:2739–2742.

7. Banas, E.M. 1969. "Multi-cloistered" NMR cells. *Appl. Spectrosc.* 23:281–282.

8. MacNamara, E., Hou, T., Fisher, G., Williams, S., and Raftery, D. 1999. Multiplex sample NMR: an approach to high-throughput NMR using a parallel coil probe. *Anal. Chim. Acta* 397:9–16.

9. Hou, T., MacNamara, E., and Raftery, D. 1999. NMR analysis of multiple samples using parallel coils: improved performance using reference deconvolution and multidimensional methods. *Anal. Chim. Acta* 400:297–305.

10. Murali, N., Miller, W.M., John, B.K., Avizonis, D.A., and Smallcombe, S.H. 2006. Spectral unraveling by space-selective Hadamard spectroscopy. *J. Magn. Reson.* 179(2):182–189.

11. Webb, A.G., and Grant, S.C. 1996. Signal-to-noise and magnetic susceptibility trade-offs in solenoidal microcoils for NMR. *J. Magn. Reson. B* 113(1):83–87.

12. Olson, D.L., Peck, T.L., Webb, A.G., Magin, R.L., and Sweedler, J.V. 1995. High-resolution microcoil ^{1}H-NMR for mass-limited, nanoliter-volume samples. *Science* 270:1967–1970.

13. Webb, A.G. 1997. Radiofrequency microcoils in magnetic resonance. *Prog. Nucl. Magn. Reson. Spectrosc.* 31(1):1–42.

14. Barbara, T. 1994. Cylindrical demagnetization fields and microprobe design in high-resolution NMR. *J. Magn. Reson. A* 109(2):265–269.

15. Hajduk, P.J., Olejniczak, E.T., and Fesik, S.W. 1997. One dimensional relaxation- and diffusion-edited NMR methods for screening compounds that bind to macromolecules. *J. Am. Chem. Soc.* 119:12257–12261.

16. Mayer, M., and Meyer, B. 1999. Characterization of ligand binding by saturation transfer difference NMR spectroscopy. *Angew. Chem. Int. Ed.* 38:1784–1788.

17. Fisher, G., Petucci, C., MacNamara, E., and Raftery, D. 1999. NMR probe for the simultaneous acquisition of multiple samples. *J. Magn. Reson.* 138(1):160–163.

18. Li, Y., Wolters, A.M., Malawey, P.V., Sweedler, J.V., and Webb, A.G. 1999. Multiple solenoidal microcoil probes for high-sensitivity, high-throughput nuclear magnetic resonance spectroscopy. *Anal. Chem.* 71:4815–4820.

19. Wu, N., Webb, A.G., Peck, T.L., and Sweedler, J.V. 1995. Online NMR detection of amino acids and peptides in microbore LC. *Anal. Chem.* 67:3101–3107.

20. Zhang, X., Sweedler, J.V., and Webb, A.G. 2001. A probe design for the acquisition of homonuclear, heteronuclear, and inverse detected NMR spectra from multiple samples. *J. Magn. Reson.* 153(2):254–258.

21. Ravi, K.C., Henry, I.D., Park, G.H.J., and Raftery, D. 2009. Design and construction of a versatile dual volume heteronuclear double resonance microcoil NMR probe. *J. Magn. Reson.* 197(2):186–192.

22. Grynbaum, M.D., Kreidler, D., Rehbein, J., Purea, A., Schuler, P., Schaal, W., Czesla, H., Webb, A., Schurig, V., and Albert, K. 2007. Hyphenation of gas chromatography to microcoil H-1 nuclear magnetic resonance spectroscopy. *Anal. Chem.* 79:2708–2713.

23. Fan, K.M., and Courtieu, J. 1980. A single-coil triple resonance probe for NMR experiments. *Rev. Sci. Instrum.* 51:887–890.

24. Li, Y., Logan, T.M., Edison, A.S., and Webb, A. 2003. Design of small volume HX and triple resonance probes for improved limits of detection in protein NMR experiments. *J. Magn. Reson.* 164(1):128–135.

25. Henry, I.D., Park, G.H.J., Kc, R., Tobias, B., and Raftery, D. 2008. Design and construction of a microcoil NMR probe for the routine analysis of 20-μL samples. *Concepts Magn. Reson. B: Magn. Reson. Eng.* 33B:1–8.

26. Cross, V.R., Hester, R.K., and Waugh, J.S. 1976. Single coil probe with transmission-line tuning for nuclear magnetic double-resonance. *Rev. Sci. Instrum.* 47:1486–1488.

27. Subramanian, R., Lam, M.M., and Webb, A.G. 1998. RF microcoil design for practical NMR of mass-limited samples. *J. Magn. Reson.* 133(1):227–231.

28. Hajduk, P.J., Meadows, R.P., and Fesik, S.W. 1997. Drug design: discovering high-affinity ligands for proteins. *Science* 278:497–499.

29. Boehm, H.J., Boehringer, M., Bur, D., Gmuender, H., Huber, W., Klaus, W., Kostrewa, D., Kuehne, H., Luebbers, T., and Meunier-Keller, N. 2000. Novel inhibitors of DNA gyrase: 3D structure based biased needle screening, hit validation by biophysical methods, and 3D guided optimization. A promising alternative to random screening. *J. Med. Chem.* 43:2664–2674.

30. Rishton, G.M. 2003. Nonleadlikeness and lead likeness in biochemical screening. *Drug Discov. Today* 8:86–96.

31. Oltersdorf, T., Elmore, S.W., Shoemaker, A.R., Armstrong, R.C., Augeri, D.J., Belli, B.A., Bruncko, M., Deckwerth, T.L., Dinges, J., Hajduk, P.J., Joseph, M.K., Kitada, S., Korsmeyer, S.J., Kunzer, A.R., Letai, A., Li, C., Mitten, M.J., Nettesheim, D.G., Ng, S., Nimmer, P.M., O'Connor, J. M., Oleksijew, A., Petros, A.M., Reed, J.C., Shen, W., Tahir, S.K., Thompson, C.B., Tomaselli, K.J., Wang, B.L., Wendt, M.D., Zhang, H.C., Fesik, S.W., and Rosenberg, S.H. 2005. An inhibitor of Bcl-2 family proteins induces regression of solid tumours. *Nature* 435:677–681.

32. Lipinski, C.A., Lombardo, F., Dominy, B.W., and Feeney, P.J. 1997. Experimental and computational approaches to estimate solubility and permeability in drug discovery and development settings. *Adv. Drug Deliv. Rev.* 23:3–25.

33. Kobayashi, M., Retra, K., Figaroa, F., Hollander, J.G., Ab, E., Heetebrij, R.J., Irth, H., and Siegal, G. 2010. Target immobilization as a strategy for NMR-based fragment screening: comparison of TINS, STD, and SPR for fragment hit identification. *J. Biomol. Screen.* 15(8):978–989.

34. Marquardsen, T., Hofmann, M., Hollander, J.G., Loch, C.M., Kiihne, S.R., Engelke, F., and Siegal, G. 2006. Development of a dual cell, flow-injection sample holder, and NMR probe for comparative ligand-binding studies. *J. Magn. Reson.* 182(1):55–65.

35. Braunlin, W.H., Strick, T.J., and Record, M.T. 1982. Equilibrium dialysis studies of polyamine binding to DNA. *Biopolymers* 21:1301–1314.

36. Vanwetswinkel, S., Heetebrij, R.J., van Duynhoven, J., Hollander, J.G., Filippov, D.V., Hajduk, P. J., and Siegal, G. 2005. TINS, Target Immobilized NMR Screening: anefficient and sensitive method for ligand discovery. *Chem. Biol.* 12:207–216.

37. Fruh, V., Zhou, Y., Chen, D., Loch, C., Ab, E., Grinkova, Y.N., Verheij, H., Sligar, S.G., Bushweller, J.H., and Siegal, G. 2010. Application of fragment-based drug discovery to membrane proteins: identification of ligands of the integral membrane enzyme DsbB. *Chem. Biol.* 17:881–891.

38. Dahmane, T., Damian, M., Mary, S., Popot, J.L., and Baneres, J.L. 2009. Amphipol-assisted in vitro folding of g protein-coupled receptors. *Biochemistry* 48:6516–6521.

39. Serrano-Vega, M.J., Magnani, F., Shibata, Y., and Tate, C.G. 2008. Conformational thermostabilization of the beta 1-adrenergic receptor in a detergent-resistant form. *Proc. Natl. Acad. Sci. USA.* 105:877–882.

40. Rasmussen, S.G.F., Choi, H.K., Rosenbaum, D.M., Kobilka, T.S., Thian, F.S., Edwards, P.C., Burghammer, M., Ratnala, V.R.P., Sanishvili, R., Fischetti, R.F., Schertler, G.F.X., William, I., Weis, W.I., and Kobilka, B.K. 2007. Crystal structure of the human beta 2 adrenergic G-proteincoupled receptor. *Nature* 450:383–387.

41. Nath, A., Atkins, W.M., and Sligar, S.G. 2007. Applications of phospholipid bilayer nanodiscs in the study of membranes and membrane proteins. *Biochemistry* 46:2059–2069.

42. Katzen, F., Fletcher, J.E., Yang, J.P., Kang, D., Peterson, T.C., Cappuccio, J.A., Blanchette, C. D., Sulchek, T., Chromy, B.A., Hoeprich, P.D., Coleman, M.A., and Kudlicki, W. 2008. Insertion of membrane proteins into discoidal membranes using a cell-free protein expression approach. *J. Proteome Res.* 7:3535–3542.

43. Leitz, A.J., Bayburt, T.H., Barnakov, A.N., Springer, B.A., and Sligar, S.G. 2006. Functional reconstitution of beta(2)-adrenergic receptors utilizing self-assembling nanodisc technology. *Biotechniques* 40:601–612.

44. Bardwell, J.C.A., Lee, J.O., Jander, G., Martin, N., Belin, D., and Beckwith, J. 1993. A pathway for disulfide bond formation in vivo. *Proc. Natl. Acad. Sci. USA.* 90:1038–1042.

45. Inaba, K., and Ito, K. 2002. Paradoxical redox properties of DsbB and DsbA in the protein disulfide-introducing reaction cascade. *EMBO J.* 21:2646–2654.

46. Stenson, T.H., and Weiss, A.A. 2002. DsbA and DsbC are required for secretion of pertussis toxin by bordetella pertussis. *Infect. Immun.* 70:2297–2303.

47. Jagusztyn-Krynicka, E.K., Rybacki, J., and Lasica, A.M. 2009. Novel strategies for antibacterial drug discovery—antitoxin drugs. *Postepy Mikrobiologii* 48:93–104.

48. Jander, G., Martin, N.L., and Beckwith, J. 1994. Two cysteines in each periplasmic domain of the membrane-protein Dsbb are required for its function in protein disulfide bond formation. *EMBO J.* 13:5121–5127.

49. Regeimbal, J., and Bardwell, J.C.A. 2002. DsbB catalyzes disulfide bond formation de novo. *J. Biol. Chem.* 277:32706–32713.

50. Kadokura, H., Bader, M., Tian, H.P., Bardwell, J.C.A., and Beckwith, J. 2000. Roles of a conserved arginine residue of DsbB in linking protein disulfidebond- formation pathway to the respiratory chain of Escherichia coli. *Proc. Natl. Acad. Sci. USA.* 97:10884–10889.

51. Inaba, K., Murakami, S., Suzuki, M., Nakagawa, A., Yamashita, E., Okada, K., and Ito, K. 2006. Crystal structure of the DsbB-DsbA complex reveals a mechanism of disulfide bond generation. *Cell* 127:789–801.

52. Zhou, Y., Cierpicki, T., Jimenez, R.H., Lukasik, S.M., Ellena, J.F., Cafiso, D.S., Kadokura, H., Beckwith, J., and Bushweller, J.H. 2008. NMR solution structure of the integral membrane enzyme DsbB: functional insights into DsbBcatalyzed disulfide bond formation. *Mol. Cell* 31:896–908.

53. Abad-Zapatero, C., and Metz, J.T. 2005. Ligand efficiency indices as guideposts for drug discovery. *Drug Discov. Today* 10:464–469.

54. Geyer, K., Codee, J.D.C., and Seeberger, P.H. 2006. Microreactors as tools for synthetic chemists—the chemists' round-bottomed flask of the 21st century?. *Chem. A Eur. J.* 12:8434–8442.

55. Kakuta, M., Jayawickrama, D.A., Wolters, A.M., Manz, A., and Sweedler, J.V. 2003. Micromixer-based time-resolved NMR: applications to ubiquitin protein conformation. *Anal. Chem.* 75:956–960.

56. Ciobanu, L., Jayawickrama, D.A., Zhang, X., Webb, A.G., and Sweedler, J.D. 2003. Measuring reaction kinetics by using multiple microcoil NMR spectroscopy. *Angew. Chem. Int. Ed.* 42:4669–4672.

57. Bart, J., Kolkman, A.J., Oosthoek–de Vries, A.J., Koch, K., Nieuwland, P.J., Janssen, H.J.W.G., Van Bentum, J.P.J.M., Ampt, K.A.M., Rutjes, F.P.J.T., Wijmenga, S.S., Gardeniers, H.J.G.E., and Kentgens, A.P.M. et al. 2009. A microfluidic highresolution NMR flow probe. *J. Am. Chem. Soc.* 131:5014–5015.

58. Gomez, M.V., Verputten, H.H.J., Diaz-Ortiz, A., Moreno, A., de la Hoz, A., and Velders, A.H. 2010. On-line monitoring of a microwave-assisted chemical reaction by nanolitre NMR-spectroscopy. *Chem. Commun.* 46:4514–4516.

59. Wensink, H., Benito-Lopez, F., Hermes, D.C., Verboom, W., Gardeniers, H.J.G.E., Reinhoudt, D. N., and van Den Berg, A. 2005. Measuring reaction kinetics in a lab-on-a-chip by microcoil NMR. *Lab Chip* 5:280–284.

60. Trumbull, J.D., Glasgow, I.K., Beebe, D.J., and Magin, R.L. 2000. Integrating microfabricated fluidic systems and NMR spectroscopy. *IEEE Trans. Biomed. Eng.* 47:3–7.

61. Massin, C., Vincent, F., Homsy, A., Ehrmann, K., Boero, G., Besse, P.A., Daridon, A., Verpoorte, E., de Rooij, N.F., and Popovic, R.S. 2003. Planar microcoil-based microfluidic NMR probes. *J. Magn. Reson.* 164(2):242–255.

62. FEMM software available from http://femm.foster-miller.net/.

63. Walton, J.H., de Ropp, J.S., Shutov, M.V., Goloshevsky, A.G., McCarthy, M.J., Smith, R.L., and Collins, S.D. 2003. A micromachined double-tuned NMR microprobe. *Anal. Chem.* 75:5030–5036.

64. Brivio, M., Fokkens, R.H., Verboom, W., Reinhoudt, D.N., Tas, N.R., Goedbloed, M., and van Den Berg, A. 2002. Integrated microfluidic system enabling (bio)chemical reactions with on-line MALDI-TOF mass spectrometry. *Anal. Chem.* 74:3972–3976.

65. Lee, M., Lee, J.-P., Rhee, H., Choo, J., Chai, Y.G., and Lee, E.K. 2003. Applicability of laser-induced Raman microscopy for *in situ* monitoring of imine formation in a glass microfluidic chip. *J. Raman Spectrosc.* 34:737–742.

66. Lee, M., Kim, H., Rhee, H., and Choo, J. 2003. Reaction monitoring of imine synthesis using Raman spectroscopy. *Bull. Korean Chem. Soc.* 24:205–208.

67. Terazima, M., Okamoto, K., and Hirota, N. 1995. Translational diffusion of transient radicals created by the photo-induced hydrogen abstraction reaction in solution; anomalous size dependence in the radical diffusion. *J. Chem. Phys.* 102:2506–2515.

68. Jones, C.J., and Larive, C.K. 2012. Could smaller really be better? Current and future trends in high-resolution microcoil NMR spectroscopy. *Anal. Bioanal. Chem.* 402:61–68.

69. Kentgens, A.P., Bart, J., van Bentum, P.J., Brinkmann, A., van Eck, E.R.H., Gardeniers, J.G.E., Janssen, J.W.G., Knijn, P., Vasa, S., and Verkuijlen, M.H.W. 2008. High-resolution liquid- and solid-state nuclear magnetic resonance of nanoliter sample volumes using microcoil detectors. *J. Chem. Phys.* 128:052202.

70. Webb, A.G. 2005. Microcoil nuclear magnetic resonance spectroscopy. *J. Pharm. Biomed. Anal.* 38:892–903.
71. van Bentum, P.J.M., Janssen, J.W.G., Kentgens, A.P.M., Bart, J., and Gardeniers, J.G.E. 2007. Stripline probes for nuclear magnetic resonance. *J. Magn. Reson.* 189(1):104–113.
72. van Bentum, P.J.M., Janssen, J.W.G., and Kentgens, A.P.M. 2004. Towards nuclear magnetic resonance micro-spectroscopy and micro-imaging. *Analyst* 129:793–803.
73. Maguire, Y., Chuang, I.L., Zhang, S.G., and Gershenfeld, N. 2007. Ultra-small-sample molecular structure detection using microslot waveguide nuclear spin resonance. *Proc. Natl. Acad. Sci. USA.* 104:9198–9203.
74. Danieli, E., Perlo, J., Blümich, B., and Casanova, F. 2010. Small magnets for portable NMR spectrometers. *Angew. Chem. Int. Edit.* 49(24):4133–4135.
75. Demas, V., Herberg, J.L., Malba, V., Bernhardt, A., Evans, L., Harvey, C., Chinn, S.C., Maxwell, R.S.A., and Reimer, J. 2007. Portable, low-cost NMR with laser-lathe lithography produced microcoils. *J. Magn. Reson.* 189(1):121–129.
76. McDowell, A., and Fukushima, E. 2008. Ultracompact NMR: 1H spectroscopy in a subkilogram magnet. *Appl. Magn. Reson.* 35(1):185–195.
77. Diekmann, J., Adams, K.L., Klunder, G.L., Evans, L., Steele, P., Vogt, C., and Herberg, J.L. 2011. Portable microcoil NMR detection coupled to capillary electrophoresis. *Anal. Chem.* 83(4):1328–1335.
78. Ardenkjær-Larsen, J.H., Fridlund, B., Gram, A., Hansson, G., Hansson, L., Lerche, M.H., Servin, R., Thaning, M., and Golman, K. 2003. Increase in signal-to-noise ratio of >10,000 times in liquid-state NMR. *Proc. Natl. Acad. Sci. USA.* 100(18):10158–10163.
79. Bowen, S., and Hilty, C. 2008. Time-resolved dynamic nuclear polarization enhanced NMR spectroscopy. *Angew. Chem. Int. Edit.* 47(28):5235–5237.
80. Zeng, H., Bowen, S., and Hilty, C. 2009. Sequentially acquired two dimensional NMR spectra from hyperpolarized sample. *J. Magn. Reson.* 199(2):159–165.

9 Quantitation of Isolated Compounds

9.1 BACKGROUND OF QUANTITATIVE NMR (qNMR)

Nuclear magnetic resonance (NMR) may be used as a quantitative technology to evaluate a compound's purity, concentration, and quantity of material. This is because signal intensity, under appropriate experimental conditions, has a proportional relationship to the number of nuclei of a particular spectral peak. Hence, NMR enables a precise measurement of the amount of compound in a sample, provided the molecule is structurally characterized. With the increase in sensitivity due to high-field superconducting magnetic fields and improved electronics, the NMR detection limits have been significantly improved enabling evaluation of smaller and smaller quantities of material.

Quantitative measurements with NMR spectroscopy (qNMR) were initially described in the 1960s by Jungnickel and Forbes [1] and Hollis [2]. In the report of Jungnickel and Forbes, the intramolecular proton ratios in 26 pure organic substances were determined, whereas the Hollis paper analyzed the amount of fractions of three analytes (aspirin, phenacetine, and caffeine) in respective mixtures. However, differences in protocol of the measurement procedure as well as the spectral processing and evaluation were responsible for the fact that quantitative investigations of identical samples in various laboratories can differ severely.

Early reports [3–6] discussed the potential use of qNMR as a main method to be used for purity assessment based on the fact that the NMR-integrated signal area is directly proportional to the number of nuclei contributing to the signal. Results reported by Maniara et al. [7] and Wells et al. [8,9] showed a 0.5% error for quantitative high-resolution ^{1}H and ^{31}P NMR measurements. This value was comparable with quantitation results from HPLC data, where HPLC is used as the standard analytical method. Because validation of a method requires the comparison of results from independent laboratories, national [10] and international comparisons [11] were conducted using over 30 laboratories at universities, research institutes, and companies. The results showed major differences in the analysis of a five-model compound mixture. This finding led to the requirement that repeatable standards are used in the setup, measurement, data processing, and evaluation procedures of quantitative NMR studies. The accepted general outline for performing qNMR studies is illustrated in Figure 9.1.

Quantitative ^{1}H and ^{13}C solution–based NMR spectroscopy has been used in pharmaceutical studies [12–18], in agriculture [8,9,19,20], in material science [21], and for military purposes [22], where purity or content determinations of substances are important issues. Use of qNMR has been accelerated by the substantial increase of the sensitivity and homogeneity of high-field NMR spectrometers as well as by modern software packages that allow accurate and precise data processing and evaluation [23]. With respect to the pharmaceutical and chemical industry, proton qNMR is most commonly applied and has been an efficient means of quantifying organic molecules. However, the implementation of qNMR in new fields of application (e.g., metabolomics, biomarker discovery, physiological pathways) brings with it more complex molecules and molecular systems, thus making the usage of ^{1}H-qNMR more challenging. In such cases, the use of other NMR active nuclei, namely ^{31}P or ^{19}F, can sometimes provide a better option.

Because the NMR signal is a linear function of the peak integration number to the number of spins (N) in the observed sample volume, quantitative NMR results can be obtained in most

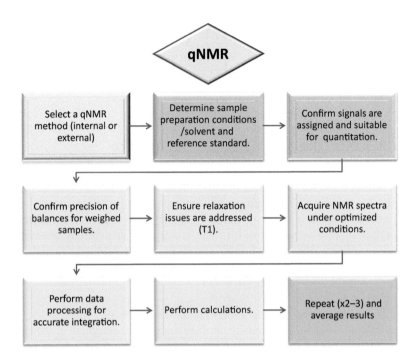

FIGURE 9.1   General outline of the qNMR method when assignment of the analyte and system suitability criteria are met.

cases, provided certain experimental conditions are met. Methods have been developed for nuclei with spin ½ because of their favorable magnetic properties. Primarily ^{1}H, ^{19}F, ^{31}P nuclei are most frequently used due to their high natural abundance. Nuclei such as ^{13}C and ^{15}N, while possible for qNMR studies, have low natural abundance that complicates the process. In addition, proton decoupling is needed to simplify the spectra and improve the signal intensity for ^{13}C, ^{15}N, and sometimes ^{31}P nuclei. However, NOE enhancements need to be suppressed to enable integration proportional to the number of nuclear spins, which adds additional layers of complication to the qNMR process. Other nuclei which possess quadrupolar or paramagnetic properties present special problems for quantification due to severe line broadening. Such cases typically will not meet the system suitability criteria for qNMR and will not be discussed further here.

Specific advantages of NMR as a quantitative method include the ability of this technology to determine structures at the atomic level. Assignment of the NMR resonances to a particular chemical structure is the first step in the quantitation process. Chemical structure is a critical part of the quantitation process since the assignment of a number of nuclei to a particular resonance is necessary to carry out accurate calculations. Also, the solved structure will further corroborate the molecular weight, which is essential to the calculation's accuracy.

A key advantage with qNMR is that there is no need for intensity calibrations since the signal area is directly proportional to the number of nuclei. In addition, calculations may be carried out with relatively short measuring times by adjusting the pulse angle, thus increasing sample throughput. Because NMR technology is nondestructive, the sample may be recovered for further use or examination. This is critical for cases such as metabolite quantitation where the sample may be needed for follow-up toxicology studies. Additionally, there are cases where prior isolation of the analyte is not feasible. In these cases, mixtures may be examined, which simplifies sample preparation and handling. Hence, the simultaneous determination of more than one analyte in a mixture is possible.

A major limitation of using qNMR spectroscopy is the high cost of the NMR technology for both purchase and maintenance. The need for accurate pulse calibration and spin relaxation also present issues. This is important when either internal or external standards are used. When considering external standards, the same instrument and same experimental conditions need to be used.

A major requirement for qNMR is that all samples meet system suitability criteria. Conditions which will be detrimental to the qNMR assay include signals that are excessively broad, cannot be accurately integrated, or accurately assigned. Suitability issues are most evident in the spectrum shown in Figure 9.2 where severe line broadening compromises assignment and integration.

It should be noted that the qNMR method has been validated. The results of the study reported by Malz and Jancke [24] defined the quantitation limits of weighed analyte and showed that the qNMR method is specific, linear, accurate, and precise within the reported system suitability requirements. The technique has also been cited and accepted in ICH and USP compendial guidances [25,26].

9.2   COMPARISON OF qNMR WITH HPLC METHODS

Liquid chromatography (LC) methods are a primary validated approach for identifying and quantitating compound purity. Developing fully characterized LC methods can be time-consuming and costly, causing the traditional method development and validation studies associated with LC development to be delayed. Nonetheless, industry and regulatory needs require selective, accurate potency determinations of drug substances from discovery through Phase I.

A major challenge with HPLC methods involves the lack of well-characterized impurity reference standards for area normalization applied to spectroscopic detection. Accurate determinations on a weight-weight basis during early stages of development requires certified reference standards of known purity. In early stages of drug development, processes are still changing and reference standards for impurities are limited, not readily available, or

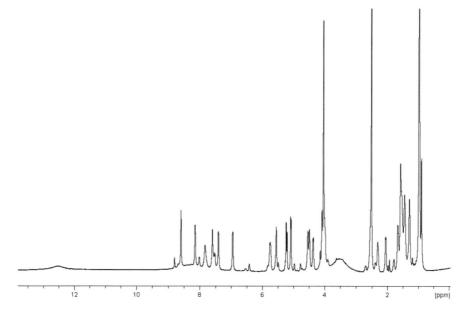

FIGURE 9.2   Example of a ^{1}H NMR spectrum (600 MHz) with severe line broadening and signal overlap. This sample does not meet system suitability requirements for qNMR studies.

have uncertain purity. Generating these standards is costly, hence, impurities are usually not well characterized.

In the absence of suitable reference materials for impurity quantitation, techniques that employ mass detectors such as the chemical luminescence detector (CLND) and the charged aerosol detector (CAD) have been used. Such detectors enable one to normalize the UV response of each impurity of interest by their molar ratios, and thus generate relative response factors without requiring isolated and purified compound-specific standards. Such detectors are limited in response and effective only with specific mobile-phase requirements.

To address the needs of rapid, selective, and accurate potency determinations without requiring full development of traditional chromatographic methods, the use of qNMR spectroscopy has been applied in early drug development stages as a mass detector. NMR has been employed as a means of providing a relative response factor (RRF) in HPLC quantitation. This has an advantage since in HPLC studies, impurities are often estimated using an area-normalization approach due to the fact that the response factor of each impurity is assumed to be identical to that of the parent compound.

Application of ^{1}H NMR spectroscopy, however, has the advantage of exploiting the universal detection of protons while not suffering from the limitations observed for CLND, CAD, and other common detectors. The determination of relative response factors using NMR has been successfully applied to several LC methods.

The calculation of RRF by ^{1}H NMR may be achieved using the following equation:

$$RRF^{UV} = \frac{Area_1^{UV}}{Area_2^{UV}} \times \frac{I_2^{NMR}}{I_1^{NMR}} \times \frac{MW_2}{MW_1} \times \frac{N_1^{H}}{N_2^{H}} \tag{9.1}$$

where:

$Area^{UV}_{1,2}$ = HPLC − UVarea count for analytes 1 and 2,
$I^{NMR}_{1,2}$ = ^{1}H NMR integral for analytes 1 and 2,
$N^{H}_{1,2}$ = number of hydrogen atoms in the integral response used for analytes 1 and 2, and
$MW_{1,2}$ = molecular weight for analytes 1 and 2.

For more information on this approach, the reader is directed to the work reported by Webster et al. [27].

9.3   SELECTING A qNMR REFERENCE STANDARD

Selection of a suitable internal standard reference material is an important part of the qNMR process. Specific criteria in qNMR reference materials have been determined to ensure accuracy in the calculated result. The critical criteria of reference materials includes high solubility in common deuterated NMR solvents since complete solubilization is imperative for accurate integration. The material must be non-hygroscopic, non-volatile, and not prone to decomposition or reactivity. These criteria are essential to ensuring accurate weighing of the analyte. The reference compound must possess chemical stability, be non-reactive, and be free of residual water, which can compromise the accuracy of integration.

The internal reference should also be selected to minimize overlap with the analyte signals. As a result of this requirement, different samples may require different references. If overlap between analyte and reference signal is unavoidable, a blank sample of the analyte may be acquired. Acquisition of the blank under identical conditions, normalization of integrals, and subtraction of integrated intensities may be used to obtain the intensity of the reference.

In order to generate accurate and precise qNMR measurements, the signal intensity ratio should be as close to 1:1 as possible. While this is recommended, primarily to avoid dynamic range issues with integration, it is not mandatory. The purity of the reference is an additional factor that needs to be determined. Commercial products are available for reference standards where purity analysis has been completed; however, if one chooses a standard that has not been assessed for purity, this test needs to be completed prior to use. For GMP studies, a Certificate of Analysis (CoA) may be required. Typical commercial standards that are certified reference materials for qNMR for ^{1}H, ^{31}P, and ^{19}F nuclei are listed in Tables 9.1, 9.2, and 9.3, respectively. The commercial reference sample list for ^{1}H NMR in Table 9.1 does not include dimethyl fumarate, which has been found to work well as a qNMR standard (Gonnella, private communication). This is due to the fact that dimethyl fumarate possesses high chemical stability and low hygroscopicity with signals appearing at 6.77 ppm for the olefinic proton and 3.74 ppm for the aliphatic methyl protons in DMSO-d_6. The ^{1}H NMR T_1 values for dimethyl fumarate in DMSO-d_6 were measured to be ~7.51 s for the olefinic peak and ~1.81 s for the aliphatic peak. Hence, care must be taken to allow a sufficient relaxation delay when the olefinic peak is selected for quantitation. The advantage of this reference in pharmaceutical studies is that it contains only two singlet ^{1}H NMR signals that appear in spectral regions, which usually do not overlap with the analyte of interest.

9.3.1 EXTERNAL VERSUS INTERNAL qNMR REFERENCE

Calculations of purities and concentrations require the presence of an accurately measured standard as a reference. As mentioned previously, there are numerous reference standards that may be considered. These standards may be used as either internal or external standards. Internal standards have an advantage in the quantitative process in that the environment is identical to that of the analyte and the samples are freshly prepared, which contributes to the spectral integrity. Because the materials are weighed into the same vial, the ratio of analyte and reference stays the same relative to the amount of solvent added. In such cases, the concentration of the solution is not critical for the quantification calculation. Because of these factors, the internal standard method has higher precision and lower uncertainties. In addition, for cases where it is necessary to check spin lattice relaxation (T_1s), these measurements may be made for both compounds using one sample.

External standards may also be used in cases where the analyte cannot be directly weighed or the volume cannot be measured with high accuracy. DE-NMR (digital ERETIC NMR) is a method of inserting a synthetic signal into an NMR spectrum using a mathematical insertion of a calculated signal with the observed FID from the sample of interest (used in place of an internal reference). To prepare an external standard, the weight must be carefully measured and the volume must be controlled to enable generation of an accurate reference concentration. Because the analyte and reference are two separate samples, the NMR acquisition and processing parameters must be identical in the NMR experiments performed [28–30].

Concentration levels must be anticipated to ensure that the receiver gain (RG) used for the calibration file can accommodate the unknown sample concentrations. Once the DE signal is inserted, the integrals of the measured samples are scaled to the known DE signal concentration (typically 10 mM).

Restrictions must be placed on the acquisition parameters and changes in the conditions of data collection cannot be made without recalibration. As with the internal reference, a constant parameter set is important. These include the pulse length, the relaxation delay, the sweep width, the carrier frequency, and the number of data points. At least two samples should be prepared and run in a similar manner. All weighed amounts of sample should be performed on a calibrated balance in a static-free environment. This will ensure accuracy in the weighing apparatus and process.

TABLE 9.1
Reference Standards for Quantitative ¹H NMR

Reference	Chemical Structure	CDCl₃		DMSOd₆		CD₃OD		CD₃CN	
		δ (ppm)	T₁ (s)	δ (ppm)	T₁ (s)	δ (ppm)	T₁ (s)	δ (ppm)	T₁ (s)
Ethylene carbonate		4.5	7.0	4.5	2.7	4.5	5.3	4.5	2.0
Benzoic acid		8.1	3.7	8.0	3.3	8.0	4.3	8.0	4.5
		7.7	4.0	7.6	3.7	7.6	4.4	7.7	2.5
		7.5	3.4	7.5	3.0	7.5	3.9	7.5	2.6
Duroquinone		2.0	3.3	2.0	3.1	2.0	4.0	2.0	5.4
Dimethyl terephthalate		8.1	3.6	8.1	2.9	8.1	4.4	8.1	4.9
		4.0	1.8	3.9	1.1	4.0	2.4	3.9	2.6
Potassium hydrogen phthalate						7.9	2.5		
1,2,4,5-Tetrachloro-3-nitro-benzene		7.8	10.7	8.5	12.6	9.4	9.6	6.8	6.4
Dimethyl sulfone		3.0	2.7	3.0	2.4	3.0	3.3	2.9	2.6
Ethyl 4 (dimethyl-amino) benzoate		7.9	3.8	7.8	2.5	7.9	3.4	7.9	5.6
		6.7	2.4	6.7	1.4	6.7	2.5	6.7	3.7
		4.3	2.8	4.2	1.9	4.3	3.3	4.3	4.1
		3.1	2.0	3.0	1.5	3.0	2.2	3.0	3.6
		1.4	2.5	1.3	2.1	1.4	2.7	1.3	3.6
Thymol		7.1	3.8	7.0	2.0	7.0	3.7	7.1	4.9
		6.8	4.5	6.7	2.2	6.6	4.4	6.7	5.7
		6.6	4.8	6.5	2.6	6.5	5.8	6.6	5.7
		3.2	4.3	3.1	2.3	3.2	4.0	3.2	5.2
		2.3	3.1	2.2	2.0	2.2	2.8	2.2	3.5
		1.3	1.9	1.1	0.9	1.2	1.8	1.2	2.4
Benzyl benzoate		5.4	4.3	5.4	2.2	5.4	2.0	5.4	1.0
		8.1	13.6	8.0	8.8	8.1	3.9	8.1	2.4

(Continued)

TABLE 9.1 (Cont.)

Reference	Chemical Structure	CDCl$_3$		DMSOd$_6$		CD$_3$OD		CD$_3$CN	
		δ (ppm)	T$_1$ (s)	δ (ppm)	T$_1$ (s)	δ (ppm)	T$_1$ (s)	δ (ppm)	T$_1$ (s)
1,3,5-Trimethoxybenzene		1.0	4.7	6.1	3.2	6.1	4.8	6.7	3.4
		2.4	2.2	3.7	1.4	3.8	2.7	4.3	2.6
1,2,4,5-Tetramethyl-benzene		7.0	6.1	6.9	4.7	6.9	5.9	6.9	7.7
		2.2	4.0	2.1	2.6	2.2	4.3	2.2	5.0
Dimethylmalonic acid				1.3	0.7	1.4	1.0	1.4	2.0
Maleic acid				6.3	3.0	6.3	3.9	6.4	2.3
Methyl 3,5-dinitrobenzoate		9.3	8.0	9.1	9.4			9.1	9.0
		9.2	6.1	8.9	7.6			9.0	8.2
		4.1	2.6	4.0	1.5			4.0	3.5

TABLE 9.2

Reference Standards for Quantitative ^{31}P NMR

Reference	Chemical Structure	CDCl$_3$		DMSOd$_6$		CD$_3$OD		CD$_3$CN	
		δ (ppm)	T$_1$(s)	δ (ppm)	T$_1$(s)	δ (ppm)	T$_1$(s)	δ (ppm)	T$_1$(s)
Triphenyl phosphate		−17.7	2.7	−17.3	1.2	−17.5	3.1	−17	4.3
Phosphonoacetic acid				14.9	1.5	17.7	2.9		

TABLE 9.3
Reference Standards for Quantitative ^{19}F NMR

Reference	Chemical Structure	CDCl₃		DMSOd₆		CD₃OD		CD₃CN	
		δ (ppm)	T₁(s)	δ (ppm)	T₁(s)	δ (ppm)	T₁(s)	δ (ppm)	T₁(s)
4,4'-Difluorobenzophenone		−105.8	2.4	−106.5	1.4	−108.1	2.8	−108.3	2.3
2,4-Dichlorobenzotrifluoride		−62.5	2.3	−61.2	1.2	−65.4	3.3	−63	2.9
2-Chloro-4-fluorotoluene		−115.8	4.4	−115.3	3.3	−117.7	4.8	−117.3	4.7

A similar approach known as ERETIC2 (**E**lectronic **RE**ference **T**o access **I**n vivo **C**oncentrations) involves the preparation of an external reference sample with a known concentration. The NMR spectrum is then measured. Analyte samples of interest in a known volume are externally referenced against it. Once again, with an external reference, there is no need to worry about overlap of analyte signal with the reference signal and no need to worry about chemical interactions.

Drawbacks of an external reference include larger error than when using an internal reference, the reference spectrum must be acquired under identical conditions, and may be susceptible to changes in volume by tube tolerances. Relaxation times of the analyte and reference sample must also be taken into account. In certain cases, differences in tuning may require adjustment. This condition may be handled with the use of the PULCON (*pul*se *l*ength–based *con*centration determination) method (see Section 9.3.2 below) [31].

9.3.2 CALIBRATION PROCEDURE

The principal methods of handling quantitative calibration for qHPLC and qNMR are described as follows:

qHPLC: Quantitative HPLC analysis is commonly applied in chromatography-based purity analysis and is particularly suitable for precious and/or mass-limited samples. The HPLC method uses the grand total measure of all signals as the basis for the 100% calculation. Reliability of data obtained by HPLC directly depends on the accuracy of the calibration. A major difficulty is obtaining and maintaining pure standards.

Calculation of the concentrations can be carried out in three different ways, all of which attempt to compensate for changes in detector's response: (1) using full standard curves constructed at each day of analysis, (2) construction of full standard curves to verify linearity over the samples' concentrations and passage through the origin and one-point recalibration on each day of analysis, and (3) use of response factors. The first approach is the ideal procedure but it takes a long time, leaving little time for the samples on each day of analysis, thus limiting sample throughput. It also uses a significant amount of standards. The second procedure involves injection of a standard of known concentration on each day of analysis which in effect verifies any change in the slope of the standard curve (i.e., change in detector's sensitivity). Although much simpler and more rapid, it has to be done carefully because there is a danger that this single point can be an outlier. The third approach, use of response factors, is also a simplification because a single reference standard is injected on each day of analysis. The response factor relative to the reference is calculated by the formula of Hart and Scott shown in Equation 9.2 [32].

$$\text{Response factor}(\text{RF}_X) = \frac{\text{peak area of analyte } (\#\mu g/ml)}{\text{peak area of reference compound } (\#\mu g/ml)} \tag{9.2}$$

The analyte concentration in the sample is calculated using Equation 9.3:

$$\text{CX}\mu g/g = \frac{A_x \times \text{total volume of extract (ml)}}{\text{RF}_X \times A_{\text{ref}} \times \text{sample weight (g)}} \tag{9.3}$$

where C_x is the concentration of the analyte, A_x is the peak area of the analyte, RFx is the response factor of the analyte, and A_{ref} is the peak area of the reference material.

qNMR: For qNMR calculations, internal calibration (IC) is the most common method and conceptually most straightforward for absolute quantitation. The process involves the addition and weighing of a certified reference to a vial containing an accurately weighed sample of the analyte.

External calibration (EC) in qNMR includes a different approach that may be conceptualized as the use of artificial signals. Signals may be obtained from standards in separate tubes devoid of analyte. This may involve concentric tubes where a capillary is inserted in a 5 mm tube or a separate tube containing the reference. The ERETIC method provides a robust approach using an external reference. Illustration of the internal vs. external reference process is given in Figure 9.3.

ERETIC (Electronic Reference To access In vivo Concentrations): ERETIC was developed in 1997 by Barantin et al. [28] to address cases where an internal standard may not be an appropriate or ideal approach in quantitative NMR studies for the determination of absolute concentrations. With ERETIC, a radio frequency (rf) reference signal is fed to the resonance circuit of the probe during the acquisition time using a free coil in the probe (heteronuclear channel). Full control of the amplitude (size), phase of the synthetic signal, and frequency (chemical shift position) enables not only chemical shift referencing but also accurate integral referencing and comparison for quantification of all components of a measured spectrum. The electronic signal is set in a free spectral range avoiding overlap with analyte signals. The synthetic signal must be calibrated separately against a real reference standard of known concentration, and it requires a rearrangement of spectrometer connections and/or filters [33]. Once an electronic signal is calibrated, analytes of interest may be studied.

ERETIC2 (Electronic Reference To access In vivo Concentrations2): An alternate approach known as ERETIC2, implemented by Bruker Biospin in Topspin software, provides a facile and efficient means of employing an external reference. With ERETIC2, a reference spectrum is acquired with a known concentration. It is imperative that the reference standard concentration be accurately prepared. The reference standard concentrations typically range from 1 to 15 mM. This reference spectrum is used to calculate the concentration of the unknown analyte using an automated calculator provided in Topspin (Bruker Biospin). Pulses need to be calibrated and the exact experimental conditions need to be duplicated in the

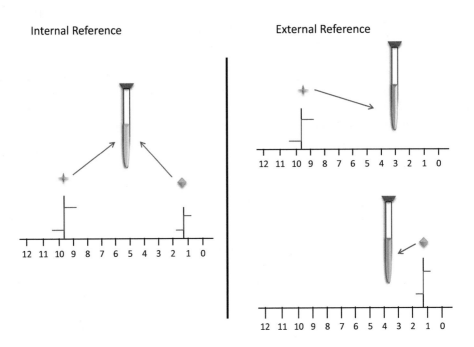

FIGURE 9.3 Illustration of an internal standard method (left) where one sample is measured containing the analyte, and reference standard (left) and an external standard method (right) where separate samples are measured containing the analyte and the reference.

reference and the unknown samples. The reference spectrum has to be acquired under identical conditions as the analyte. The number of scans and receiver gain may be varied or kept the same. The concentration of the analyte may be obtained using Equation 9.4.

$$[\text{concentration of analyte}] = k_{\text{ERETICref}}/k_{\text{analyte}} \times [\text{ERETIC}_{\text{ref}}] \times I_{\text{analyte}}/I_{\text{ERETICref}}$$

(9.4)

where:

k = number of spins
I_{analyte} = integration of analyte
$I_{\text{ERETIC reference}}$ = integration of reference
$[\text{ERETIC}_{\text{ref}}]$ = concentration of reference
$[\text{analyte}]$ = concentration of analyte

As previously noted, with an external reference there is no concern about signal overlap or chemical interactions with the analyte. With the ERETIC2 application, sample purity and concentration may be calculated. If volume of the analyte is known in preparing the sample, the amount (mass) of analyte in the NMR tube may be calculated. Since separate tubes are used for the NMR samples of analyte and reference, the measurements are susceptible to potential changes in volume by tube tolerances; hence, it is important to use tubes of the same quality.

Cases where the solvent composition is different from that of the reference (usually pronounced in aqueous biological systems), compensation for sample differences in tuning may be addressed using PULCON.

With PULCON, for a selected probe where the sample in the NMR tube fills the active volume of the coil in the z direction, the unknown concentration C_U of sample U can be calculated from the known concentration C_R of a reference sample R with Equation 9.5

$$C_U = f_T \cdot C_R \frac{I_U T_U \theta^U{}_{360} n_R}{I_R T_R \theta^R{}_{360} n_U}$$

(9.5)

where I stands for the resonance integrals, T for the sample temperature in Kelvin, θ_{360} for the 360° RF pulse, and n for the number of scans used for the measurements of samples U and R, respectively. The factor f_T accounts for variations in signal intensities caused by different experimental schemes used in the measurement of the samples U (unknown) and R (reference). The factor $f_T = 1$ when measurements of U and R are one-pulse experiments with the signal acquisition immediately following the RF pulse. For multiple scans per spectrum, the relaxation period between scans must be long enough to exclude T_1 (longitudinal relaxation times) differences for I_U and I_R. Equation 9.5 is valid when the measurements are done with the same NMR probe, which is tuned and matched to the same amplifier delivering the same power. The acquisition and transformation parameters can be adapted to the individual samples. It is best to use the same receiver gain in the two measurements, otherwise the scaling introduced by different receiver gains has to be determined separately. Furthermore, window functions applied during Fourier transformation and scaling factors during processing data should be consistent to ensure proper peak integrals [31].

9.4   INSTRUMENT SETTINGS, SOLVENT SELECTION, RAPID SPECTRAL ANALYSIS

Quantitative NMR measurements require a compromise between time and accuracy. Samples that are submitted for analysis and are run "as fast as possible" may give rise to errors in measurement if appropriate experimental settings are not employed. In solvent selection, NMR experiments conducted in non-volatile solvents, such as DMSO-d_6, have an advantage in ensuring good solubilization of both analyte and internal standard. In addition, evaporation of solvent from micro-tube environments that can lead to precipitation of the analyte and/or reference standard during data acquisition is less likely to occur when using solvents such as DMSO. Hence, instrument settings and solvent selection and other experimental factors are important components to enable rapid spectral analysis.

The qNMR experiments should be carried out under conditions where sufficient signal-to-noise is present for accurate integration. For samples of small organic molecules with 5–10 mg in about 700 μl of solution, about 256 to 500 scans is sufficient per sample, depending on concentration. The number of points for ^{1}H NMR may be set to 65,536. If the pulse width is set to 10% of the calibrated 90° pulse width and the relaxation delay set to 2 s, errors are usually ± 2%. Total acquisition time ranges from 20 to 40 minutes. Sample temperature may be kept constant at 300 K and the sample is run non-spinning. Fourier transform is carried out with line broadening of 0.05 Hz. Sweep width for ^{1}H may be set up to 20 ppm and the center carrier frequency set at 6 ppm. For tube acquisitions, the samples may be acquired in automation using a sample changer.

The tube quality and NMR hardware are important when an external reference is used. The same NMR probe and high-quality NMR tubes are recommended to avoid differences due to tube variation.

The active length of the RF coil and the inner diameter of the NMR tube are critical in the measurement of the volume of a sample. Hence, quantitative NMR measurements involving external samples should be performed on the same spectrometer probe since quantitation between different probes is not directly possible.

Differences of volume are critical when using external referencing. Because of this, appropriate NMR tubes are also required since variation of the inner diameter will affect the measured result. Other tube specifications including concentricity and camber are not as critical (see Figure 9.4).

Relaxation effects (T_1 and T_2) also have a profound effect on the outcome of the experiments. A fully relaxed nuclear spin responds to a radio-frequency pulse with signal area that is directly proportional to the number of spins in the sample. The applied radiofrequency pulse

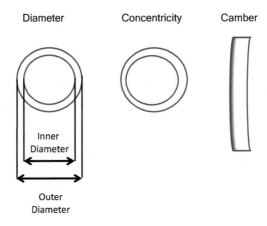

FIGURE 9.4 Illustration of NMR-tube specifications. When using an external reference in qNMR studies, the inner diameter supersedes camber and concentricity in affecting measured results.

excites the nuclei of interest in the sample. However, accurate signal intensities are dependent on pulse lengths and bandwidth coverage such that spins are being excited uniformly.

The T_1 relaxation is critical in qNMR measurements since long T_1s can attenuate the NMR signal if a 90° pulse is used and the recycle delay is less than $5 \times T_1$. T_2 relaxation is also an important consideration since short T_2s can contribute to intensity losses due to missing signal at the beginning of the FID and/or broad signals, which will introduce integration errors. Quadrupolar relaxation is a key contributor to T_2 signal loss. Other contributions to signal loss include chemical exchange where line broadening or reduced signal intensity can occur. Relaxation by aniosotropic motion (CSA), spin rotation, scalar interactions (*J*-coupling) can also cause relaxation of NMR signals. Finally ionic solutions (e.g., aqueous buffered solutions) may introduce resistive losses induced by ionic inductance in high Q-coils. This condition can also change probe tuning as well as pulse calibration.

Understanding of the excitation profile or bandwidth is essential. One pre-requisite for the linearity of integration is the maintenance of equally relaxed nuclei during the experiment. Information about the range of the T_1 relaxation times may be obtained from inversion recovery experiments. A sample with highly coupled, fast-relaxing protons ($T_1 < 1$ s) together with almost non-coupled nuclei ($T_1 \sim 2$ s) shows the linearity of integration is better than 1% when using short pulse angles (30°) and pulse repetition rates of about 3 times the longest T_1 in the sample.

Proper pulse calibration is essential to achieve accuracy in calculation. If a 90° pulse is used, a relaxation delay of 5 times the spin-lattice relaxation time (T_1) is required to achieve full relaxation of the signal prior to application of a second pulse. Because multiple scans are often required to achieve good signal-to-noise, experiments may become excessively long per sample. To enable data acquisition in a timely manner, it has been determined that use of a 30° pulse with relaxation delays of 5 s or a 10° pulse with a relaxation delay of 2 s will provide good accuracy. Because the 90° pulse gives the largest signal, mis-calibration will result in smaller error. For studies of small organic molecules, a 10° pulse with a 2 s delay, approximately a $\pm 2\%$ error, is typically observed (Gonnella, private communication).

Solvents and solubility play a critical role in experiment accuracy. Analyte and reference samples must be devoid of volatile solvent, otherwise this can compromise the weighing of the analyte or reference. Weight measurements are essential to accurate calculation. In addition, rapid evaporation of solvent in the NMR tube could induce precipitation in the NMR tube, which will adversely affect the resonance integration. Even when there appears to be promising solubility, there may be samples that do not fully dissolve in some solvents. Hence additional dilution and heating may be required to obtain a fully dissolved sample. DMSO is a preferred solvent since it enables full solubility of many analytes and is often the solvent of choice in performing qNMR studies. For other solvents such as chloroform or acetonitrile, it is important to ensure that all compounds are completely dissolved. Cloudy or occluded samples are not acceptable. Discrepancies in dissolution may be determined by visual inspection. Differences in good versus poor dissolution are displayed in Figure 9.5. However, for some cases, the differences may require careful magnified examination.

Care must be taken to ensure accurate integration. Baseline correction is a critical step in the integration process. After phasing the spectra, baseline correction may be achieved with an automated command. Integrations of ^{1}H spectra should never include ^{13}C satellites. ^{13}C satellites are part of every carbon-based molecule. Since ^{13}C has a natural abundance of 1.1%, the main signal in ^{1}H NMR represents 98.9% of the total signal. In cases where ^{13}C satellites overlap with the main signal of other resonances, the integration of main signals becomes compromised. Decoupling ^{13}C may be possible; however, ^{13}C decoupling can heat the sample, which becomes problematic particularly with long acquisition times where decoupling sidebands may be introduced.

FIGURE 9.5 Visual display of (a) acceptable dissolution for qNMR studies and (b) dissolution that does not meet suitability requirements for qNMR.

9.5 TYPES OF qNMR CALCULATIONS

The types of qNMR experiments in cases where external references are used include (a) calculations of analyte concentration, (b) mass of compound, and (c) sample purity. These experiments are especially useful for the evaluation of analytes isolated using chromatographic or SPE methods.

9.5.1 CONCENTRATION (ERETIC2)

To determine concentration of the analyte signal using an external reference (see Equation 9.4), the requirements are that the concentration and integrations of the reference sample must be accurately measured. A reference sample with a known concentration is needed. Samples of interest may then be externally referenced against it. With use of ERETIC2, there is no need to worry about overlap of sample and reference signals. There is no need to worry about chemical interactions. If identical conditions cannot be used (particularly with respect to tuning and matching the sample), compensation for sample differences in tuning may be achieved using PULCON. The reference spectrum should be freshly prepared and has to be acquired under identical conditions as the analyte.

9.5.2 MASS OF COMPOUND (INTERNAL STANDARD AND ERETIC)

For samples where an internal standard has a known molecular weight and mass and where the analyte has known molecular weight, then the mass of the analyte may be calculated using Equation 9.6.

$$\text{mass}_a = \frac{\text{mass}_{\text{ref}} * I_a * \text{MW}_a}{I_{\text{ref}} * \text{MW}_{\text{ref}}}$$

(9.6)

where:

    $mass_a$ = mass of analyte
    $mass_{ref}$ = mass of reference
    I_a = integration of analyte
    I_{ref} = integration of reference
    MW_a = molecular weight of analyte
    MW_{ref} = molecular weight of reference

With qNMR, where an external reference is used, once the concentration of the analyte (C_a) is calculated (see Equation 9.4). The mass of the compound may be calculated using Equation 9.7 provided the analyte volume (V_a) is known.

$$M_a = V_a \times C_a \tag{9.7}$$

where:

    M_a is mass of analyte
    V_a is volume of analyte
    C_a is concentration of analyte

9.5.3   PURITY

In cases where the amount of compound is determined, the purity may be calculated using Equation 9.8. When a significant amount of analyte sample is available, the analyte's weight should be measured *via* a high accuracy balance that measures to at least five decimal places. Otherwise, it may be computed from calculation of concentration and known volume as previously described.

$$P_a = (I_a/I_{Std})(N_{Std}/N_a)(M_a/M_{Std})(m_{Std}/m_a)(P_{Std})100\% \tag{9.8}$$

where:

    P_a = % purity of the analyte
    M_a and M_{Std} = molar masses of analyte and standard, respectively
    m_a = weighed mass of analyte
    m_{Std} and P_{Std} = weighed mass and purity of standard
    N_a and N_{Std} = number of spins of analyte and standard, respectively
    I_a and I_{Std} = integrated signal areas of analyte and standard, respectively

Template for calculating purity is shown in Figure 9.6 with dimethyl fumarate as the internal standard. In this example, the purity of dimethyl fumarate was obtained from NMR integral analysis in DMSO-d_6.

9.6   APPLICATIONS

Application of qNMR to characterize pharmaceutical and biomedical compounds demonstrates the value and importance of qNMR for determining purity, concentration, and mass of the analyte [7,25,34–42]. Typically, ^{1}H-NMR is used as the quantitation method of choice. This allows the technique to operate as a "universal" detector. Unlike molar absorptive differences in UV detection and ionization efficiency differences in mass spectrometry detection, protons in NMR spectrometry respond quantitatively from molecule to molecule assuming relaxation is taken into account.

Purity Calculation Template:

Sample name:

Sample X of 2   Dimethyl Fumarate Purity = 99.9%

$$\frac{\dfrac{\textbf{D.DDDDD}}{\#H}}{\dfrac{\textbf{E.EEEEEE}}{\#H} \times 144.13} \times \textbf{A.AA} \times \frac{\textbf{C.CC} \times 0.999}{\textbf{B.BB}} \times 100\% = \textbf{F.FF}\%$$

Average purity: **G.GG** %

	Mwt	Mass (mg)	Proton Integration
Sample name	**A.AA**	**B.BB**	**D.DDDDD**
Dimethyl Fumarate	144.13	**C.CC**	**E.EEEEEE**

FIGURE 9.6 Template showing the key measured values (A–E) that are required for qNMR calculations. The purity of the reference standard (99.9%) is important in the calculation and needs to be determined.

Metabolites: Structure elucidation and quantitation of drug metabolites is an important part of drug development. Identification of metabolites from active pharmaceutical ingredients (APIs) or related toxicological compounds usually occurs using high-resolution LC-MS/MS. However, some metabolite structures can only be tentatively assigned based on MS/MS fragmentation. NMR provides a powerful technology not only for structure elucidation at the atomic level but also for sample quantitation. Cases where purity, concentration, and mass of compound in can be calculated using qNMR have been aptly demonstrated in metabolism studies.

One example involves metabolism of compound A (Figure 9.7), an agonist of a potential therapeutic target, for controlling inflammatory autoimmunity. The resultant metabolite was isolated, purified, and submitted for NMR analysis. MS analysis of the metabolite (data not shown) was consistent with hydroxylation of compound A (mwt 552.6) with the expected site of hydroxylation on either the methyl compound B or methylene position compound C of the ethyl sulfone moiety (Figure 9.7). Because attempts to synthesize the metabolite were unsuccessful, the metabolite was biosynthesized from dog-liver microsomes. A total yield of 500 μg was estimated based on small-scale turnover experiments.

Structural characterization of the parent agonist using NMR spectroscopy was consistent with compound A. ^{1}H chemical shift (δ) assignments were made for carbon and proton signals (referenced to dimethylsulfoxide-d_6 at $\delta 2.50$ ppm for ^{1}H and 39.5 ppm for ^{13}C) (Gonnella and M. Cerny, private communication). Once the structure was established, the purity, mass, and concentration were determined using qNMR (ERETIC2).

Proton NMR spectra were run for the external standard dimethylfumarate in DMSO-d_6 using the same conditions used for the analyte (compound B). The concentration of the reference standard was calculated from weighed mass and measured volume used in preparing the sample. A signal generated from the external reference standard was used to calculate the concentration of the analyte using the Bruker Biospin ERETIC2 software module.

The concentration of the analyte was calculated using integrations from three different ^{1}H resonances in the NMR spectrum (Figure 9.8).

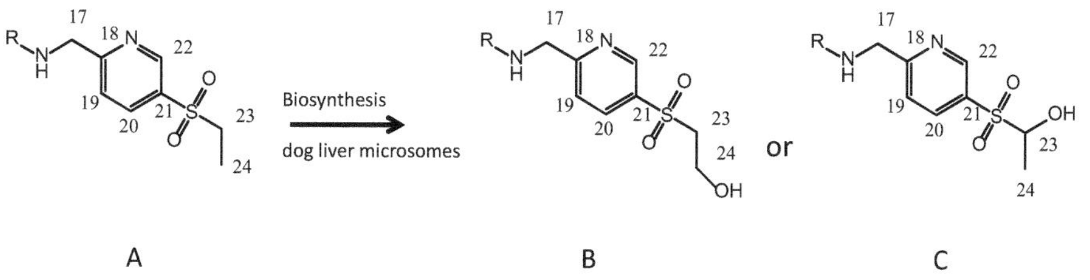

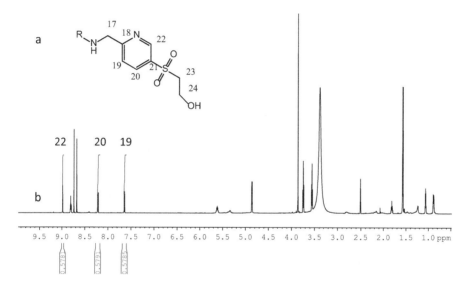

FIGURE 9.7 Incubation of pharmaceutical agonist A with dog-liver microsomes yielded a hydroxylated metabolite. NMR characterization of the isolated metabolite showed hydroxylation occurred at C-24 (compound B) and not C-23 (compound C).

FIGURE 9.8 ^{1}H NMR spectrum (600 MHz) showing the integrated intensities of analyte B used in the qNMR calculation.

Calculated concentration results showed good agreement among all four peaks (see Table 9.4). The average concentration was found to be 21.557 mM. Given the known volume, concentration, and molecular weight, the average amount of compound B present in the sample was calculated to be 416.9 µg.

Purity of the metabolite was calculated using Equation 9.9. Purity of the external reference standard (99.9%) was determined using NMR analysis.

$$x = \frac{\text{Conc. of Std.}}{\text{Conc. of Analyte}} \times \frac{\text{Integration analyte peak}}{\text{\#protons analyte peak}}$$
$$\times \frac{\text{\#protons Std. peak}}{\text{Integration Std. peak}} \times 0.999 \times 100\% \tag{9.9}$$

The integration of the internal standard peak was normalized to the number of protons in creating the ERETIC reference. The aliphatic protons of dimethylfumarate were used as the reference peak.

TABLE 9.4

Concentration and Purity of Metabolite Compound B (10° pulse; 20 s delay)

Sample	Mass	Molecular weight	Volume	Concentration	Integration	# protons	Purity %
Dimethyl Fumarate	5.37 mg	144.13	1.0 mL	37.258 mM	6	6	
Cpd. B	**416.9 µg**	552.61	35 µL	**21.557 mM**			**99.89%**
(metabolite)	(avg)*			(avg)*			(avg)*
Peak 19	416.9 µg*			21.555 mM*	0.5785	1	99.89%*
Peak 20	417.3 µg*			21.576 mM*	0.5791	1	99.90%*
Peak 22	416.6 µg*			21.540 mM*	0.5781	1	99.89%*

* = calculated values

$$1 = \frac{\#\text{protons Std. peak}}{\text{Integration Std. peak}} \tag{9.10}$$

The average NMR purity result for this sample was calculated to be 99.89% (see Table 9.4).

Overall structure elucidation of metabolite compound B, along with the concentration, amount of compound (mass), and purity, was achieved with NMR spectroscopy. This study demonstrates the power of qNMR in characterizing and quantitating compounds such as metabolites that cannot be synthetically prepared. Additionally, the use of an external reference enables the analyte sample to be used for follow-up toxicity studies.

The second example highlights the approach where traditional biosynthesis and isolation of desired metabolites from a biological matrix did not yield sufficient quantity and purity suitable for structure elucidation by NMR spectroscopy [43]. Production of metabolites *via* classical-liver microsomal incubations was carried out, and isolation achieved using supercritical fluid chromatography (SFC) followed by liquid chromatography-solid-phase extraction-nuclear magnetic resonance (LC-SPE-NMR). This approach was used to investigate the biosynthesis of 4-hydroxyestrone glucuronides from human-liver microsomes (Figure 9.9). The isolated metabolite was unequivocally characterized and quantified by NMR spectroscopy.

The full chemical shift assignment and structure characterization by NMR spectroscopy established compound C as the isolated metabolite. Key HMBC correlations between H-6/C-4 and H-19/C-4 showed the glucuronidation occurred at the OH on C4 rather than C-3. (Figure 9.10). Further relative chair stereochemistry was determined based upon ROESY data showing correlations between H-22/H-20, H-21/H-23, H-21/H-19, H-19/H-23.

With the structure of compound C in hand, it was possible to calculate the concentration and mass of the isolated compound with qNMR and ERETIC2 (see Table 9.5). In this study, dimethyl fumarate was used as an external reference prepared with a concentration of 7.64 mM and 4.68 mM. Reference peaks used the aliphatic methyl at 3.75 ppm and the olefinic protons at 6.97 ppm. NMR spectra were collected with and without solvent suppression.

The molar concentration of the metabolite in DMSO-d_6 was found to be about 4.03 mM. The amount of 4-hydroxyestrone glucuronide in the NMR tube was calculated to be ~60 µg. Hence, 4-hydroxy-estrone glucuronide was successfully biosynthesized, unequivocally characterized using LC-SPE-NMR, and subsequently quantified using qNMR (ERETIC2).

The third case study highlighting the importance of qNMR in the investigation of drug metabolites was reported by Walker et al [44]. In this study, biosynthetic methods were used to produce drug metabolites needed to study their activity and toxicity. The structures and concentrations of five metabolites were elucidated using NMR and qNMR technology.

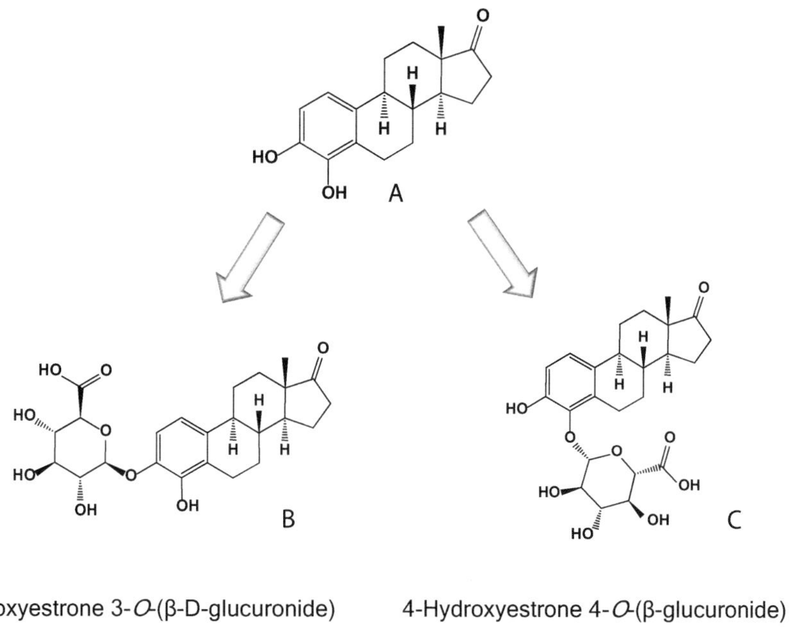

4-Hydroxyestrone 3-*O*-(β-D-glucuronide)   4-Hydroxyestrone 4-*O*-(β-glucuronide)

FIGURE 9.9   Chemical structures of 4-hydroxyesterone (A) and possible glucuronidation products B and C from incubation with human liver microsomes.

The types of metabolites produced in these studies consisted of one glucuronide and four compounds yielding oxidative products. These seven metabolites are shown in Figure 9.11.

A closer look at one of the metabolism studies (Figure 9.11, compound B) provides insight into the rationale and subsequent isolation, purification, and characterization process, and showcases the power of NMR technology in solving important pharmacological problems. Compound B, a phosphodiesterase inhibitor [45], produced metabolites upon incubation in dog-liver microsomes with similar activity as the parent compound. The observed metabolites included two hydroxyl metabolites with the same molecular weight (m/z 390) (shown in Figure 9.11), as well as one doubly hydroxylated metabolite (m/z 406), and two arising from oxidative opening of the azetidine ring (m/z 406 and m/z 392, retention) (not shown).

Because the two singly hydroxylated metabolites (m/z 390) were found to have activity at the target enzyme, a large-scale biosynthesis using dog-liver microsomes was carried out to prepare significant quantities for structural characterization and quantitation. The work-up included the addition of CH_3CN to produce a precipitate followed by HPLC and fraction collection. The MS evaluation showed the protonated molecular ion m/z 390 was consistent with hydroxy metabolites for two fractions. Collection vessels containing the two hydrolysis products were evaporated by vacuum centrifugation and prepared for NMR analysis.

The two metabolites of were structurally characterized by NMR spectroscopy and the sites of oxidation were determined. The numbered metabolites structures B1 and B2 are illustrated in Figure 9.11.

The concentrations of metabolites B1 and B2 were determined using qNMR. One dimensional ^{1}H NMR spectra of compounds B, B1, and B2 are shown in Figure 9.12. In the ^{1}H spectrum of compound B, the resonance at 2.19 ppm is assigned as the methylene H25 (Figure 9.12A). The ^{1}H chemical shifts of H24 and H26 were only observed as shallow broad peaks at 3.75 ppm.

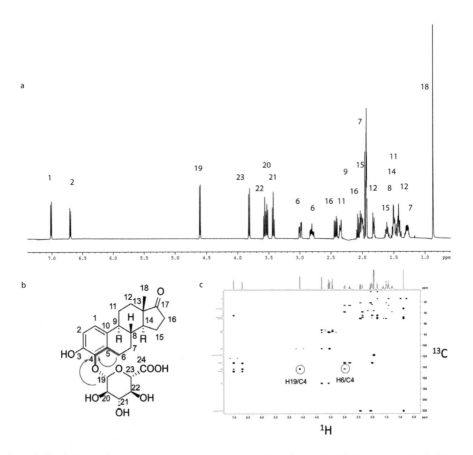

FIGURE 9.10 NMR structure characterization of metabolite. (a) ^{1}H NMR spectrum (600 MHz) of metabolite B showing hydrogen atom assignments, (b) numbered chemical structure of compound B, (c) ^{1}H,^{13}C HMBC spectrum showing connectivity between H19 and C4, H6 and C4 that establishes (4-OH) as the site of glucuronidation.

TABLE 9.5

Use of qNMR and ERETIC2 to Calculate Metabolite (Compound C) Concentration

Data Set	Dimethyl fumarate Conc. (mM)	Reference Peak (ppm)	NMR Spectra	Eretic2 Conc. (mM)
1	7.64	3.75	1D ^{1}H NMR	4.00
2	External Reference 1	6.97	1D ^{1}H NMR	4.02
3	4.68	3.75	WET ^{1}H NMR	4.06
4	External Reference 2	6.97	WET ^{1}H NMR	4.02

Average 4.03*

* Isolated ~30 μL of 4.0 mM 4-hydroxyestrone glucuronide (~60 μg) in DMSO-d$_6$

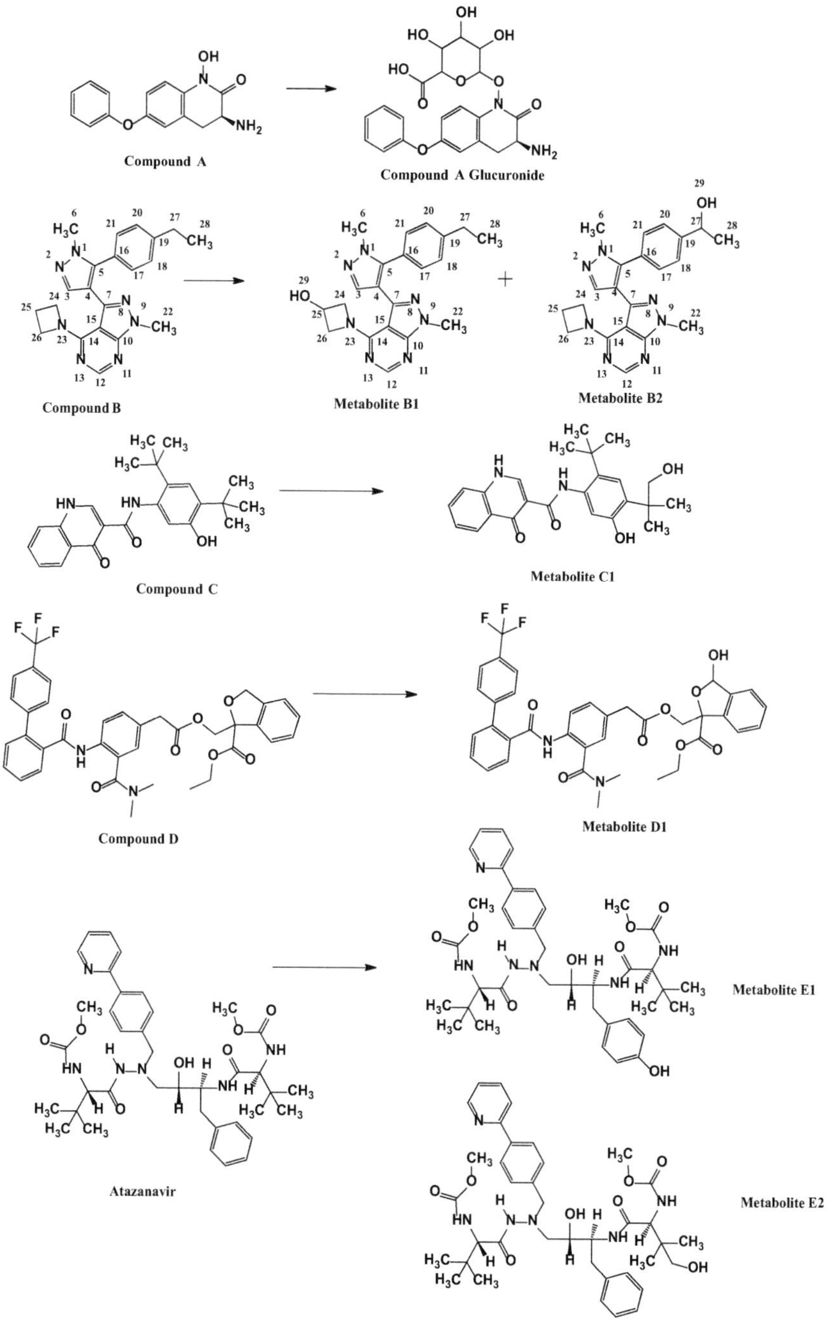

FIGURE 9.11 Example compounds and their metabolites. Numbering scheme is shown for compound B and their metabolites B1 and B2. (From GS Walker, JN Bauman, TF Ryder, EB Smith, DK Spracklin, and RS Obach (2014) Biosynthesis of Drug Metabolites and Quantitation Using NMR Spectroscopy for Use in Pharmacologic and Drug Metabolism Studies, *Drug Metab Dispos*, 42(10): 1627–39.)

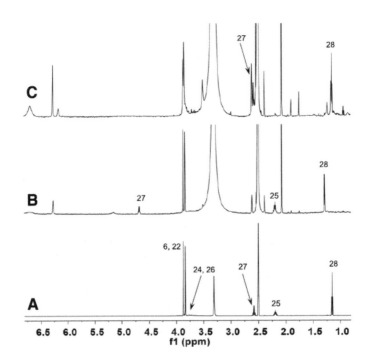

FIGURE 9.12 NMR spectra of compound B and its metabolites B1 and B2. (A) Aliphatic portion of the ^{1}H spectrum of compound B. (B) Aliphatic portion of the ^{1}H spectrum of metabolite B2. (C) Aliphatic portion of the ^{1}H spectrum of metabolite B1. (From GS Walker, JN Bauman, TF Ryder, EB Smith, DK Spracklin, and RS Obach (2014) Biosynthesis of Drug Metabolites and Quantitation Using NMR Spectroscopy for Use in Pharmacologic and Drug Metabolism Studies, *Drug Metab Dispos*, 42(10): 1627–39.)

For the ^{1}H spectrum of metabolite B1, the ^{1}H resonances of H-24, H-25, and H-26 (Figure 9.12C) were not observed. However, all other ^{1}H resonances in the metabolite B1 spectrum were present and unmodified from those observed in the parent compound B. Because of this, the exact site of oxidation for metabolite B1 could only be implied to occur on C-24, C-25, or C-26. However, if the site of oxidation was on either the C-24 or C-26, the resulting metabolite would be a carbanolamine and would likely be unstable. Because the metabolite appeared to be sufficiently stable to isolate, the oxidation was assigned to the C-25 carbon. The ambiguity in the structural assignment had no impact on the determination of concentration of the metabolite B1.

The chemical structure of metabolite B2 was definitively identified based on ^{1}H NMR data. Compound B contains an ethyl group on the phenyl ring. In the ^{1}H NMR spectrum of the parent compound, the methyl H-28 of the ethyl is assigned as a triplet at 1.15 ppm and the methylene protons H-27 are assigned as a quartet at 2.58 ppm. For the spectrum of the isolated metabolite B2, however, resonances have changed (Figure. 9.12B). The methyl resonance was now a doublet at 1.28 (^{3}H, J = 6.5 Hz), and the methylene changed to a methine and was assigned to the quartet at 4.67 (^{1}H, J = 6.5 Hz).

Because there was an abundance of peaks in which the number of hydrogens were known, calculation of the concentration could be readily determined. The concentration of metabolite B1 was calculated to be 0.048 mM and the concentration of metabolite B2 was 0.92 mM using ERETIC (artificial signal insertion for calculation of concentration).

The stock solutions of the two metabolites were then used to determine the target enzyme IC_{50}s of the metabolites that were found to be 10 and 42 nM for B1 and B2, respectively, as compared with 2 nM for the parent compound B.

Concentrations of the all the reported metabolites (shown in Figure 9.11) ranged from 0.048 to 8.3 mM. Because of the non-destructive nature of NMR evaluation, these isolated metabolites with defined concentrations could then be used as standards in pharmacology assays, assays to assess toxic properties, and studies to determine absorption, distribution, metabolism, and excretion of these new chemical entities.

Natural Products: A study involving the quantification of the shellfish toxin, okadaic acid (OA), was reported using ERETIC2 based on PULCON methodology [46]. The diarrheic shellfish toxin, OA, [47] is a lipophilic marine toxin produced by toxic dinoflagellates Prorocentrum lima [48] and Dinophysis spp [49]. The toxin can cause diarrhea and vomiting in humans when consumed [50].

OA was isolated from a the toxic dinoflagellate Prorocentrum lima using liquid–liquid partitioning and requiring multiple-column chromatography steps [48]. The purity analysis of OA was performed with qNMR. Purities were detected in the range of ~95%. The five ^{1}H NMR resonances appearing at 5.78 (H-14), 5.52 (H-15), 5.36 (H-41), 5.30 (H-9), and 5.06 (H-41) ppm of OA were used for quantitation. The numbered chemical structure of OA and 1D ^{1}H NMR spectrum are shown in Figure 9.13.

A slight fluctuation of the baseline was reported even after spectral baseline correction. This effect was detected using either an internal or external reference, which can compromise signal integration. Also noted was the observation that lower concentration values were

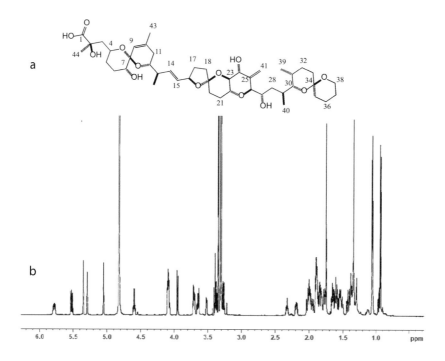

FIGURE 9.13 (a) Numbered structure of okadaic acid and (b) ^{1}H NMR spectrum dissolved in CD$_3$OD (800 MHz). (Adapted from Watanabe, R., Sugai, C., Yamazaki, T., Matsushima, R., Uchida, H., Matsumiya, M., Takatsu, A., Suzuki, T. (2016) Quantitative Nuclear Magnetic Resonance Spectroscopy Based on PULCON Methodology: Application to Quantification of Invaluable Marine Toxin, Okadaic Acid. Toxins, 8, 294; © 2016 by the authors; licensee MDPI, Basel, Switzerland.)

observed for signal 1 compared to the other signals; therefore, signal 1 was excluded from calculations for quantitation (Table 9.6).

In the reported study, comparison of OA concentrations was made between the PULCON method and the internal standard method. Because it is generally found that qNMR using an internal standard method yields higher precision and accuracy than using an external standard, the precision and accuracy of PULCON vs. an internal standard method was investigated.

Each method was performed in duplicate. The reference standard (internal and external) was 1,4 bis(trimethylsilyl) benzene (1,4-BTMSB-d$_4$). In both methods, the concentrations obtained from selected signals were averaged in each tube and then the average concentration in each tube was averaged.

The OA concentration calculated by the internal standard was 352.8 + 7.5 µg/g while the concentration quantified by PULCON gave 355.4 + 0.1 µg/g (Table 9.6). Because quantification of OA by PULCON with 1,4-BTMSB-d$_4$ external standard gave a result equivalent to the internal standard, it was concluded that the PULCON method can be reliably applied to the quantification of invaluable marine toxins without any introduction of an internal standard.

Fluorine (^{19}F) qNMR: For the most part, ^{1}H NMR may be considered the primary technique for qNMR studies; however, Martino et al. also showed qNMR can be an effective tool using ^{19}F and ^{31}P NMR as well [51]. An example where ^{19}F qNMR was successfully used to quantitate the content of fluorinated pharmaceuticals was described in a paper by Okaru et al. [52]. In this study, ^{19}F NMR spectra were obtained in dimethylsulfoxide-d$_6$ or aqueous buffer, and trifluoroacetic acid was used as the internal reference. Thirteen fluorine-containing pharmaceuticals spanning various pharmacological classes were studied with HPLC and/or NMR

TABLE 9.6
Quantitation Results Obtained by Internal Standard Method and PULCON

Quantitation (µg/g)

Method	Tube	Signal	Integral region	Run 1	Run2	Run3	Average	Std. dev	RSD (%)
Internal standard method	1	1	5.85–5.72	344.3	344.08	340.38	**342.92**	2.20	0.64
		2	5.57–5.49	346.23	345.58	344.36	**345.39**	0.95	0.28
		3	5.38–5.25	345.23	349.27	354.62	**349.78**	4.61	1.32
		4	5.09–5.03	367.78	332.26	345.23	**349.78**	17.98	5.16
	2	1	5.85–5.72	346.39	344.29	349.36	**346.68**	2.55	0.74
		2	5.57–5.49	362.29	360.85	357.01	**360.05**	2.73	0.76
		3	5.38–5.25	365.55	361.48	355.51	**360.85**	5.05	1.40
		4	5.09–5.03	361.77	355.68	355.76	**357.74**	3.49	0.98
PULCON	1	1	5.86–5.72	344.81	333.91	340.59	**339.77**	5.50	1.62
		2	5.58–5.48	359.29	355.49	355.08	**356.62**	2.32	0.65
		3	5.41–5.24	363.20	360.73	338.95	**354.29**	13.34	3.77
		4	5.10–5.01	371.00	363.92	364.74	**366.55**	3.87	1.06
	2	1	5.86–5.72	339.47	351.09	342.76	**344.44**	5.99	1.74
		2	5.58–5.48	351.92	360.04	357.88	**356.61**	4.21	1.18
		3	5.41–5.24	353.67	359.63	342.35	**351.88**	8.78	2.49
		4	5.10–5.01	359.73	365.08	366.31	**363.71**	3.50	0.96

to determine compound content. Purities of fluorometholone, flumazenil, flunitrazepam, fluta-mide, fluvastatin, and fluprednidene acetate were determined using ^{19}F qNMR and HPLC. The qNMR studies were carried out in DMSO-d$_6$ and the chemical structures of the analytes and purity results for qNMR and HPLC are shown in Table 9.7.

TABLE 9.7

Purities for DMSO-d$_6$ Soluble Fluorinated Pharmaceuticals

Structure	Compound Name	^{19}F NMR Purity (% w/w)	Purity (% w/w) HPLC
	Fluorometholone	95.9	100.0
	Flumazenil	94.9	99.8
	Flunitrazepam (Rohypnol)	98.8	99.8
	Flutamide	98.2	n.d
	Fluvastatin sodium hydrate	83.0	84.6
	Fluprednidene acetate	98.1	99.7

n.d. = not determined

The reliability of the data obtained by the ^{19}F qNMR spectroscopic method was tested by comparison with data obtained using HPLC method. The data shows that similarity of assay values using NMR spectroscopy versus HPLC was <5%. Because small differences (<5%) were found for the fluorinated pharmaceuticals determined through the qNMR method compared with HPLC method, the two methods were considered equally applicable in the analysis of fluorinated pharmaceuticals.

In a similar manner, compounds such as fluphenazine, flucytosine, fluconazole, flurbiprofen, fluoxetine, fludarabine, and flurazepam had purities determined in an aqueous buffered environment using ^{19}F qNMR (see Table 9.8). In this study, comparison with the HPLC method was not available.

Only one inconsistency was reported for the Flurazepam study in the lithium carbonate buffer since the ^{19}F NMR spectrum contained two distinct signals instead of one signal. This observation may be attributed to hydrolytic decomposition of the compound in acidic water solution. The phenomenon was previously found in ^{19}F NMR spectra where a 44:56 ratio between Flurazepam and its ring-opened form was observed after 24 h equilibrium [53]. In the reported study, the combined integral of both signals was used for quantification although follow-up analysis using LC/MS/MS would be required for further clarification. Combined signal integration may also be used in the qNMR investigation of rotamers or cases of severe overlap.

For fluorine containing compounds, ^{19}F qNMR offers the advantage of high natural abundance and less risk of signal overlap since ^{19}F NMR has a broad chemical shift range (~500 ppm) [51]; therefore, fluorine signals are less prone to resonance overlap or interference from homonuclear coupling [54]. Another major advantage is that NMR solvents typically contain no fluorine atoms, hence, solvent suppression is not necessary. This makes the technique uniquely suitable for quantification of fluorine containing compounds, especially in aqueous environments.

A possible disadvantage of the ^{19}F NMR large chemical shift range is the inability to obtain uniform excitation over the entire chemical shift range. This problem may be addressed by carrying out two measurements: one over the entire range to identify the optimal sweep width and then a second experiment for quantification in the optimized range. Additionally, the internal reference standard should be selected to have a similar chemical shift as the analyte signal of interest.

Phosphorous (^{31}P) qNMR: The use of ^{31}P NMR for quantification of the active ingredient in a formulation may be achieved with the use of a single standard reference material. This is in contrast to chromatographic methods. As discussed previously, quantitation with chromatography requires a separate standard for each target analyte to accurately determine the composition of a mixture. The effort required to establish good chromatographic methods and relative response factors is significant and limitations can introduce sizable errors in chemical analysis. This is not required in qNMR analysis, which directly measures the relative numbers of nuclei that are observed; hence, ^{31}P qNMR can present a significant advantage in formulation analysis.

Selecting a reference for ^{31}P NMR is an important process in qNMR and depends on the type of chemical structure. For example, the use of sodium phosphate as the ^{31}P NMR standard may lead to a possible underestimation of purity arising from the nuclear Overhauser effect (NOE) on the analyte peak. Sodium phosphate contains no protons, and so it cannot exhibit an NOE, whereas peaks from the analyte may have strong proton-phosphorous coupling that can contribute to an NOE effect. Even if the NOE is mimimised with inverse-gated decoupling, complete elimination of coupling effects cannot be guaranteed and may create a disparity in measured values. For this reason, references with similar NOE to the analyte may be preferred.

In qNMR, the choice of the nucleus may be critical to the accuracy in the execution of the experiment. For example, ^{1}H NMR, which is usually the method of choice for qNMR studies, may not be a suitable method for the study of formulations. Typically, formulations contain multiple excipients in high concentration that can cause interference in signal integration of the

TABLE 9.8

Purities for buffer-soluble fluorinated pharmaceuticals

Structure	Compound name	Purity (% w/w) 19F qNMR
	Fluphenazine dihydrochloride	88.8
	Fluoxetine HCl	84.0
	Fludarabine phosphate	89.1
	Flurazepam monohydrochloride	86.5
	5-Fluorocytosine	82.8
	Fluconazole	86.2
	Sodium flurbiprofen dihydrate	89.3

analyte. This is because formulation compounds usually are high in concentration relative to the analyte and proton-rich. An abundance of protons from multiple species can mask the signals of interest. In cases of severe overlap of proton signals, the use of ^{31}P NMR is preferable since few formulation constituents contain phosphorus. Hence, observation of the ^{31}P nucleus will pose little risk of masking the active ingredient resonance.

Phosphorous-31 qNMR spectroscopy has been used as a preferred method for quantification of agrochemicals and associated formulations. One major reason is due to the fact that a large number of agrochemicals contain phosphorus nuclei, which have properties conducive for use in qNMR studies.

An example of the use of ^{31}P qNMR involves a method developed to investigate the decomposition of malathion in various formulations during storage. In this study, malathion (Figure 9.14) in various formulations was subjected to storage at approximately 54°C (±2°C). The samples were first analyzed for content of the active ingredient and the impurities O,O,O-trimethylphosphorothioate, O,O,S-trimethylphosphorodithioate, malaoxon, and iso-malathion. This mixture of malathion and impurities was subsequently quantitated by ^{31}P qNMR.

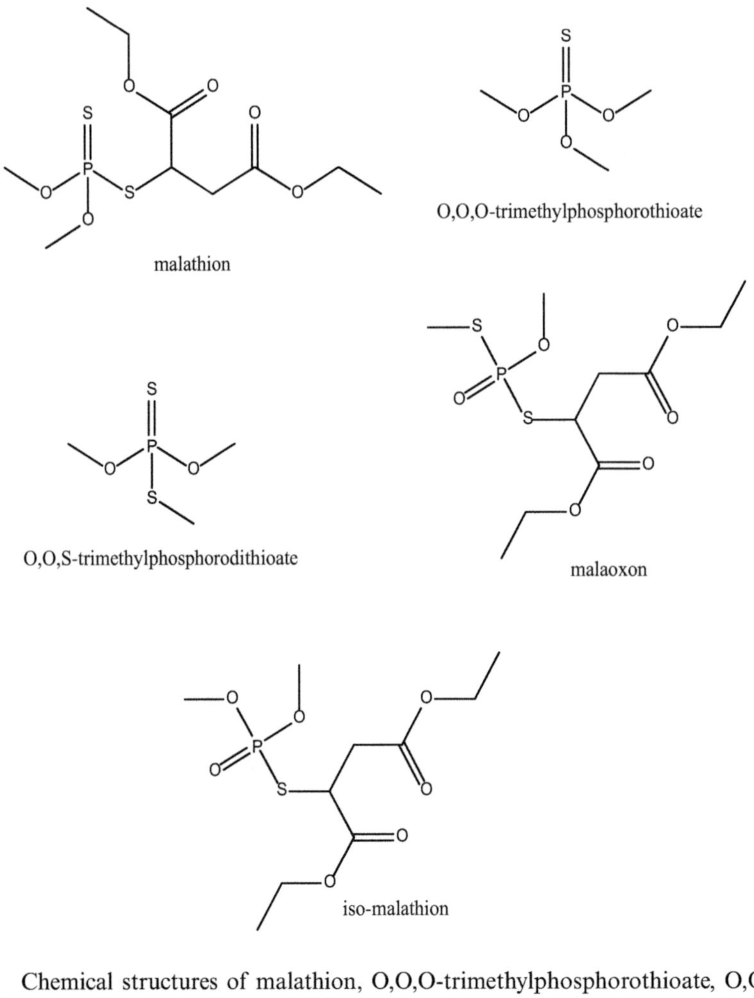

FIGURE 9.14 Chemical structures of malathion, O,O,O-trimethylphosphorothioate, O,O,S-trimethyl-phosphorodithioate, malaoxon, and iso-malathion.

Figure 9.15 shows the ^{31}P spectra of an emulsifiable concentrate formulation of malathion immediately following formulation preparation and after storage at 54°C for 14 days [55]. The spectrum shows NMR resonances free from signal overlap. Application of ^{31}P qNMR allowed the quantation of malathion and impurities O,O,O-trimethylphosphorothioate, O,O,S-trimethylphosphorodithioate, malaoxon, and isomalathion [56]. In this study, a portion of the test formulation is dissolved in deuterated chloroform and the molar ratios between compounds are determined by ^{31}P-NMR spectroscopy. Components of the mixture such as isomalathion were then calculated from the malathion content as determined by gas chromatography.

The decline in malathion content and generation of isomalathion during 14 days of storage at 54°C depended on the formulation. For example, during 14 days of storage at 54°C, the average final level of isomalathion (expressed as % of active ingredient) in the emulsifiable concentrate formulation was reported to be 0.11%. Other degradants in the formulation were similarly measured [56].

Overall, quantitative NMR (qNMR) spectroscopy can provide an accurate and precise method that can complement and/or corroborate to the HPLC method. The technique sometimes does not require isolation of the analyte. Unlike HPLC, no expensive chemical reference substances are necessary. A major advantage for NMR is that the technology enables structural information of analytes to be determined. Because the NMR technology is non-destructive, the analyte samples may be used in follow-up studies of the compound's activity or toxic properties, when an external reference is used. The ability to use multiple nuclei (e.g., ^{1}H, ^{19}F, and ^{31}P) expands the range and versatility of these investigations. NMR

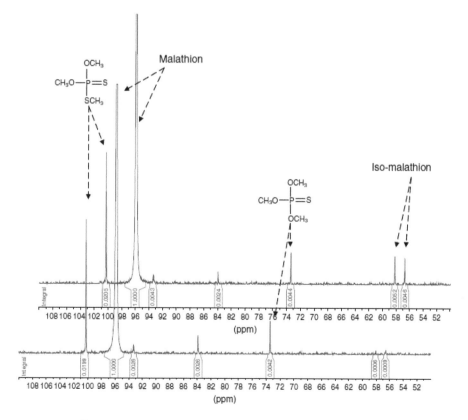

FIGURE 9.15 The ^{31}P NMR spectra of a freshly prepared malathion emulsion concentrate formulation standard (lower trace) and the same formulation stored for 14 days at 54°C (upper trace).

can be less time-consuming when relaxation issues and purification are not needed. When system suitability is met, the qNMR assay is easy to perform and more specific leading to high reproducibility. Combined with ERETIC2/PULCON, qNMR provides a powerful means of determining purity, concentration, and mass of a compound, thereby providing a potent technology for quantifying challenging analytes.

REFERENCES

1. Jungnickel, J.L., and Forbes, J.W. 1963. Quantitative measurement of hydrogen types by intergrated nuclear magnetic resonance intensities. *Anal. Chem.* 35:938–942.
2. Hollis, D.P. 1963. Quantitative analysis of aspirin, phenacetin, and caffeine mixtures by nuclear magnetic resonance spectrometry. *Anal. Chem.* 35:1682–1684.
3. Jancke, H. 1998. NMR as primary analytical measuring method. *Chem. Tech. Lab.* 46:720–722.
4. King, B. 2000. The practical realization of the traceability of chemical measurements standards. *Accredit. Qual. Assur.* 5:429–436.
5. King, B. 2000. Metrology in chemistry: part II. future requirements in Europe. *Accredit. Qual. Assur.* 5:266–271.
6. Milton, M.J., and Quinn, T.J. 2001. Primary methods for the measurement of amount of substance. *Metrologia.* 38:289–296.
7. Maniara, G., Rajamoorthi, K., Rajan, S., and Stokton, G.W. 1998. Method performance and validation for quantitative analysis by ^{1}H and ^{31}P NMR spectroscopy. Applications to analytical standards and agricultural chemicals. *Anal. Chem.* 70:4921–4928.
8. Wells, R.J., and Cheung, J. 2001. *The Chemistry Preprint Server CPS: Analchem/0103002.* http://preprint.chemweb.com.
9. Wells, R.J., Hook, J.M., Al-Deen, T.S., and Hibbert, D.B. 2002. Quantitative nuclear magnetic resonance (QNMR) spectroscopy for assessing the purity of technical grade agrochemicals: 2,4-dichlorophenoxy-acetic acid (2,4-D) and sodium 2,2-dichloropropionate (dalapon sodium). *J. Agric. Food Chem.* 50:3366–3374.
10. Jancke, H., and Malz, F. 1999. *Abschlussbericht NMR-1.* Berlin.
11. Jancke, H., and Malz, F. 1999. *International Comparison CCQM-4.* Berlin: Final Report.
12. Turczan, J.W., and Medwick, T. 1977. Qualitative and quantitative analysis of amygdalin using NMR spectroscopy. *Anal. Lett.* 10:581–590.
13. Cockerill, A.F., Davis, G.L.O., Harrison, R.G., and Rackham, D.M. 1974. NMR determination of the enantiomer composition of penicillamine using an optically active europium shift reagent. *Org. Magn. Reson.* 6:669–670.
14. Holzgrabe, U., Diehl, B.W., and Wawer, I. 1998. NMR spectroscopy in pharmacy. *J. Pharmaceut. Biomed.* 17:557–616.
15. Vailaya, A., Wang, T., Chen, Y., and Huffman, M. 2001. Quantitative analysis of dimethyl titanocene by iodometric titration, gas chromatography and NMR. *J. Pharmaceut. Biomed.* 25:577–588.
16. Forshed, J., Andersson, F.O., and Jacobsson, S.P. 2002. NMR and Bayesian regularized neural network regression for impurity determination of 4-aminophenol. *J. Pharmaceut. Biomed.* 29:495–505.
17. Meshitsuka, S., Morio, Y., Nagashima, H., and Teshima, R. 2001. ^{1}H-NMR studies of cerebrospinal fluid: endogenous ethanol in patients with cervical myelopathy. *Clin. Chim. Acta.* 312:25–30.
18. Sahrbacher, U., Pehlke-Rimpf, A., Rohr, G., Eggert-Kruse, W., and Kalbitzer, H.R. 2002. High resolution proton magnetic resonance spectroscopy of human cervical mucus. *J. Pharmaceut. Biomed.* 28:827–840.
19. Xia, Z., Akim, L.G., and Argyropoulos, D.S. 2001. Quantitative ^{13}C NMR analysis of lignins with internal standards. *J. Agric. Food Chem.* 49:3573–3578.
20. Igarashi, T., Aursand, M., Hirata, Y., Gribbestad, I.S., Wada, S., and Nonaka, M. 2000. Nondestructive quantitative determination of docosahexaenoic acid and n−3 fatty acids in fish oils by high-resolution ^{1}H nuclear magnetic resonance spectroscopy. *JAOCS.* 77:737–748.
21. Papke, N., and Karger-Kocsis, J. 2001. C-13- and H-1-NMR analysis of a nitrile rubber with and without glycidylmethacrylate grating. *Eur. Polym. J.* 37:547–557.
22. Henderson, T.J. 2002. Quantitative NMR spectroscopy using coaxial inserts containing a reference standard: purity determinations for military nerve agents. *Anal. Chem.* 74:191–198.

23. Derome, A.E. 1987. *Modern NMR Techniques for Chemistry Research.* Oxford: Pergamon Press.

24. Malz, F., and Jancke, H. 2005. Validation of quantitative NMR. *J. Pharm. Biomed. Anal.* 38:813–823.

25. Holzgrabe, U., Deubner, R., Schollmayer, C., and Waibel, B. 2005. Quantitative NMR spectroscopy–applications in drug analysis. *J. Pharm. Biomed. Anal.* 38:806–812.

26. Branch, S.K. 2005. Guidelines from the International Conference on Harmonisation (ICH). *J. Pharm. Biomed. Anal.* 38:798–805.

27. Webster, G.K., Marsdena, I., Pommereninga, C.A., Tyrakowski, C.M., and Tobias, B. 2009. Determination of relative response factors for chromatographic investigations using NMR spectrometry. *J. Pharm. Biomed. Anal.* 49:1261–1265.

28. Barantin, L., Le Pape, A., and Akoka, S. 1997. A new method for absolute quantitation of MRS metabolites. *Magn. Reson. Med.* 38:179–182.

29. Akoka, S., Barantin, L., and Trierweiler, M. 1999. Concentration measurement by proton NMR using the ERETIC method. *Anal. Chem.* 71:2554–2557.

30. Remaud, G.S., Silvestre, V., and Akoka, S. 2005. Traceability in quantitative NMR using an electronic signal as working standard. *Accredit. Qual. Assur.* 10:415–420.

31. Wider, G., and Dreier, L. 2006. Measuring protein concentrations by NMR spectroscopy. *J. Am. Chem. Soc.* 128(8):2571–2576.

32. Hart, D.J., and Scott, K.J. 1995. Development and evaluation of an HPLC method for the analysis of carotenoids in foods, and the measurement of the carotenoid content of vegetables and fruits commonly consumed in the UK. *Food Chem.* 54:101–111.

33. Holzgrabe, U. 2010. Quantitative NMR spectroscopy in pharmaceutical applications. *Prog. Nucl. Magn. Reson. Spectrosc.* 57:229–240.

34. Diehl, B. 2008. NMR applications for polymer characterization. In *NMR Spectroscopy in Pharmaceutical Analysis, Elsevier, Amsterdam*, eds. U. Holzgrabe, I. Wawer, and B. Diehl, 157–180. Oxford and Amsterdam: Linacre House and Elsevier B.V.

35. Jones, C. 2005. NMR assays for carbohydrate-based vaccines. *J. Pharm. Biomed. Anal.* 38:840–850.

36. Pauli, G.F., Jaki, B.U., and Lankin, D.C. 2005. Quantitative ^{1}H NMR: development and potential of a method for natural products analysis. *J. Nat. Prod.* 68:133–149.

37. Pauli, G.F., Jaki, B.U., and Lankin, D.C. 2007. A routine experimental protocol for qHNMR illustrated with taxol. *J. Nat. Prod.* 70:589–595.

38. Kellenbach, E., Sanders, K., Zomer, G., and Overbeeke, P.L. 2008. The use of proton NMR as an alternative for the amino acid analysis as identity test for peptides. *Pharmaeuropa Sci. Notes.* 2008 (1):1–7.

39. Wells, R.J., Cheung, J., and Hook, J.M. 2008. The use of qNMR for the analysis of agrochemicals. In *NMR Spectroscopy in Pharmaceutical Analysis, Elsevier, Amsterdam*, eds. U. Holzgrabe, I. Wawer, and B. Diehl, 291–315. Oxford and Amsterdam: Linacre House and Elsevier B.V.

40. Nord, L.I., Vaag, P., and Duus, J.Ø. 2004. Quantification of organic and amino acids in beer by ^{1}H NMR spectroscopy. *Anal. Chem.* 76:4790–4798.

41. Consonni, R., Cagliani, L.R., Benevelli, F., Spraul, M., Humpfer, E., and Stocchero, M. 2008. NMR and chemometric methods: a powerful combination for characterization of balsamic and traditional balsamic vinegar of modena. *Anal. Chim. Acta.* 611:31–40.

42. Almeida, C., Duarte, I.F., Barros, A., Rodrigues, J., Spraul, M., and Gil, A.M. 2006. Composition of beer by ^{1}H NMR spectroscopy: effects of brewing site and date of production. *J. Agric. Food Chem.* 54:700–706.

43. Liu, P., Temrikar, Z., Lee, H., Gao, A., Gonnella, N., Xin, D., Taub, M.E., and Teitelbaum, A. 2018. Purification of biosynthesized drug metabolites by supercritical fluid chromatography. *Drug Metab. Pharmacokinet.* 33:S78.

44. Walker, G.S., Bauman, J.N., Ryder, T.F., Smith, E.B., Spracklin, D.K., and Obach, R.S. 2014. Biosynthesis of drug metabolites and quantitation using NMR spectroscopy for use in pharmacologic and drug metabolism studies. *Drug Metab. Dispos.* 42(10):1627–1639.

45. Helal, C.J., Chappie, T.A., and Humphrey, J.M. 2012. inventors, Pfizer, assignee. Pyrazolo[3,4-D] pyrimidine compounds and their use as PDE2 inhibitors and/or CYP3A4 inhibitors. European Patent Office Application WO2012168817 A1. 2012 Dec 13.

46. Watanabe, R., Sugai, C., Yamazaki, T., Matsushima, R., Uchida, H., Matsumiya, M., Takatsu, A., and Suzuki, T. 2016. Quantitative nuclear magnetic resonance spectroscopy based on PULCON methodology: application to quantification of invaluable marine toxin, okadaic acid. *Toxins.* 8:294.

47. Tachibana, K., Scheuer, P.J., Tsukitani, Y., Kikuchi, H., Van Engen, D., Clardy, J., Gopichand, Y., and Schmitz, F.J. 1981. Okadaic acid, a cytotoxic polyether from two marine sponges of the genus Halichondria. *J. Am. Chem. Soc.* 103(9):2469–2471.

48. Nagahama, Y., Murray, S., Tomaru, A., and Fukuyo, Y. 2011. Species boundaries in the toxic dinoflagellate prorocentrum lima (Dinophyceae, Prorocentrales), based on morphological and phylogenetic characters. *J. Phycol.* 47:178–189.

49. Yasumoto, T., Oshima, Y., Sugawara, W., Fukuyo, Y., Oguri, H., Igarashi, T., and Fujita, N. 1980. Identification of Dinophysis fortii as the causative organism of diarrhetic shellfish poisoning. *Bull. Jpn. Soc. Sci. Fish.* 46:1405–1411.

50. Yasumoto, T., Oshima, Y., and Yamagichi, M. 1978. Occurrence of a new type of shellfish poisoning in the Tohoku district. *Bull. Jpn. Soc. Sci. Fish.* 44:1249–1255.

51. Martino, R., Gilard, V., Desmoulin, F., and Malet-Martino, M. 2006. Interest of fluorine-19 nuclear magnetic resonance spectroscopy in the detection, identification and quantification of metabolites of anticancer and antifungal fluoropyrimidine drugs in human biofluids. *Chemotherapy.* 52(5):215–219.

52. Okaru, A.O., Brunner, T.S., Ackermann, S.M., Kuballa, T., Walch, S.G., Kohl-Himmelseher, M., and Lachenmeier, D.W. 2017. Application of ^{19}F NMR spectroscopy for content determination of fluorinated pharmaceuticals. *J. Anal. Methods Chem.* 2017:1–7.

53. Holzgrabe, U., Wawer, I., and Diehl, B. 1999. *NMR Spectroscopy in Drug Development and Analysis.* Weinheim: Wiley Press.

54. Dalvit, C., Fagerness, P.E., Hadden, D.T.A., Sarver, R.W., and Stockman, B.J. 2003. Fluorine-NMR experiments for high-throughput screening: theoretical aspects, practical considerations, and range of applicability. *J. Am. Chem. Soc.* 125(25):7696–7703.

55. Wells, R.J., Cheung, J., and Hook, J.M. 2008. *The Use of qNMR for the Aanalysis of Agrochemicals, NMR Spectroscopy in Pharmaceutical Analysis*, 291–315. Oxford and Amsterdam: Linacre House and Elsevier B.V.

56. FAO. Specifications and evaluations for agricultural Pesticides Malathion (S-1,2-bis(ethoxycarbonyl) ethyl O,O-dimethylphosphorodithioate), 1–77.

10 QM/DFT Chemical Shift Prediction

10.1 COMPUTATIONAL METHODS

Nuclear magnetic resonance (NMR) spectroscopy is a powerful analytic tool used in organic chemistry for assigning the identities and structures of molecules at the atomic level [1]. Over the years considerable effort has been made to assist in the NMR assignment process. These efforts resulted in the development of knowledge-based approaches to predict NMR chemical shifts and coupling constants. The knowledge-based approaches that account for the effects of neighboring chemical groups provide guidance in predicting chemical shifts and, in some cases, coupling constants. Such prediction programs are commercially available and extensively used in the organic chemistry community as a guide in structural assignment [2,3]. These knowledge-based programs, however, have limited utility and often fall short when addressing challenging or unusual structures.

Quantum chemical methods to predict NMR spectra provide a higher level of accuracy, although such methods are not as widely used by the practicing structural chemist. This is because, until recently, automated programs for performing such high-level chemical shift predictions have not been commercially available [4].

Determination of isotropic shielding constants, and hence isotropic chemical shifts, using quantum chemistry and density functional theory (DFT), has become a practical technology in structural chemistry. The process involves use of the gauge independent atomic orbital (GIAO) method [5] to address invariant magnetic properties. The method was further refined by Pulay and co-workers [6,7] and combined with determination of accurate molecular geometries that were key components to development of a robust process. The term "GIAO/NMR" is used to describe GIAO techniques for structure elucidation. These approaches have provided an important path forward in the development of chemical shift prediction programs for structure elucidation of organic molecules.

The ability to calculate NMR chemical shifts *ab initio* is a very important advancement in quantum chemistry. Development of DFT methods to calculate NMR spectra has proven very useful in the assignment of NMR chemical shifts or to verify resonance assignments. The objective in using DFT methods is to calculate absolute shielding constants for every relevant atom, which may then be converted (via the use of a reference term) to chemical shifts. These predicted chemical shifts can then be compared with experimental data. Good agreement results in high confidence in a proposed structure.

DFT methods may be selected to enable NMR calculations of a variety of organic compounds. Numerous theoretical methods have been investigated, yielding varying degrees of accuracy. Often selection of a particular method depends on an assessment of what is most convenient in terms of quality and computational cost [8,9]. Once shielding constants are obtained, there are various reference approaches that may be applied, which can also affect the accuracy of the predicted chemical shifts relative to experimental data.

Computational prediction of chemical shifts has seen a marked increase in accuracy and computational affordability, especially as applied to ^1H, ^{13}C, and ^{15}N nuclei. These improvements came from advancements in computational techniques, processes, and computational power. As a result, reliable chemical shift calculations using quantum chemistry and DFT may now be routinely accessible to organic chemists.

The history of NMR calculations can be found in numerous publications, review articles, and books [10–15]. In-depth discussion of the theoretical foundations of computational NMR has also been extensively reported [16–23]. The purpose of this chapter is to outline an accurate, automated, and robust method for computing isotropic chemical shifts. Benchmarking and identification of a key method, use of systematic error correction terms, referencing methods, and examples are described as well.

10.1.1 METHOD EVALUATION

There are a variety of quantum mechanical calculation methods that account for electron correlation effects to varying degrees. While the Hartree Fock method treats electronic interactions in an average manner, it does not include specific correlation effects. Methods, such as coupled-cluster theory, that capture correlation effects are computationally less feasible, except for very small molecules. However, DFT enables correlation in an approximate manner providing good results at reasonable computational cost. Therefore, in the absence of extensive computational resources and highly specialized experience, DFT calculations have been used for routine determination of computed chemical shifts.

As part of an investigation applicable to drug-like molecules, benchmarking of computational methods used in NMR structure elucidation was reported, where investigators used a molecule that contains functional groups commonly encountered in pharmaceutically relevant compounds. Compound **1** (Figure 10.1) was selected for this purpose [24,25]. The

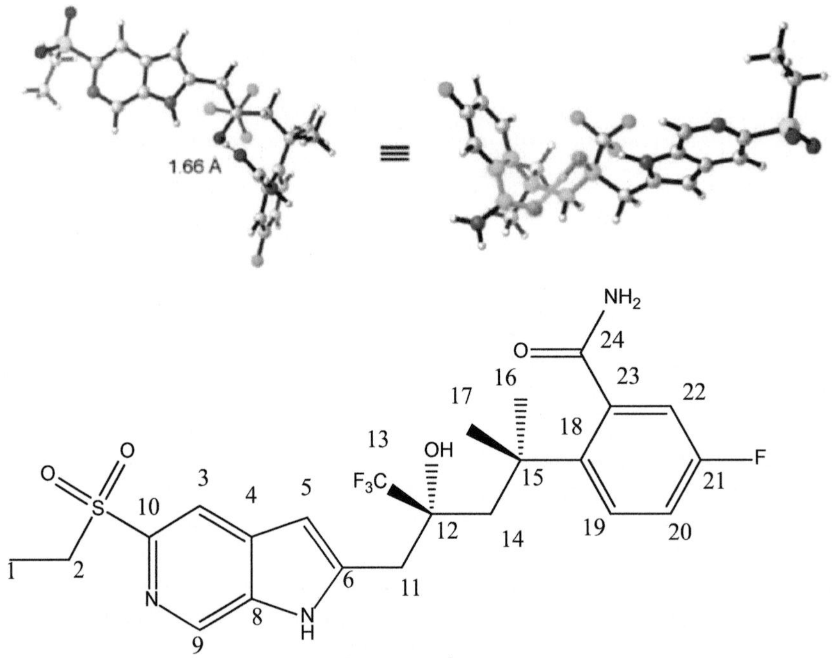

FIGURE 10.1 Top: The lowest energy conformer of compound 1 ((R)-2-(4-((5-(ethylsulfonyl)-1H-pyrrolo[2,3-c]pyridin-2-yl)methyl)-5,5,5-trifluoro-4-hydroxy-2-methylpentan-2-yl)-5-fluorobenzamide) at the B3LYP/cc-pVDZ level of theory with CPCM solvation in dimethyl sulfoxide. Bottom: The numbered chemical structure of Compound 1. (Reprinted (adapted) with permission from Xin et al. [4]. Copyright (2017) American Chemical Society.)

functionals that were selected for testing are listed in Table 10.1 These functionals include B3LYP [26,27], ωB97XD [28], M06-2X [29], WC04 [30], and B3LYP-D3 with Becke–Johnson damping [31]. The Pople-type basis sets include 6-31+G(d,p) [32,33] and 6-311++G(2d,p), the Ahlrichs-type basis set includes def2-TZVP [34], the correlation-consistent cc-pVDZ basis set of Dunning [35], and the polarization-consistent pcS-2 basis set of Jensen [36]. The WC04 functional and the pcS family of basis sets were designed to improve accuracy in computing magnetic shielding constants. However, these methods are not appropriate for calculation of free energies; therefore, single-point GIAO calculations employing these methods were performed using optimized geometries, while frequency corrections were performed at the B3LYP/6-311++G(2d,p) level. The accuracy of each functional/basis set combination was compared with root-mean-square deviation (RMSD) values.

The lowest energy conformer and numbered chemical structure of compound **1** are shown in Figure 10.1. The global minimum conformer shows a stabilizing intramolecular hydrogen bond of 1.66 Å between the hydrogen of the alcohol group on C12 and the carbonyl oxygen on C24, resulting in a pseudo nine-membered ring with a boat–boat conformation.

The performance of the computational methods for the prediction of ^{13}C NMR chemical shifts in compound **1** is summarized in Table 10.1, in which RMSD values are displayed in ppm. A set of 25 conformers was used as input; however, convergence failures or the presence of imaginary frequencies prevented the use of all 25 conformers with some functional basis set combinations. The difference between the best performing method (ωB97XD/cc-pVDZ; RMSD = 3.20 ppm) and the worst performing method (M06-2X/pcS-2//B3LYP/6-311++G(2d,p); RMSD = 24.8 ppm) is more than 20 ppm, which shows that the choice of calculation method is critical to produce acceptable accuracy. Notably, ωB97XD performed better than the other density functionals for sulfone and trifluoromethyl groups. The use of B3LYP-D3 resulted in marginal improvement in the RMSD values with respect to standard B3LYP and no significant qualitative changes in the optimized geometries or difference plots were observed. The study found B3LYP/cc-pVDZ yielded comparable accuracy with other methods examined and was achieved at low computational cost [4].

The reported benchmarking assessment showed that B3LYP/cc-pVDZ is reliable and robust in ^{13}C chemical shift prediction. The method balances accuracy with computational time. An alternate method reported by Toomsalu and Burk showed that the aug-cc-pVDZ basis set produces accurate results for ^{13}C chemical shifts with a preference for use of PBE1PBE method and the continuous set of gauge transformation (CSGT) schemes [37]. The results of this approach were found to be parallel with the cc-pVDZ basis set analysis findings for ^{13}C NMR predictions.

TABLE 10.1

Root Mean Square Deviations (ppm) of ^{13}C Chemical Shift Calculations at Various Levels of Theory

Functional	Basis set				
	cc-pVDZ	6-31+G(d,p)	6-311++G(2d,p)	def2-TZVP	pcS-2[a]
B3LYP	4.06	4.00	8.01	8.83	15.2
B3LYP-D3	3.60	3.86	7.78	8.71	15.2
ωB97XD	3.20	3.28	5.99	–	15.1
M06-2X	14.02	11.36	–	–	24.8
WC04[b]	3.24	4.61	3.41	3.83	6.10

[a,b] Single point GIAO calculations using B3LYP/6-311++G(2d,p) optimized geometries and frequencies.

10.1.2 SYSTEMATIC ERRORS

It is widely reported that NMR chemical shift calculations, even when performed using high level computational methods, can produce average errors. These have been reported to be ≥ 0.4 ppm for ^{1}H chemical shifts, ≥ 10 ppm for ^{13}C chemical shifts, and ≥ 16 ppm $\geq$ for ^{15}N [38–41]. Unfortunately, for interpretation of computed chemical shifts relative to determining or verifying chemical structure, these errors when not understood or identified can be misleading.

Various sources of systematic error have been reported for the observed deviations, which include limitations in theory and solvation effects. A major contribution to the errors encountered in chemical shift calculations may be attributed to heavy-atom effects. The calculation of isotropic shielding constants and chemical shifts for carbon atoms attached to halogens and other atoms of the third row or greater are subject to significant errors, causing the computed shifts to be deshielded relative to experiment [22,23,42–47]. This is also true for chemical shifts for heavy atoms themselves; however, for small organic molecules, described in this chapter, the focus will remain on effects for ^{1}H, ^{13}C, and ^{15}N. The observed errors have been attributed to spin orbital contributions from relativistic effects and electron correlation effects. Common DFT methods do not address compensation for these effects [22,23,42–44,46–49].

For ^{13}C NMR chemical shift prediction, the heavy-atom effect is proportional to the number and size of heavy atoms attached to a carbon, and the deviation from experiment is systematic in nature [43]. These errors are evident in chemical shift calculations regardless of whether a molecular reference, such as tetramethylsilane (TMS), or a linear scaling reference is used. Fortunately, this effect is highly localized and has been reported in the case of ^{13}C to show no significant effect on chemical shifts of neighboring nuclei or adjacent carbon-bound protons [43]. It should be noted that for protons directly attached to heavy atoms, the resulting error can be greater than the entire proton chemical shift scale [23]; hence, such nuclei are typically eliminated from structure elucidation studies and will not be addressed further here. Heavy-atom effects in relevant organic molecules have been observed for carbon atoms attached to silicon, phosphorus, and sulfur, although relativity and electron correlation are significantly observed for halogens, in particular chlorine, bromine, and iodine [42,50]. A common reference for carbon-13 chemical shift prediction is in TMS. The heavy-atom effect for silicon in TMS affects its use as a reference for ^{13}C computed chemical shifts and this effect has been systematically investigated (see Section 10.3.1).

An example of systematic error effects is illustrated in the computational studies performed on compound **1**. Plots of the differences between experimental and computed ^{13}C chemical shifts for compound **1** at the B3LYP/cc-pVDZ level of theory are shown in Figure 10.2. In this example, the computed isotropic shielding constants were converted to chemical shifts using a TMS reference. The bar graph shows that sulfone carbons C2 and C10 are predicted to be more deshielded relative to experimental data due to heavy-atom effects of sulfur. Likewise, computed ^{13}C chemical shifts in CF$_3$ display similar deviations from experimental data at the B3LYP/cc-pVDZ level. The predicted chemical shifts of carbons C2, C10, and C13, along with C15 in the flexible alkyl chain show relatively worse agreement than the rest of the molecule. Also the larger chemical shift deviations of ^{13}C prediction from experiment (δ_{exp}-δ_{calc}) are localized alpha to the sulfone, bonded to three fluorine atoms or in a quaternary environment. As noted previously, these deviations are due to limitations of theory, solvation modeling, or the combination of both.

The data obtained from a set of 69 pharmaceutically relevant compounds were examined with B3LYP/cc-pVDZ, and consistent errors were reported for several functional groups and atom types. Because these deviations are systematic, a correction factor can be employed to compensate for the discrepancies between experimental data and calculated

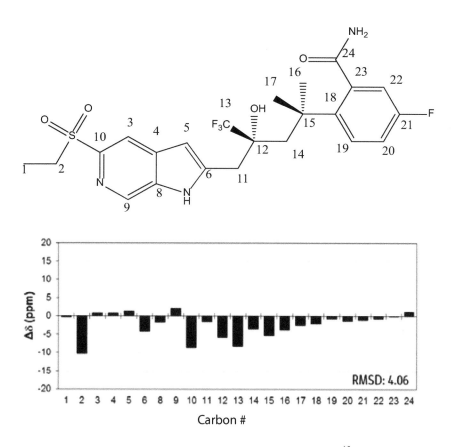

FIGURE 10.2 Plots of the differences between the calculated and experimental ^{13}C NMR chemical shifts of Compound 1. Shielding constants were computed at the B3LYP/cc-pVDZ level of theory and converted to chemical shifts referenced by TMS. RMSD values are given in ppm. Reprinted (adapted) with permission from Xin et al. [4]. Copyright (2017) American Chemical Society.

results. Hence application of systematic error corrections enables a method to be selected that can preserve the balance between computed data and accuracy.

Systematic investigation of halogen-bearing aromatics enabled calculation of correction terms for halogen-bearing carbons. Correction terms for C−F, C−Cl, and C−Br were calculated with B3LYP/cc-pVDZ, while for C−I with B3LYP/lanl2dz. As shown in Figure 10.3, the errors are clustered within a narrow range of values and thus are amenable to application of an appropriate correction term.

Functional groups and atom types that showed consistent deviation from experimental data included aromatic carboxylic acids, carbons alpha to sulfone groups, carbons alpha to sulfur in thienopyrimidines, tertiary carbons alpha to phosphine oxides, CF$_3$, aminothiazoles, and quaternary carbons. Although these errors result from a complex set of interactions (heavy atom, solvation, limitation of theory, and so on), the inherent errors appear systematic and thus may be compensated by a suitable postcomputational addition factor. A summary of correction terms for ^{13}C is listed in Table 10.2.

For compound **1** (Figure 10.1), several functionalities were shown to be systematically deficient in the predicted NMR chemical shifts (sulfones, trifluoromethyl, and quaternary carbon). When experimentally determined systematic error corrections were applied for predictions using B3LYP/cc-pVDZ, the RMSD becomes 2.29 ppm, showing a significant and reproducible

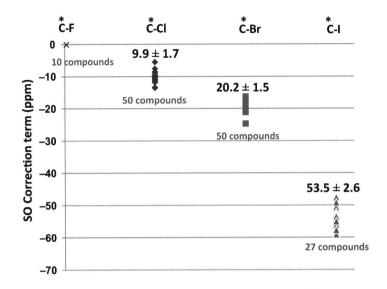

FIGURE 10.3 Graph shows ^{13}C chemical shift deviations (ppm) for halogenated aromatic compounds. Correction terms were derived from the mean calculated differences.

TABLE 10.2
Correction Terms for Selected Functional Groups

Carbon-X Group	Mean Spin-orbit (SO) Correction (ppm)	Number of Compounds	Maximum Deviation Without Correction (ppm)
Cl	+9.9	>50	−13.5
Br	+20.2	>50	−24.7
COOH	−4.0	38	6.8
Sulfone groups	+8.9	25	−16.2
CF$_3$	+6.8	6	−7.5
Quaternary	+8.3	15	−12.2
Aminothiazole*[a]	+9.3	12	−15.0
Thieno[3,2-d]pyrimidine C4[b]	+7.7	10	−8.1
Thieno[3,2-d]pyrimidine C1[b]	+4.9	10	−5.3
t-butylphosphine oxide*[c]	+12.2	8	−12.6

[a] All data collected at the B3LYP/cc-pVDZ level of theory using the TMS reference.
[b]
[c]

improvement and enhancement in the accuracy of the prediction, without incurring additional computational cost. Demonstration of an applied systematic error correction term may be found in the monobromophakellin study [39] (Figure 10.4). Applying +20.2 ppm correction, the deviation from the bromine-bearing carbon at C4 was brought to within 5 ppm of the experimental value.

For ^{15}N NMR chemical shift predictions at the B3LYP/cc-pVDZ level of theory, a systematic error was identified and calculated for primary amines. This deviation showed an average of +16.5 ppm (based on 20 primary amines (Figure 10.5). This may be attributed to solvent effects, in particular hydrogen-bonding interactions with solvent molecules such as DMSO, which are not considered in the calculations [51].

Application of these correction factors places all the predicted chemical shifts obtained at the B3LYP/cc-pVDZ level of theory to be within ±5 ppm of experimental values for ^{13}C,

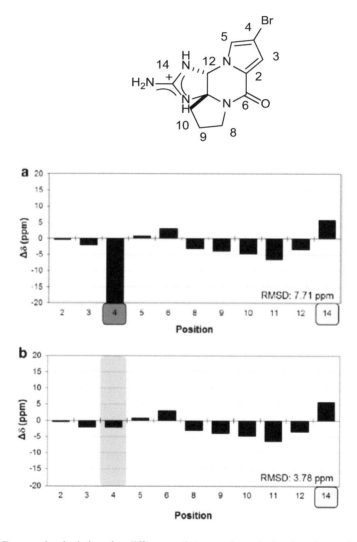

FIGURE 10.4 Bar graphs depicting the differences between the calculated and experimental ^{13}C NMR chemical shifts of monobromophakellin, before (a) and after (b) applying systematic error corrections for Br (+20.2 ppm). Shielding constants were computed at the B3LYP/cc-pVDZ level of theory and converted to chemical shifts referenced to TMS. RMSD values are given in ppm.

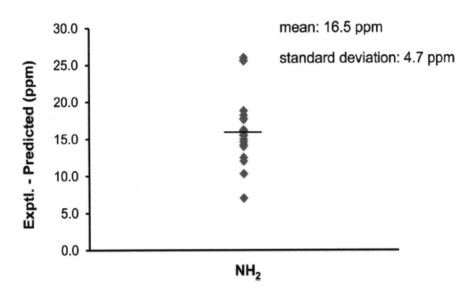

FIGURE 10.5   Graph shows ^{15}N chemical shift deviations (ppm) for primary amines. Correction term was derived from the mean calculated difference. (Xin et al. [55]) (Reproduced with permission of The Royal Society of Chemistry.)

which had been statistically determined to be the ppm range for structural confidence. For ^{15}N, the correction factor puts the chemical shifts within a ± 10 ppm distribution from experiment, which statically favors the correct structure (see Section 10.2).

10.2   STATISTICAL DISTRIBUTION OF CHEMICAL SHIFT DIFFERENCES (^{1}H, ^{13}C, AND ^{15}N NMR)

Comparison was made between TMS and the Linear Regression method [4] relative to the chemical shift prediction accuracy. Figure 10.6 shows a pair of histograms that illustrates the distribution of differences between the experimental chemical shifts and calculated chemical shifts referenced to TMS and those referenced with linear scaling; determined for a set of 51 organic molecules (766 chemical shifts). Because systematic errors introduced by third row and higher elements in the periodic table can adversely affect linear correlations derived mainly from, and intended for, compounds lacking these heavier elements, known systematic errors were excluded from the histograms in Figure 10.6. Specifically, 9 chemical shifts of halogen-substituted carbons, 17 chemical shifts of carbons bonded to sulfones, 10 chemical shifts of carbons bonded to sulfur in aminothiazole heterocycles, and 3 chemical shifts of carbons bonded to phosphorus were excluded from the analysis. When TMS is used as a reference (Figure 10.6, top), the distribution is approximately normal with a bias toward negative differences.

For many chemical shift predictions using TMS as a reference, B3LYP/cc-pVDZ overestimates the chemical shift with respect to experiment by a mean of -1.5 ppm, which provided confirmation of a systematic deviation at this level of theory. For ease of analysis, the ranges of chemical shift differences were separated into zones based on standard deviations from the mean of a Gaussian distribution. With a standard deviation of 2.7 ppm, the TMS reference data shows that 95% of all calculated chemical shifts are expected to be within the

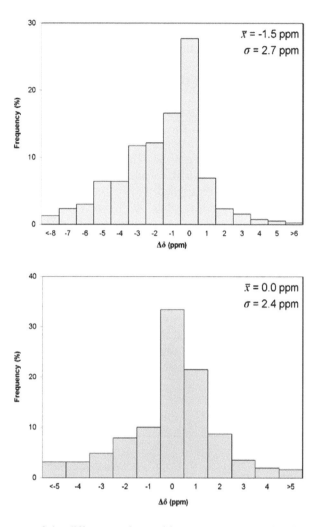

FIGURE 10.6 Histogram of the differences observed between experimental and calculated ^{13}C chemical shifts for a set of 51 organic molecules at the B3LYP/cc-pVDZ level of theory in DMSO. Top graph shows TMS reference ($\delta_{expt} - \delta_{calc}$); bottom graph shows Linear Scaled reference (δ = (intercept − isotropic magnetic shielding)/slope). (Reprinted (adapted) with permission from Xin et al. [4]. Copyright (2017) American Chemical Society.)

range of −6.9 to +3.9 ppm of the correct assignment. About 5% of computed chemical shifts differ from the corresponding experimental shift in the ranges of −9.6 to −6.9 or +3.9 to +6.6. Less than 0.3% of computed chemical shifts will result in a difference of less than −9.6 ppm or greater than 6.6 ppm. The statistical analysis gives a foundational basis for an acceptable fit to experimental data for the B3LYP/cc-pVDZ method.

The distribution of errors after linear scaling (Figure 10.6, bottom) is Gaussian with a mean of 0.0 ppm and a standard deviation of 2.4 ppm. The data show that 95% of differences between experimental and computed chemical shifts are expected to be within ±4.8 ppm from the average value. This deviation from experiment plays a prominent role in establishing confidence in the differences between experiment and prediction, especially when distinction among multiple isomers is required.

For ^{15}N NMR studies, comparison between calculated chemical shifts from linear scaling conversion with experimental values showed that the deviations are <10 ppm. Figure 10.7 shows the histogram that illustrates the distribution of calculated errors ($\delta_{exp} - \delta_{calc}$).

Two models, Student-t distribution and normal distribution, were employed to fit the experimental distribution. The results of the fitting with the Stats module of Scipy20 show μ (mean) = 0 ppm (due to linear scaling), σ (standard deviation) = 4.77 ppm, v (degrees of freedom) = 647.2 for Student-t distribution and μ = 0 ppm, σ = 4.78 ppm for normal distribution. The high degrees of freedom in the Student-t distribution fitting support a normal distribution to describe the calculated error distribution. Therefore, a normal distribution with μ = 0 ppm, σ = 4.78 ppm may be utilized. This work provided a quantitative statistical analysis of error distributions for ^{15}N NMR calculation in the literature. The results indicate that when the linear scaling terms are applied, 95% of calculated chemical shifts are within 9.56 ppm (2σ) of the experimental values while 99.7% are within 14.34 ppm (3σ). As previously noted, for NH$_2$ there is a systematic error, which may be addressed through application of a −16.5 ppm correction that consistently places the predicted nitrogen chemical shift well within the 2σ range of experimental values. Compared with ^{13}C NMR calculations, the σ value is greater than the average results (σ = 2–3 ppm) for ^{13}C NMR [8,52,53]. However, given the fact that the chemical shift range for ^{15}N NMR is substantially larger than ^{13}C NMR, the relative calculation accuracy for ^{15}N is comparable to ^{13}C. Moreover, compared with ^{13}C NMR, the chemical shift of ^{15}N varies much more drastically in different chemical environments. The shift of ^{15}N peaks caused by subtle changes in chemical structure is usually more distinguishable and, therefore, more advantageous in determining the correct structure. Thus, ^{15}N NMR calculation using a linear reference is complementary to ^{13}C NMR prediction

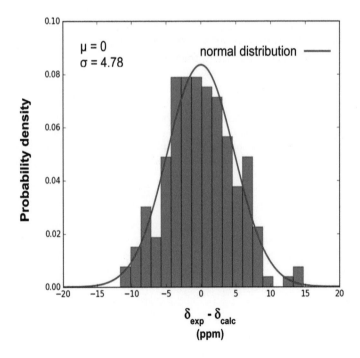

FIGURE 10.7 Normal distribution of ^{15}N calculation errors without inclusion of NH$_2$ chemical shifts. The mean is μ and the standard deviation σ = 4.78 ppm. ((Xin et al. [55]). Reproduced with permission of The Royal Society of Chemistry.)

and can sometimes provide an alternative to ^{13}C NMR prediction in structural elucidation of organic molecules.

10.3   NMR REFERENCING

10.3.1   TMS

In ^{1}H and ^{13}C NMR spectroscopy, the accepted experimental reference is TMS. All experimental chemical shifts are referenced to TMS at 0 ppm by convention. Calculated chemical shifts also require use of a reference term. For experimental spectra obtained at room temperature in a designated solvent, the computed shielding constants may be generated using a Boltzmann–weighted average of the Gibbs free energies of approximately 30 conformers (where energy duplicates are removed). To this end, calculated shielding constants for TMS in the same designated solvent may be used to convert analyte shielding constants for ^{1}H and ^{13}C to chemical shifts.

Systematic studies using B3LYP/cc–pVDZ/GIAO//B3LYP/cc–pVDZ level of theory have been reported with implicit solvation conductor-like polarizable continuum model (CPCM), where reference molecules were calculated at the same level of theory. With TMS, only one conformation is possible, hence shielding constants in a variety of solvents can be rapidly produced. The ^{13}C reference correction terms for TMS were calculated in four different solvents [4]. The results were as follows:

a) For DMSO-d_6, the isotropic shielding constants were corrected by subtracting the shielding constants of the analyte from 192.6 ppm.
b) For CDCl$_3$, the isotropic shielding constants were corrected by subtracting the shielding constants of the analyte from 192.5 ppm.
c) For acetonitrile-d_3, the isotropic shielding constants were corrected by subtracting the shielding constants of the analyte from 192.6 ppm.
d) For methanol-d_3, the isotropic shielding constants were corrected by subtracting the shielding constants of the analyte from 192.7 ppm.

The ^{1}H reference correction terms for TMS were calculated in the same four solvents [54]. The results were as follows:

a) For DMSO-d_6, the isotropic shielding constants were corrected by subtracting the shielding constants of the analyte from 31.30 ppm.
b) For CDCl$_3$, the isotropic shielding constants were corrected by subtracting the shielding constants of the analyte from 31.30 ppm.
c) For acetonitrile-d_3, the isotropic shielding constants were corrected by subtracting the shielding constants of the analyte from 31.30 ppm.
d) For methanol-d_3, the isotropic shielding constants were corrected by subtracting the shielding constants of the analyte from 31.29 ppm.

These values may be used as reference terms for ^{13}C and ^{1}H nuclei when calculations of the analyte are performed with the B3LYP-cc-pVDZ method.

10.3.2   NITROMETHANE/LNH$_3$

When referencing ^{15}N data, two commonly used references may be considered. They include nitromethane or liquid ammonia. Extensive systematic investigation of ^{15}N at the B3LYP/ ccpVDZ level of theory was carried out in DMSO solvent alone. The choice of DMSO has been mainly due to the fact that poor nitrogen sensitivity (0.39% natural abundance) necessitates the use of high concentrations. DMSO effectively solubilizes most organic compounds, making it the solvent of choice even when indirect detection NMR experiments are used in collecting data.

The ^{15}N reference correction terms for nitromethane and liquid ammonia were developed using generalized linear scaling. The scaling terms were developed using 182 nitrogen atom chemical shifts derived from 63 diverse nitrogen types [55].

10.3.3   LINEAR SCALING

Linear scaling may be carried out using calculations of shielding constants from within a single molecule as well as with shielding constants from a large sampling of diverse atom types for a particular nucleus. For linear scaling within the molecule, chemical shift prediction sometimes shows a better match with the experimental data yielding a false high probability for the incorrect structure.

A comparison of linear scaling within the molecule, with generalized linear scaling, showed how linear scaling within the molecule significantly reduced the MAE (mean absolute error) for the incorrect structure while slightly increasing the MAE for the correct structures. This is due to the fact that the slopes and intercepts are different for differing isomers, which can lead to an incorrect prediction of the wrong isomer. Hence real errors caused by the incorrectness of the structure might be suppressed or even completely concealed by linear scaling within the molecule [4].

The Gaussian distribution of errors after linear scaling (Figure 10.6 bottom) by necessity has a mean of 0.0 ppm. As noted previously, the reported data for generalized linear scaling showed that 95% of differences between experimental and computed chemical shifts at the B3LYP/cc-pVDZ level of theory are within ±4.8 ppm for ^{13}C. This value has a prominent role in establishing the reliability of the ^{13}C chemical shift prediction method. A linear fit for calculated ^{13}C isotropic shielding constants plotted against corresponding experimental chemical shifts was carried out using a set of 51 small-to-medium-sized organic molecules. A tight correlation was evident by the R^2 value of 0.9979 that indicates almost no random error associated with B3LYP/cc-pVDZ in DMSO. This tight correlation is a direct result of excluding carbon atoms that introduce systematic errors. Deviation of the slope from unity is a prime indicator of systematic error in the method. Typically this slope should be within the range of 0.95−1.05 for a reliable method [6,11].

The linear scaling ^{13}C reference correction terms for B3LYP/cc-pVDZ were calculated for four solvents [4]. The isotropic chemical shifts were calculated by subtracting the intercept from the shielding constant and dividing by the slope. The (slope/intercept) values obtained in four different solvents are as follows:

a) For DMSO-d$_6$, the isotropic shielding constants were corrected by (188.57 − x/0.9759).
b) For CDCl$_3$, the isotropic shielding constants were corrected by (189.29 − x/0.9710).
c) For acetonitrile-d$_3$, the isotropic shielding constants were corrected by (189.66 − x/0.9705).
d) For methanol-d$_3$, the isotropic shielding constants were corrected by (190.59 − x/0.9766).

where x = computed shielding constant for Carbon.

The linear scaling ^{1}H reference correction terms for B3LYP/cc-pVDZ were also calculated for four solvents [54]. The (slope/intercept) values obtained for ^{1}H in four different solvents are as follows:

a) For DMSO-d_6, the isotropic shielding constants were corrected by $(31.31 - x/0.9726)$.

b) For CDCl$_3$, the isotropic shielding constants were corrected by $(31.498 - x/1.0089)$.

c) For acetonitrile-d_3, the isotropic shielding constants were corrected by $(31.36 - x/0.9872)$.

d) For methanol-d_3, the isotropic shielding constants were corrected by $(31.41 - x/0.9923)$.

where x = computed shielding constant for proton.

Either TMS or linear regression may be used as chemical shift references with comparable overall conclusions. Since linear scaling consistently performs better than TMS, linear scaling is often selected when such correction terms are available.

For ^{15}N NMR referencing, generalized linear scaling has been shown to enhance the accuracy of chemical shift calculations [55]. While an alternative approach sometimes used with ^{13}C NMR chemical shift calculations is to employ empirical linear scaling within each molecule, often this approach is not suitable for ^{15}N referencing due to the limited number of nitrogen atoms in each molecule.

Linear scaling analysis was reported for ^{15}N NMR datasets, which showed that the inclusion of NH$_2$ chemical shifts led to inferior linear correlation in the analysis. This was due to the systematic error associated with primary amines. A linear scaling analysis without NH$_2$ resonances was performed (Figure 10.8) at the B3LYp/ccpVDZ level of theory yielding an excellent linear correlation ($R^2 = 0.9971$).

The results show that, with the exception of NH$_2$, ^{15}N NMR calculation using generalized linear scaling at the B3LYP/cc-pVDZ level of theory performs well for various types of nitrogen atoms with little random error. The data in numerous reported case studies indicated that the linear scaling terms are accurate and widely applicable in converting the calculated isotropic shielding constants to ^{15}N chemical shifts. The expanded application of generalized linear regression referencing may be attributed to the diversity of nitrogen types in the dataset [55]. Conversion based on Equations 10.1 and 10.2 relative to nitromethane and liquid

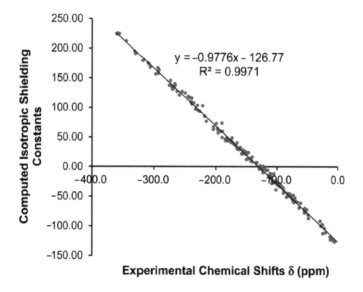

FIGURE 10.8 Linear correlation between experimental chemical shifts and calculated shielding constants for 182 ^{15}N chemical shifts. (Xin et al. [55]). (Reproduced with permission of The Royal Society of Chemistry.)

ammonia, respectively, yields calculated chemical shifts with high accuracy for all nitrogen atoms excluding primary amines.

$$\delta = -(\sigma + 126.77)/0.9776 \tag{10.1}$$

$$\delta = 380.23 - (\sigma + 126.77)/0.9776 \tag{10.2}$$

10.4 AUTOMATED PROGRAM DEVELOPMENT

10.4.1 H*i*PAS HOLISTIC *IN-SILICO* PREDICTION APPLICATION SOFTWARE

The H*i*PAS method is a process used to carry out NMR chemical shift prediction [14]. The process involving chemical shift prediction, which is tedious and highly labor intensive, has been fully automated, making this a practical procedure. The method begins with the proposal of possible chemical structures. These structures may be regioisomers, or compounds with various oxidation states, protonation states, and in some cases stereochemistry. The next step involves carrying out a gas phase conformational search. The search is typically carried out using LowModeMD or systematic conformational search methods [56]. Each conformer is then subjected to geometry optimization and frequency calculation. Although several software packages have been developed for quantum mechanical electronic structure calculations (e.g., Gaussian, GAMESS, Jaguar, and Spartan), Gaussian was selected because of its wide use as a software package for NMR computation. NMR shielding constants were then calculated for every relevant atom in every conformer of every structure variation. These calculations were carried out at the B3LYP/cc-pVDZ level of theory. DFT GIAOs (Gauge-Independent Atomic Orbitals) are used in the calculation of shielding constants followed by calculation of a Boltzmann average weighted by energy. The Boltzmann averaged shielding constants may then be automatically converted to chemical shifts using an applied reference term. Calculated chemical shifts are subtracted from experimental chemical shifts and the differences are displayed in a bar graph where the zero line represents the experimental chemical shift and the bar represents deviation from experiment, so the smaller the bar the better the agreement between experiment and prediction. For example, with ^{13}C NMR data, if the deviations in the bar graph fall between ±5ppm, there is a high probability that the structure is correct. A schematic diagram of the H*i*PAS process is presented in Figure 10.9.

10.4.2 DP4, DP4+, AND D*i*CE DIASTEROMERIC *IN-SILICO* CHIRAL ELUCIDATION

Determining stereochemistry of diastereomers by NMR spectroscopy represents a significant challenge in organic chemistry. Stereochemistry is an integral part of a compound's physical and biological properties. In the pharmaceutical industry, stereochemistry can be a critical component in defining a molecule's activity. Examples include compounds such as Zoloft, an antidepressant for major depressive disorders, obsessive-compulsive disorder, panic disorder, and social anxiety disorder. This drug has two chiral centers, yet only the (S,S)-diastereomer has significant activity and is marketed [57]. Penicillin, an antibiotic originally isolated from mold, is used in the treatment of pathogenic bacterial infections such as tetanus, typhoid fever, diphtheria, syphilis, and leprosy. The compound has three stereo centers yielding eight possible stereoisomers, however, only the (2S,5R,6R)-diastereomer has significant antibacterial activity. Another antibiotic, chloramphenicol, has been used in the treatment of meningitis, cholera, and typhoid fever. For this compound, only the (R,R)-diastereomer is found to be active [58–60]. Structures of Zoloft, penicillin, and chloramphenicol are shown in Figure 10.10.

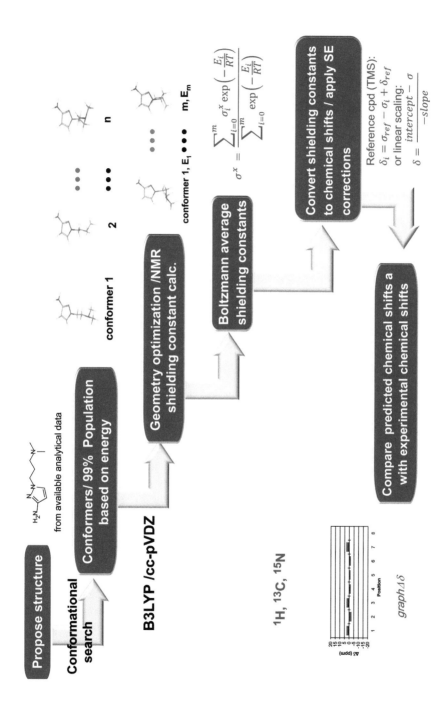

FIGURE 10.9 Schematic representation of the H*i*PAS chemical shift prediction process.

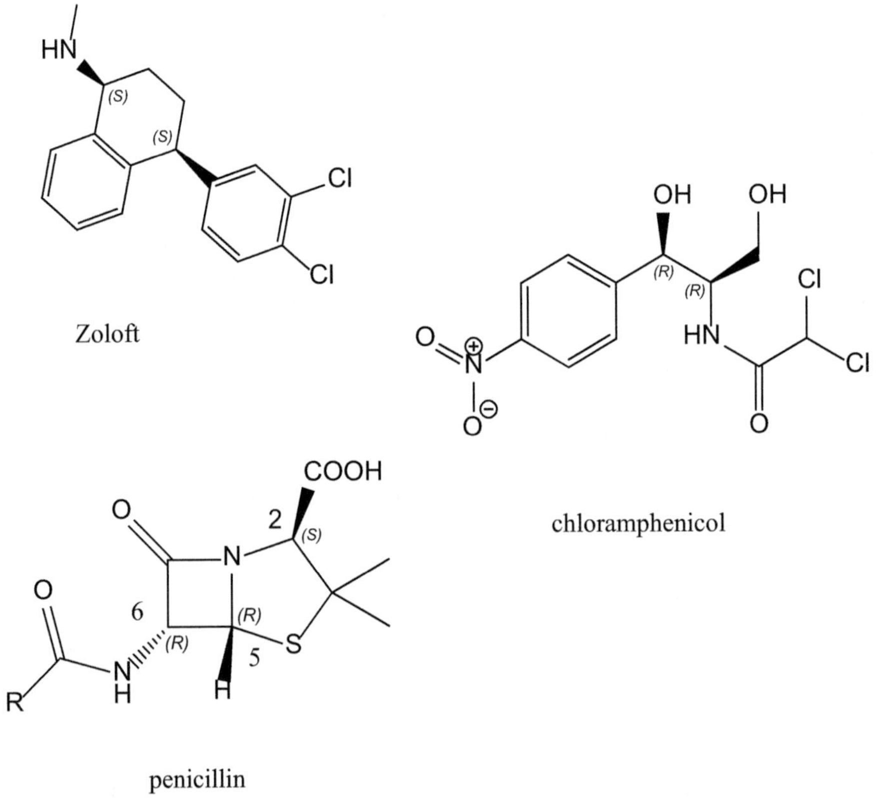

FIGURE 10.10 Chemical structures of Zoloft, penicillin, and chloramphenicol with the active stereochemistry are shown.

Although single-crystal X-ray can unequivocally establish the stereochemistry of diaster-eomers, the technology requires sufficient quantity, purity, and solubility of the material as well as the time and resources to grow a diffraction-quality crystal. NMR spectroscopy has several advantages over single-crystal X-ray crystallography. First, generating a diffraction-quality crystal is not needed. In addition, NMR technology has the ability to analyze compound mixtures as well as samples where only microgram quantities of material are available. For liquid chromatography-solid-phase extraction (LC–SPE) NMR studies, often only microgram quantities of material are available or samples may be the mixtures of com-pound of interest with minor impurities.

NMR spectroscopy has been able to solve the stereochemistry of diastereomers of rigid sys-tems with well-defined conformational orientation but faces severe challenges with compounds having high flexibility and rotatable bonds. Chemical synthesis in combination with NMR spec-troscopy has been used to solve stereochemistry of diastereomeric structures; however, a significant amount of time and resources needs to be employed in synthesizing multiple stereoisomers. This approach is not practical when a compound has multiple chiral centers. The nuclear Overhauser effect (NOE), which is dependent on the through-space distance (<4 Å for small molecules) separating the cross-relaxing nuclei, has been used to determine the relative and, in some cases, the absolute stereochemistry for rigid diastereotopic chemical scaffolds, but this approach is not definitive for highly flexible systems.

Synthetic efforts, where highly flexible organic molecules are converted to rigid structures, have been used with NOE studies to provide chiral elucidation of diastereomers; however, such chemical modifications may be challenging as well as requiring synthetic resources [61]. Another approach in stereochemistry elucidation of flexible molecules involves use of NMR methods to measure long-range heteronuclear coupling constants. While these studies have been used to assign relative stereochemistry, additional comparisons with NOE data, degradation products, and synthetic controls have been required [62]. Overall, significant challenges are associated with the use of NMR spectroscopy in the elucidation of the stereochemistry of diastereomers. For more detailed discussion of the NOE effect in structure elucidation, the reader is referred to discussion in Chapters 2.6 and 6.3 and references therein.

To address the challenge of determining stereochemistry of diastereomers when only one set of chemical shifts is available, Smith and Goodman [52] reported a method that invokes the use of probabilities from differences between GIAO NMR calculations and NMR experimental data. The probability analysis was developed using ^{13}C or ^{1}H chemical shifts. This process has resulted in the ability to make stereochemical assignments of diastereomers with high confidence.

Smith and Goodman used NMR shifts calculated with computationally inexpensive single-point calculations on molecular mechanics geometries without time-consuming *ab initio* geometry optimization.

The program, named DP4, is based on Bayesian probability theory. A percentage probability $P_{N\%}(i)$ of each candidate structure, i being the correct structure is based on NMR chemical shift prediction of nucleus N, and can be further calculated with Equation 10.3. In Equation 10.3, F is the cumulative distribution function of either a normal distribution with mean μ and standard deviation σ or a t distribution with an additional degrees of freedom parameter v.

$$P_{N\%}(i|\delta_1, \delta_2, \ldots, \delta_n) = \frac{\prod_{k=1}^{n}\left(1 - F^v\left(\frac{\left|(\delta^i_{calc,k} - \delta^i_{exp,k})\right| - \mu}{\sigma}\right)\right)}{\sum_{j=1}^{m}\left[\prod_{k=1}^{n}\left(1 - F^v\left(\frac{\left|(\delta^i_{calc,k} - \delta^i_{exp,k})\right| - \mu}{\sigma}\right)\right)\right]} \times 100\% \tag{10.3}$$

The probability measure was shown to enable assignment of stereochemistry or structure of 21 natural products that had been misassigned in the literature or that required extensive synthesis of diastereomers to establish their stereochemistry.

Two other approaches to enhance the DP4 method have been reported. The first approach that improves the performance of the DP4 method is named DP4+. The main differences of DP4+ from the original DP4 method are the inclusion of unscaled data (combined with scaled probabilities) and the use of higher levels of theory in the NMR calculation procedure. In addition, combined ^{1}H and ^{13}C probabilities are employed to achieve a higher level of accuracy.

To determine whether one nucleus is more discriminating than the other (since proton was suggested to allow a better differentiation among stereoisomers) [63,64], studies using only ^{1}H or ^{13}C data alone were found to significantly reduce the performance of DP4+ even when using scaled or unscaled DP4+ probabilities. For instance, with the exclusive use of scaled ^{1}H NMR shifts or ^{13}C NMR shifts, (26%) and (23%), respectively, produced predictions where the structures were incorrectly assigned. Using both ^{1}H and ^{13}C data significantly improved the overall performance of scaled DP4+, reducing to 6% the number of bad examples. Similar behavior was observed with unscaled chemical shifts. Overall the performance of the method significantly improved upon the inclusion of ^{1}H and ^{13}C data that are scaled and unscaled with only 6% showing modest confidence.

A second probability theory process called D*i*CE (Diastereomeric *in-silico* Chiral Elucidation) was developed to be compatible with the previously reported chemical shift prediction approach (H*i*PAS) performed at the B3LYP/cc-pVDZ level of theory.

While there are several probability methods for calculation of stereochemistry of diastereomers, this section will focus on applications of D*i*CE. These calculations used geometry optimization with a CPCM solvent model for each conformer. Generalized linear scaling terms obtained in four different solvents contributed to the improved accuracy in chemical shift prediction for ^{1}H and ^{13}C nuclei. This process was also extended to ^{15}N in DMSO. Statistical distribution was evaluated for individual solvents and generalized for combined solvents. Hence the probability theory algorithm was developed for individual solvents as well as combined solvent data. A combined total of 1,745, 1,133, and 182 chemical shift differences, for carbon (^{13}C), proton (^{1}H), and nitrogen (^{15}N), respectively, was used in the statistical distribution calculations. Besides stereochemistry of diastereomers, this approach was also found to be powerful in identifying regiochemistry for cases where chemical shift prediction yields ambiguous or inconclusive results.

The D*i*CE application was further improved with the use of combined D*i*CE. This approach enables the combination of multiple probabilities from different nuclei. Combined D*i*CE uses the product of probabilities of multiple nuclei for an individual isomer divided by the sum of the products of probabilities for each of the possible isomers. An illustration of the equation for two possible isomers and probabilities for three nuclei (^{1}H, ^{13}C, and ^{15}N) is given in Equation 10.4

$$P_{1C\%} \times P_{1H\%} \times P_{1N\%} = \prod p_{1CHN\%} \quad P_{2C\%} \times P_{2H\%} \times P_{2N\%} = \prod p_{2CHN\%}$$

$$[\prod p_{1CHN\%} \div (\prod p_{1CHN\%} + \prod p_{2CHN\%})] \times 100\% = P_{\text{combined probability for isomer 1}} \quad (10.4)$$

where

$P_{1C\%}$ = percentage probability of ^{13}C for isomer 1
$P_{1H\%}$ = percentage probability of ^{1}H for isomer 1
$P_{1N\%}$ = percentage probability of ^{15}N for isomer 1
$P_{2C\%}$ = percentage probability of ^{13}C for isomer 2
$P_{2H\%}$ = percentage probability of ^{1}H for isomer 2
$P_{2N\%}$ = percentage probability of ^{15}N for isomer 2

The probabilities for the correct and incorrect diastereomers are comparable to signal and noise. Ideally, noise, if completely random, will be canceled out by averaging, whereas signal would become more prominent. Use of combined D*i*CE was reported for 40 challenging diastereomers that failed with DP4. The study resulted in an 87% success rate for the 40 very difficult cases [54].

10.5 APPLICATIONS

10.5.1 REGIOISOMERS

The H*i*PAS application has been demonstrated to be powerful for resolving the structures of regioisomers. A particularly difficult case where experimental data fall short was the nevirapine hydrolysis product (NHP) [4]. In this case, there were three possible sites where it was believed hydrolysis could occur leading to three possible products [Figure 10.11].

The dilemma was that experimental data were inconclusive. NMR data supported structure C (Figure 10.11) due to a COSY cross peak between an NH exchangeable proton and a proton on the cyclopropyl ring. Mass spectrometry data, however, showed loss of +44 consistent with the presence of a carboxylic acid supporting structure A. A fourth possibility, structure D, was subsequently proposed based on cleavage of the amide bond and a subsequent Smiles rearrangement [Figure 10.12], however, experimental data alone was not conclusive.

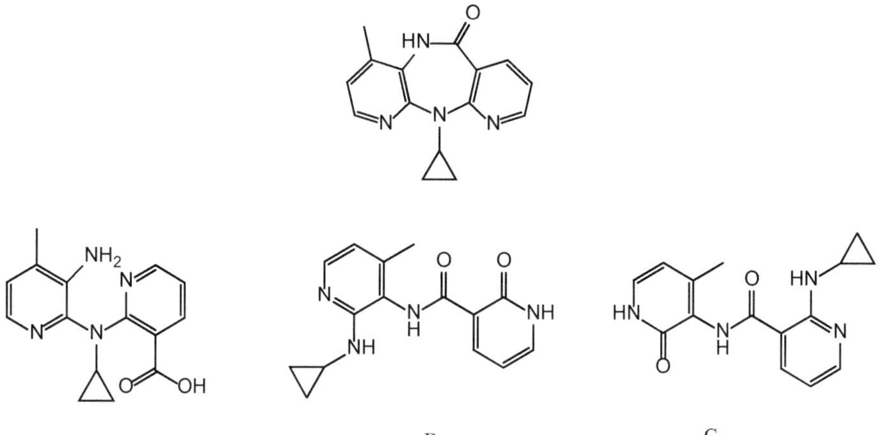

FIGURE 10.11 Chemical structures of nevirapine (top) and three proposed hydrolysis products (bottom).

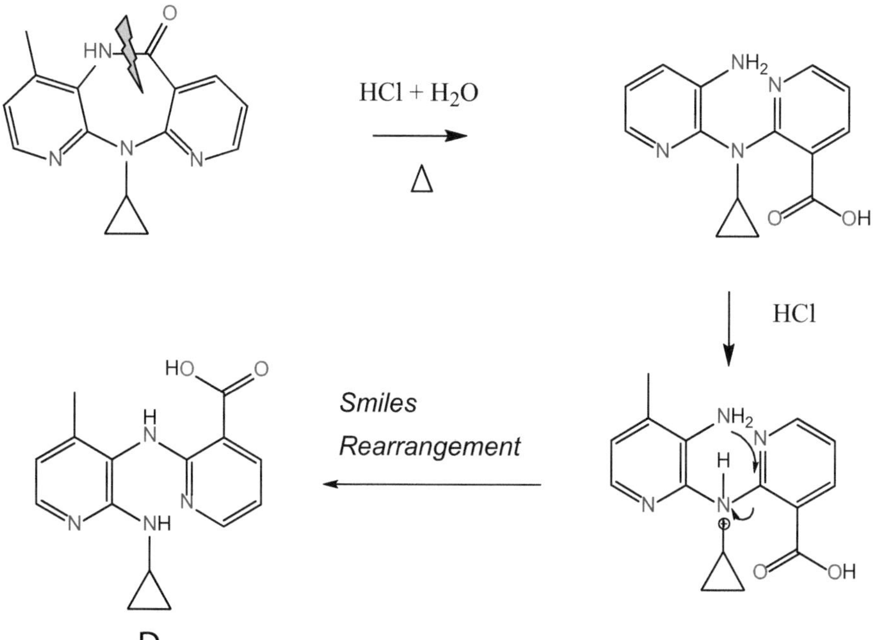

FIGURE 10.12 Mechanism showing hydrolysis of nevirapine followed by a Smiles rearrangement to yield structure D.

Application of ^{13}C H*i*PAS to the four possible regioisomers resulted in confirmation of structure D [Figure 10.13]. The bar graph representation shows deviation of the predicted chemical shifts from experimental chemical shifts. When the deviations are within 2σ of the mean statistical distribution, the structure has a high probability of being correct as shown for structure D.

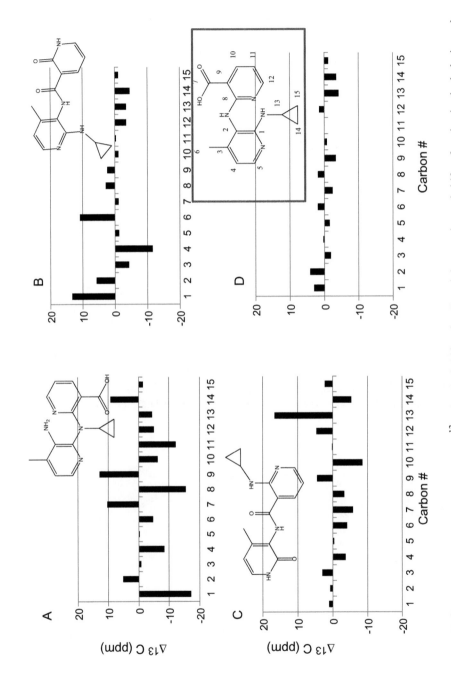

FIGURE 10.13 Plots of the differences between the computed ^{13}C chemical shifts of A–D and the experimental shifts of nevirapine hydrolysis product. Computed at the B3LYP/cc-pVDZ level of theory with CPCM solvation in dimethyl sulfoxide with linear scaling reference.

10.5.2 TAUTOMERS

The H*i*PAS program was also found to be highly sensitive to the existence of preferred tautomers in solution. This may be illustrated in the example of an isomer of NHP. Compound C was synthesized and characterized by NMR spectroscopy in dimethyl sulfoxide-d_6. NMR data showed a preference in the tautomeric equilibrium between the pyridinone and pyridinol. Figure 10.14 shows the two possible tautomers.

Comparison of the computed chemical shifts of pyridinone and pyridinol tautomer with the available experimental shifts showed a significant difference in C1−C5 of the ring, favoring the pyridinone (Figure 10.14). Hence NMR chemical shift computations at the B3LYP/cc-pVDZ level of theory were consistent with the stable preferred tautomer. The same preferred tautomeric result was obtained with either TMS or a linear scaling reference. The results showing tautomer comparison using a linear scaling reference is shown in Figure 10.15.

This result was further corroborated by computations at the B3LYP/ccpVDZ level of theory predicting that the pyridinone is more stable than the pyridinol tautomer by 5.8 kcal/mol [4].

10.5.3 NATURAL PRODUCTS

Pyrrole-imidazole alkaloids (PIAs) are a family of natural products with high nitrogen content that are isolated from sponge [65]. The phakellins are substructures within the PIA family and those moieties can be found in more complex larger alkaloids such as dimeric PIAs [66]. NMR characterization of a PIA natural product monobromophakellin was carried out and the structure was published as neutral compound **a** in Figure 10.16, with full carbon, proton, and nitrogen chemical shifts reported [67]. The ^{13}C NMR prediction calculations for monobromophakellin, however, showed significant differences between experimental and computed ^{13}C chemical shifts for compound **a** while the prediction results favored protonated species **b** (Figure 10.16).

Bar graphs generated using a TMS reference for compounds **a** and **b** are shown in Figure 10.17a–c. The graph shows deviations are within three standard deviations from the mean for

ΔG = 0.0   ΔG = 5.8

Tautomeric equilibrium of NHP isomer. Relative Gibbs free energies are in kcal/mol.

FIGURE 10.14 Chemical structures of tautomeric equilibrium for an isomer of the nevirapine hydrolysis product.

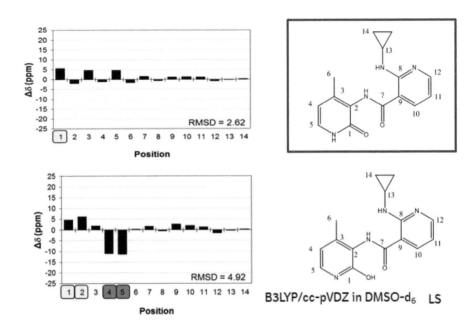

FIGURE 10.15 Bar graphs of the differences between the computed ^{13}C chemical shifts of pyridinone and pyridinol and the experimental shifts of the NHP isomer. Computed at the B3LYP/cc-pVDZ level of theory with CPCM solvation in dimethyl sulfoxide with linear scaling reference. (Reprinted (adapted) with permission from Xin et al. [4]. Copyright (2017) American Chemical Society).

monobromephakellin

a

protonated monobromephakellin

b

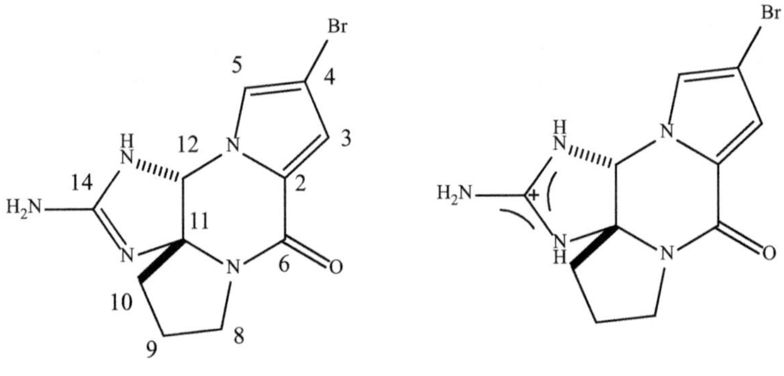

FIGURE 10.16 Numbered chemical structure of (a) monobromephakellin and (b) protonated monobromephakellin.

compound **b** (Figure 10.17 b), while a significant deviation of −12.2 ppm was observed at C11 for compound **a** (Figure 10.17 c).

The data, plotted in Figure 10.17 a–c, also show application of the systematic error correction term (−20.2 ppm) for Br compensated for the large deviation between prediction and experiment

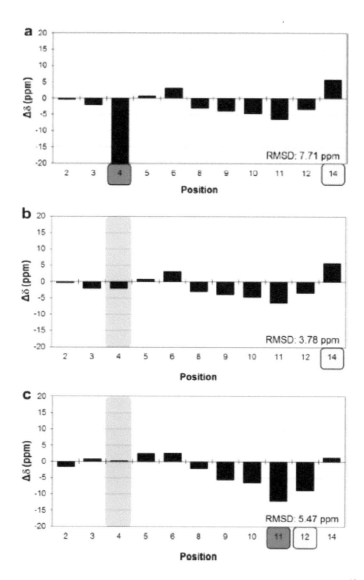

FIGURE 10.17 Bar graphs of the differences between the calculated and experimental ^{13}C NMR chemical shifts of protonated monobromephakellin before (a) and after (b) applying systematic error corrections for Br (+20.2 ppm) and monobromephakellin after applying Br corrections (c). Shielding constants were computed at the B3LYP/cc-pVDZ level of theory and converted to chemical shifts referenced to TMS. RMSD values are given in ppm. (Reprinted (adapted) with permission from Xin et al. [4]. Copyright (2017) American Chemical Society).

for compound **b** (Figure 10.17 b) as well as compound **a** (Figure 10.17c) at carbon 4 in structures **a** and **b** (Figure 10.16). A red box over a position number in graphs a and c of Figure 10.17 serves as a literal "red flag" by indicating a difference that is greater than three standard deviations away from the mean of −1.5 ppm, an observation that has a low probability of 0.3%. A yellow box indicates that the observed difference falls between two and three standard deviations away from the mean, which has a probability of ~5%. These calculations support a protonated monobromophakellin (Figure 10.17b) which shows better agreement with experimental data than neutral monobromophakellin (Figure 10.17 c).

These results and conclusions were also consistent with a ^{15}N chemical shift prediction study, where the protonated structure of monobromophakellin was found to agree with experimental data [55]. The experimental ^{15}N chemical shifts of monobromophakellin in DMSO were compared with ^{15}N NMR calculated chemical shifts to explore the ability to identify protonation states as distinguished from alternative tautomers of monobromophakellin. The results in Figure 10.18 indicate the protonated structure **d** offers the best match between ^{15}N experimental data and calculated chemical shifts. The data for structure **d** shows all the deviations are within 9.56 ppm (2σ) after applying the systematic error correction of −16.5 ppm for NH$_2$. All of the tautomers of the free base a–c showed significant deviations (>20 ppm) for at least one nitrogen atom. In each of the incorrect structures the biggest deviation occurs on the nitrogen with the wrong number of attached hydrogen atoms.

Therefore both ^{13}C and ^{15}N calculations were shown to support a protonated structure of monobromophakellin.

10.5.4 ASSIGNMENT AMBIGUITIES

The ^{13}C H*i*PAS calculations have a distinct advantage of resolving assignment ambiguities. This can be particularly problematic in cases where there are numerous impurities, severe overlap of cross peaks in HMBC spectra or insufficient signal for unequivocal assignments. An example of a case where chemical shift assignments were resolved through chemical shift prediction is shown in Figure 10.19. In this case, the sample in question had numerous impurities, which complicated the assignment. In addition, the through-bond HMBC correlations for each of the carbons in question (C2 and C5) were inconclusive. Chemical shift predictions were carried out, which showed huge deviations between prediction and experiment for C2 and C5. When the experimental assignments were switched, excellent agreement between the experimental and predicted values was observed. The clarified carbon assignments became evident upon more careful examination of the compromised spectra [68].

10.5.5 DECOMPOSITION PRODUCTS

Compound stability is a major criterion in the development of pharmaceutical products. The integrity of formulation components must be studied and monitored to ensure the system composition is not adversely modified. Such studies invoke the testing of formulation composition and sometimes the high-performance liquid chromatography separation and isolation of degradation products. Although as previously discussed, use of LC-SPE-NMR can result in isolated samples with sufficient quantity and purity for full structural characterization, there are instances where sufficient material for rigorous purification may not be available. In such cases when mass spectrometry can produce a chemical formula of the major component, NMR analysis of mixtures may be carried out that can provide structural evaluation of proposed compounds.

An example of this situation was found in a degradation study of a steroid where excipients had interacted with the active ingredient [68]. Two key ingredients in the formulation included hydrochloric acid and propylene glycol. The mass spectrometry evaluation showed a mass of m/z 399.2 and a chemical formula of $C_{24}H_{30}O_4$. The NMR spectra of the isolated compound contained numerous impurities. Figure 10.20 shows the 1D ^{1}H NMR spectrum with numerous impurities present, which severely complicates the assignment and analysis of the data.

Examination of the NMR data resulted in a proposed structure (A), however, an alternate proposed structure (B) invoked the formation of an eight-membered ring. The ^{13}C chemical

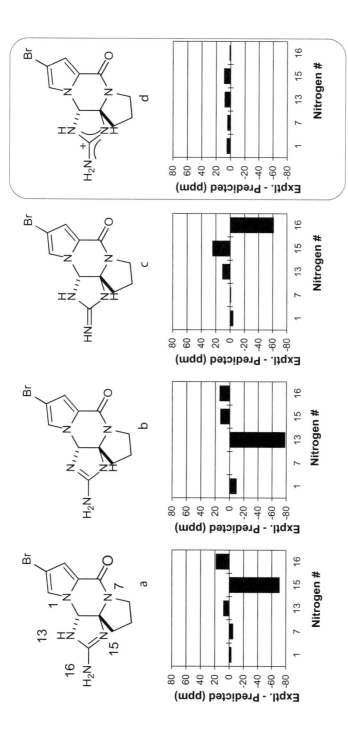

FIGURE 10.18 Bar graph depicting difference between experiment and predicted ^{15}N chemical shifts for monobromephakellin tautomers a–c and protonated species d. Systematic error correction of −16.5 ppm was applied to primary amine N16. (Xin et al. [55]) Reproduced with permission of The Royal Society of Chemistry.

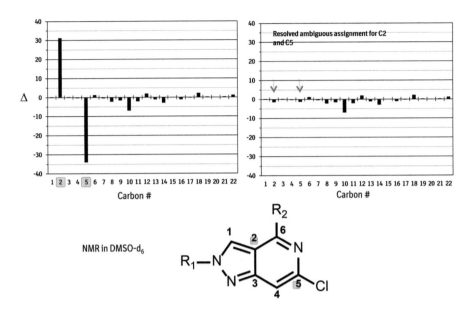

FIGURE 10.19 Bar graph depicting clarification of mis-assignments for carbons 2 and 5. These data illustrate the power of chemical shift predictions in resolving assignment ambiguities [68].

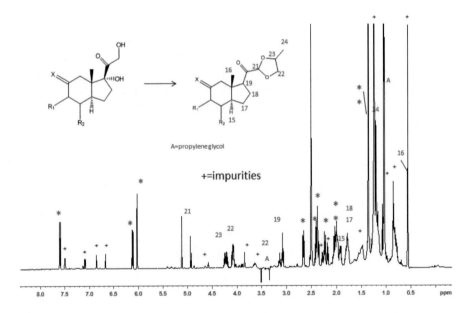

FIGURE 10.20 Proton NMR spectrum of an isolated steroid degradant where the sample contained numerous impurities introducing spectral complexity [68].

shift predictions at the B3LYP/cc-pVDZ level of theory for both structures when compared with experimental data are shown in Figure 10.21.

The results show that structure A is consistent with the ^{13}C NMR experimental results, whereas structure B shows large statistical deviations. Although the predictions support

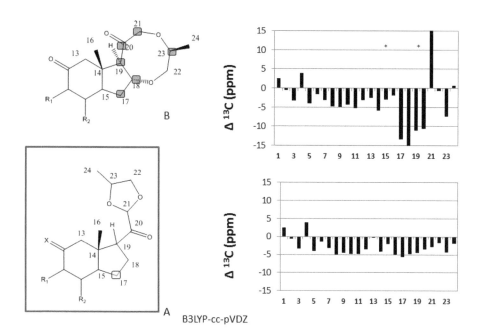

FIGURE 10.21 Bar graph showing chemical shift prediction at the B3LYP/cc-pVDZ level of theory, which supports formation of the dioxolane [68].

structure A, additional information was desired to further corroborate the results. This corroboration was found in a paper published by Herzog et al. [69], where it was shown that steroids in the presence of a mild acid can be transformed into an acetaldehyde upon loss of water. The acetaldehyde may then react with alcohols or in the present case dioxolane to produce ketals or acetals. (Figure 10.22). This information taken together with the chemical shift predictions and the experimental NMR assignments supported the proposed structure A (Figure 10.21).

10.5.6 DIASTEREOMERS

Use of probability theory along with high-level chemical shift prediction at the B3LYP/cc-pVDZ level of theory was shown to be highly effective in determining stereochemistry of disatereomers [52]. In some difficult cases, however, the probabilities of individual nuclei may give inconclusive or ambiguous results. To address such situations, the use of combined probabilities was successfully employed. Combinations of ^{1}H and ^{13}C DiCE probabilities were found to be improved over those of a single nucleus. Further improvement in accuracy of the DiCE probability calculation was reported by including ^{15}N DiCE calculations for combined analysis.

This was illustrated using palau'amine [70] (Figure 10.23), a nitrogen-rich polycyclic dimeric PIA natural product, which was investigated with multi-nuclei DiCE calculations. Historically, the structure of palau'amine had been very difficult to solve and was revised several times, particularly with respect to the relative configurations of the eight stereogenic centers. The stereochemistry at C12, C17, and C20 appeared to be especially ambiguous and had the most disagreements in the literature, until the final structural revision [70] and total synthesis of this natural product [71].

The performance of combined DiCE was tested using, ^{1}H, ^{13}C, and ^{15}N DiCE calculations, performed for eight possible isomers generated by varying the stereochemistry at positions 12,

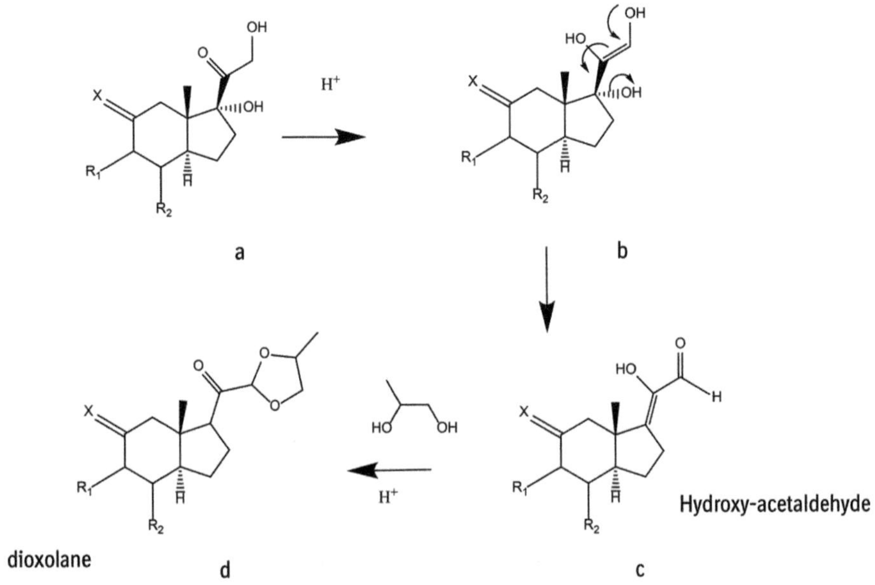

FIGURE 10.22 Mechanism for the formation of a dioxolane species from steroids when placed in the presence of propylene glycol and acid [68].

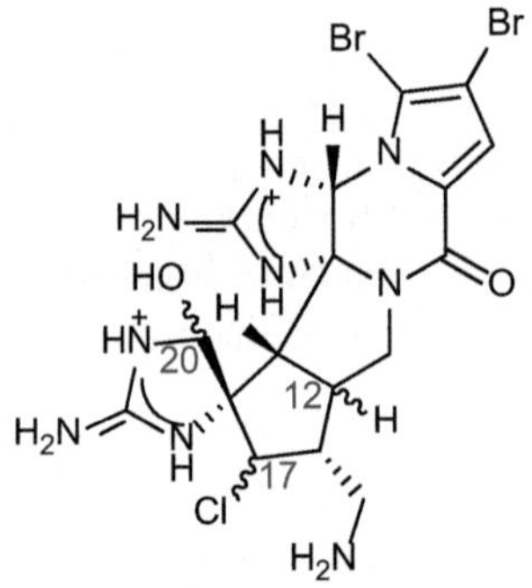

Dibromopalau'amine

FIGURE 10.23 Chemical structure of dibromopalau'amine with key stereo center carbons 12, 17, and 20 labeled.

17, and 20 in dibromopalau'amine (Figure 10.23). Experimental ^{1}H, ^{13}C, and ^{15}N data were available in the literature for this palau'amine derivative making this study possible [71].

Of the eight diastereomers, none of the single nucleus D*i*CE calculation could reliably assign the relative stereochemistry for dibromopalau'amine. In fact, only ^{13}C D*i*CE yielded the highest probability for the correct isomer, while ^{1}H and ^{15}N D*i*CE ranked the correct isomer as the second highest probability structure. Combining ^{13}C and ^{1}H D*i*CE could lead to a higher tendency for the correct isomer 18a (12R, 17R, 20R), but even so, the probability value was only 76%, indicating ambiguity in the prediction. However, after the inclusion of ^{15}N D*i*CE, the probability for the correct diastereomer was >98%

TABLE 10.3
^{13}C, ^{1}H, and ^{15}N DiCE probabilities for dibromopalau'amine

Isomer	Stereochemistry (12, 17, 20)	^{13}C^a	^{1}H	^{15}N^a	^{1}H + ^{13}C	^{13}C + ^{1}H + ^{15}N
18a	(R, R, R)	66.17	30.20	16.24	76.87	98.74
18b	(S, R, R)	6.86	5.97	0.02	1.58	0.00
18c	(R, S, R)	8.65	36.09	1.27	12.01	1.21
18d	(S, S, R)	0.01	1.63	82.06	0.00	0.00
18e	(R, R, S)	9.50	26.11	0.07	9.54	0.05
18f	(S, R, S)	0.00	0.00	0.00	0.00	0.00
18g	(R, S, S)	8.81	0.00	0.33	0.00	0.00
18h	(S, S, S)	0.00	0.00	0.00	0.00	0.00

^{a} Atoms with known systematic errors were excluded from DiCE calculations.

(Table 10.3). This improved performance may be attributed to the independence of the DiCE probabilities for different nuclei.

Because the probabilities for the correct and incorrect diastereomers may be comparable to signal and noise, the, random events would be canceled, while the nonrandom signal would be additive and become more prominent in this process. Therefore, the incorporation of the ^{15}N DiCE calculation was shown to be highly beneficial for the structure elucidation of the nitrogen-containing compounds.

Overall, a practical approach to high-level NMR chemical shift prediction has been reported. Use of B3LYP/cc-pVDZ affords a robust and accurate method for chemical shift prediction with low computational cost. Because B3LYP remains the most commonly known and employed functional for computational experts and novices alike, it was a preferred method of choice. The cc-pVDZ basis set provides a balance between accuracy and computational cost.

Linear scaling reference terms for four solvents, and systematic error corrections for ten functional groups were identified, calculated, and reported. Overall, the methods described herein have proven to be robust and broadly applicable for a wide range of organic compounds.

In addition, a probability theory algorithm (DiCE) for ^{1}H, ^{13}C, and ^{15}N NMR data is described that improves the accuracy in the prediction of the stereochemistry and regiochemistry in cases where only one set of experimental data is available. The algorithm was generated using NMR chemical shift calculations at the B3LYP/cc-pVDZ level of theory using geometry optimization and the CPCM solvent model. The generalized linear scaling terms calculated for four different solvents had a major impact on the predicted accuracy. The improved chemical shift predictions enabled a higher level of precision in calculating the statistical distribution, standard deviation, and degrees of freedom. DiCE was found to be highly successful at making accurate predictions in very challenging diastereomer and regioisomer cases, where the chemical shift prediction comparisons were ambiguous or inconclusive.

10.6 SUMMARY

Over the past 10 years, major improvements in the performance of LC-NMR have been realized. The addition of post-column SPE, advances in probe technology including cryogenic probes and microcoil probes, improved solvent suppression pulse sequences, and shielded magnets with better homogeneity have all contributed to rapid advancements in this technology.

Application of LC-NMR to problems in pharmaceutical development has had a major impact on structure elucidation studies. LC–NMR has been successfully applied to determine the structures of degradation products, impurities, mixtures of compounds, and metabolites. Use of stop-flow techniques with LC-NMR experiments has been a critical means of identifying unstable compounds and studying conformational kinetics.

The integration of SPE as an intermediate step between the LC unit and the NMR spectrometer has vastly improved the power of the hyphenated technique in trace analysis applications. Online postcolumn enrichment of chromatographic peaks by SPE dramatically reduces the NMR acquisition times by allowing repeated injections to be trapped onto the same cartridge or different cartridges. Because protonated solvents can be easily removed with a drying procedure, solvents and buffers may be freely chosen for maximizing chromatographic separation without compromising NMR spectral quality. The compound of interest may then be eluted from an SPE cartridge, using deuterated organic solvent, which helps to reduce dynamic range issues.

When combined with cryogenically cooled microcapillary probes, the sensitivity of the NMR signal increases about 10-fold over the conventional room temperature probes, enabling full structure characterization at the microgram level. Heteronuclear experiments with concentrations previously only possible in a limited number of cases, have now become standard experiments. The availability of HSQC and HMBC experiments and microcoil/cryogenic technology opens the possibility of using LC-SPE NMR for the structural elucidation of complete unknowns on a microgram scale.

To enable significant future downscaling beyond the current capabilities, improved performance of the LC-NMR interface and improved SPE cartridge retention need to be addressed. In addition, the active volume of the NMR flow cell or capillary tube will have to shrink along with the corresponding detection coils in order not to lose filling factor. As the size of the NMR probes become more efficient with respect to mass sensitivity, techniques such as CE or CEC (capillary electrochromatography) may be interfaced more successfully with NMR spectroscopy.

Because of the tremendous advances in LC-SPE-NMR technology as well as NMR software and hardware, it is now possible to not only isolate and structurally characterize challenging compounds, but carryout quantitative analysis of these compounds. With quantitative NMR (qNMR) and use of external references, determination of an isolated sample's purity, concentration, and mass is possible. This becomes important when investigating isolated metabolites degradants or impurities on a microgram scale. Nondestructive quantitation of a sample becomes critical when conducting follow-up of a compound's toxic properties. This is especially important when the isolated material presents a major synthetic challenge.

The application of computational chemistry in solving chemical structure is another approach that augments the LC-SPE-NMR process. In particular, the use of quantum chemistry and DFT, in the *ab initio* prediction of chemical shifts, offers a powerful supplement in structure elucidation. This is very important with LC-SPE-NMR analyses due to the fact that frequently the structures isolated from this approach are unusual or unstable. Often only partial NMR data are available, hence the comparison of predicted chemical shifts from a proposed structure with experimental chemical shifts from an unknown sample may be used as a primarily a guide that can accelerate the structure elucidation process. A match between prediction and experiment, when used in conjunction with a totality of chemical and experimental data can corroborate a structure with a high level of confidence.

Overall, the current state of the art in LC–NMR has shown proven utility in a variety of applications. When combined with SPE and cryotechnology, LC-NMR is an extremely valuable tool for mass-limited samples enabling structure elucidation without the need for laborious scale-up, serial isolation, and purification procedures. Application of qNMR enables quantitation on isolated microgram quantities of materials that previously would have been a major challenge. Finally, the combined use of LC-SPE-NMR with QM/DFT chemical shift prediction offers a robust and reliable process for accurate structure determination that has been automated for the practicing structural chemist.

REFERENCES

1. Claridge, T.D.W. 2008. *High-Resolution NMR Techniques in Organic Chemistry*, 2nd edition, Vol. 27, New York: Elsevier.
2. Moore, K.W., Li, R., Pelczer, I., and Rabitz, H. 2012. NMR landscapes for chemical shift prediction. *The Journal of Physical Chemistry. A.* 116 (36):9142–9157.
3. Abraham, R.J., and Mobili, M. 2008. *Modelling 1H NMR Spectra of Organic Compounds: Theory and Applications*, Chichester: John Wiley & Sons.
4. Xin, D., Sader, C.A., Chaudhary, O., Jones, P.-J., Wagner, K., Tautermann, C., Yang, Z., Busacca, C.A., Saraceno, R.A., Fandrick, K.R., Gonnella, N.C., Horspool, K., Hansen, G., and Senanayake, C.H. 2017. Development of a 13C NMR chemical shift prediction procedure using B3LYP/cc-pVDZ and empirically derived systematic error correction terms: a computational small molecule structure elucidation method. *J. Org. Chem.* 82:5135−5145.
5. Ditchfield, R. 1974. Self-consistent perturbation theory of diamagnetism I. A gauge-invariant LCAO method for N.M.R. chemical shifts. *Mol. Phys.* 27:789–807.
6. Wolinski, K., Hinton, J.F., and Pulay, P. 1990. Efficient implementation of the gauge-independent atomic orbital method for NMR chemical shift calculations. *J. Am. Chem. Soc.* 112 (23):8251–8260.
7. Rauhut, G., Puyear, S., Wolinski, K., and Pulay, P.J. 1996. Comparison of NMR chemical shieldings calculated from Hartree-Fock and density functional wave functions using gauge-including atomic orbitals. *Phys. Chem.* 100 (63):10–16.
8. Lodewyk, M.W., Siebert, M.R., and Tantillo, D.J. 2012. Computational prediction of ^{1}H and ^{13}C chemical shifts: a useful tool for natural product, mechanistic and synthetic organic chemistry. *Chem. Rev.* 112:1839–1862.
9. Tantillo, D.J. 2013. Walking in the woods with quantum chemistry–applications of quantum chemical calculations in natural products research. *Nat. Prod. Rep.* 30:1079–1086.
10. Schleyer, P.V.R., and Maerker, C. 1995. Exact structures of carbocations established by combined computational and experimental methods. *Pure Appl. Chem.* 67:755–760.
11. Siehl, H.-U. 2008. The interplay between experiment and theory: computational NMR spectroscopy of carbocations. *Adv. Phys. Org. Chem.* 42:125–165.
12. Buehl, M., and Schleyer, P.V.R. 1992. Application and evaluation of ab initio chemical shift calculations for boranes and carboranes. How reliable are "accurate" experimental structures?. *J. Am. Chem. Soc.* 114 (2):477–491.
13. Buehl, M. 1998. *In Encyclopedia of Computational Chemistry*, eds. P.V.R. Schleyer, P.R. Schreiner, N.L. Allinger, T. Clark, J. Gasteiger, P. Kollmann, and H.F. Schaefer, III. Chichester: Wiley 1835.
14. Kaupp, M., Buehl, M., and Malkin, V.G., eds. 2004. *Calculation of NMR and EPR Parameters.* Verlag: Wiley-VCH.
15. Harding, M.E., Gauss, J., and Schleyer, P.V.R. 2011. Why benchmark-quality computations are needed to reproduce 1-adamantyl cation nmr chemical shifts accurately. *J. Phys. Chem. A.* 115:2340–2344.
16. Alkorta, I., and Elguero, J. 2010. Computational NMR spectroscopy. In *Computational Spectroscopy: Methods, Experiments and Applications*, J. Grunenberg ed., Weinheim, Germany: Wiley-VCH, 37.
17. Casabianca, L.B., and de Dios, A.C. 2008. Ab initio calculations of NMR chemical shifts. *J. Chem. Phys.* 128 (5):052201/1–052201/10.
18. Mulder, F.A.A., and Filatov, M. 2010. NMR chemical shift data and ab initio shielding calculations: emerging tools for protein structure determination. *Chem. Soc. Rev.* 39:578–590.
19. Petrovic, A.G., Navarro-Vazquez, A., and Alonso-Gomez, J.L. 2010. From relative to absolute configuration of complex natural products. Interplay between NMR, ECD, VCD, and ORD assisted by ab initio calculations. *Curr. Org. Chem.* 14:1612–1628.
20. Barone, V., Improta, R., and Rega, N. 2008. Quantum mechanical computations and spectroscopy: from small rigid molecules in the gas phase to large flexible molecules in solution. *Acc. Chem. Res.* 41:605–616.

21. Bifulco, G., Dambruoso, P., Gomez-Paloma, L., and Riccio, R. 2007. Determination of relative configuration in organic compounds by NMR spectroscopy and computational methods. *Chem. Rev.* 107:3744–3779.

22. Bagno, A., and Saielli, G. 2007. Computational NMR spectroscopy: reversing the information flow. *Theor. Chem. Acc.* 117:603–619.

23. Helgaker, T., Jaszunski, M., and Ruud, K. 1999. Ab initio methods for the calculation of NMR shielding and indirect spin-spin coupling constants. *Chem. Rev.* 99:293–352.

24. Reeves, J.T., Fandrick, D.R., Tan, Z., Song, J.J., Rodriguez, S., Qu, B., Kim, S., Niemeier, O., Li, Z., Byrne, D., Campbell, S., Chitroda, A., DeCroos, P., Fachinger, T., Fuchs, V., Gonnella, N.C., Grinberg, N., Haddad, N., Jäger, B., Lee, H., Lorenz, J.C., Ma, S., Narayanan, B.A., Nummy, L.J., Premasiri, A., Roschangar, F., Sarvestani, M., Shen, S., Spinelli, E., Sun, X., Varsolona, R.J., Yee, N., Brenner, M., and Senanayake, C.H. 2013. Development of a large scale asymmetric synthesis of the glucocorticoid agonist BI 653048 BS H₃PO₄. *J. Org. Chem.* 78:3616–3635.

25. Betageri, R., Bosanac, T., Burke, M.J., Harcken, C., Kim, S., Kuzmich, D., Lee, T.W.H., Li, Z., Liu, P., Lord, J., Razavi, H., Reeves, J.T., and Thomson, D. PCT. int. appl. WO/2009/149139, 2009.

26. Becke, A.D.J. 1993. Density-functional thermochemistry. III. The role of exact exchange. *Chem. Phys.* 98:5648–5652.

27. Lee, C., Yang, W., and Parr, R.G. 1988. Development of the Colle-Salvetti correlation-energy formula into a functional of the electron density. *Phys. Rev. B: Condens. Matter Mater. Phys.* 37:785–789.

28. Chai, J.D., and Head-Gordon, M. 2008. Long-range corrected hybrid density functionals with damped atom-atom dispersion corrections. *Phys. Chem. Chem. Phys.* 10:6615–6620.

29. Zhao, Y., and Truhlar, D.G. 2008. The M06 suite of density functionals for main group thermochemistry, thermochemical kinetics, noncovalent interactions, excited states, and transition elements: two new functionals and systematic testing of four M06-class functionals and 12 other functionals. *Theor. Chem. Acc.* 120:215–241.

30. Wiitala, K.W., Hoye, T.R., and Cramer, C.J. 2006. Hybrid density functional methods empirically optimized for the computation of 13C and 1H chemical shifts in chloroform solution *J. Chem. Theory Comput.* 2:1085–1092.

31. Grimme, S., Ehrlich, S., and Goerigk, L. 2011. Effect of the damping function in dispersion corrected density functional theory. *J. Comput. Chem.* 32:1456–1465.

32. Hehre, W.J., Ditchfield, R., and Pople, J.A. 1972. Self-consistent molecular orbital methods. XII. further extensions of gaussian-type basis sets for use in molecular orbital studies of organic molecules. *J. Chem. Phys.* 56:2257–2261.

33. Francl, M.M., Pietro, W.J., Hehre, W.J., Binkley, J.S., Gordon, M.S., DeFrees, D.J., and Pople, J.A. 1982. Self-consistent molecular orbital methods. XXIII. A polarization-type basis set for second-row elements. *J. Chem. Phys.* 77:3654–3665.

34. Weigend, F., and Ahlrichs, R. 2005. Balanced basis sets of split valence, triple zeta valence and quadruple zeta valence quality for H to Rn: design and assessment of accuracy. *Phys. Chem. Chem. Phys.* 7:3297–3305.

35. Dunning, T.H. 1989. Gaussian basis sets for use in correlated molecular calculations. I. The atoms boron through neon and hydrogen. *J. Chem. Phys.* 90:1007–1023.

36. Jensen, F.J. 2008. Basis set convergence of nuclear magnetic shielding constants calculated by density functional methods. *Chem. Theory Comput.* 4:719–727.

37. Tormena, C.F., and Da Silva, G.V.J. 2004. Chemical shifts calculations on aromatic systems: a comparison of models and basis sets. *Chem. Phys. Lett.* 398:466–470.

38. Sarotti, A.M., and Pellegrinet, S.C. 2009. A multi-standard approach for GIAO 13C NMR calculations. *J. Org. Chem.* 74:7254–7260.

39. Jain, R.J., Bally, T., and Rablen, P.R. 2009. Calculating accurate proton chemical shifts of organic molecules with density functional methods and modest basis sets. *J. Org. Chem.* 74:4017–4023.

40. Forsyth, D.A., and Sebag, A.G. 1997. Computed 13C nmr chemical shifts via empirically scaled giao shieldings and molecular mechanics geometries. Conformation and configuration from 13C shifts. *J. Am. Chem. Soc.* 119 (40):9483–9494.

41. Chesnut, D.B. 1997. On the calculation of hydrogen NMR chemical shielding. *Chem. Phys.* 214:73–79.

42. Dybiec, K., and Gryff-Keller, A. 2009. Remarks on GIAO-DFT predictions of 13C chemical shifts. *Magn. Reson. Chem.* 47:63–66.

43. d'Antuono, P., Botek, E., and Champagne, B. 2006. Theoretical investigation on 1H and 13C NMR chemical shifts of small alkanes and chloroalkanes. *J. Chem. Phys.* 2006 (125): 144309/1-144309/12.

44. Bagno, A., Rastrelli, F., and Saielli, G. 2003. Predicting 13C NMR spectra by DFT calculations. *J. Phys. Chem. A.* 107:9964–9973.

45. Giesen, D.J., and Zumbulyadis, N. 2002. A hybrid quantum mechanical and empirical model for the prediction of isotropic 13C shielding constants of organic molecules. *Phys. Chem. Chem. Phys.* 4:5498–5507.

46. Autschbach, J., and Zheng, S. 2009. Relativistic computations of NMR parameters from first principles: theory and applications. *S. Annu. Rep. NMR Spectrosc.* 67:1–95.

47. Autschbach, J., and Ziegler, T. 2002. Relativistic computation of NMR shieldings and spin-spin coupling constants. In *Encyclopedia of Nuclear Magnetic Resonance*, eds. D.M. Grant and R. K. Harris. Vol. 9, Chichester: John Wiley & Sons, 306.

48. Kaupp, M. 2004. Interpretation of NMR chemical shifts. In *Calculation of NMR and EPR Parameters*, eds. M. Kaupp, M. B€uhl, and V.G. Malkin. Verlag: Wiley-VCH, 123.

49. Neto, A.C., Ducati, L.C., Rittner, R., Tormena, C.F., Contreras, R.H., and Frenking, G. 2009. Heavy halogen atom effect on 13C NMR chemical shifts in monohalo derivatives of cyclohexane and pyran. Experimental and theoretical Study. *J. Chem. Theory Comput.* 5:2222–2228.

50. Rozhenko, A.B., and Trachevsky, V.V. 2009. Specificity of 13C NMR shielding calculations in thiocarbonyl compounds. *Phosphorus, Sulfur Silicon Relat. Elem.* 184:1386–1405.

51. Facelli, J.C., Pugmire, R.J., and Grant, D.M. 1996. Effects of hydrogen bonding in the calculation of ^{15}N chemical shift tensors: benzamide. *J. Am. Chem. Soc.* 118 (23):5488–5489.

52. Smith, S.G., and Goodman, J.M. 2010. assigning stereochemistry to single diastereoisomers by GIAO NMR Calculation: the DP4 probability. *J. Am. Chem. Soc.* 132 (37):12946–12959.

53. Grimblat, N., Zanardi, M.M., and Sarotti, A.M. 2015. Beyond DP4: an improved probability for the stereochemical assignment of isomeric compounds using quantum chemical calculations of NMR shifts. *J. Org. Chem.* 80:12526–12534.

54. Xin, D., Jones, P.-J., and Gonnella, N.C. 2018. DiCE: diastereomeric in silico chiral elucidation, expanded DP4 probability theory method for diastereomer and structural assignment. *J. Org. Chem.* 83:5035−5043.

55. Xin, D., Sader, C.A., Fischer, U., Wagner, K., Jones, P.-J., Xing, M., Fandrick, K.R., and Gonnella, N.C. 2017. Systematic investigation of DFT-GIAO 15N NMR chemical shift prediction using B3LYP/cc-pVDZ: application to studies of regioisomers, tautomers, protonation states and N-oxides. *Org. Biomol. Chem.* 15:928–936.

56. CCG reference Molecular Operating Environment (MOE), 2014.09; Chemical Computing Group Inc.: Montreal, QC, Canada, 2015.

57. McConathy, J., and Owens, M.J. 2003. Stereochemistry in drug action. prim care companion. *J. Clin. Psychiatry.* 05:70−73.

58. Hutt, A.G., and O'Grady, J. 1996. Drug chirality: a consideration of the significance of the stereochemistry of antimicrobial agents. *J. Antimicrob. Chemother.* 37:7−32.

59. Young, D.W., Morecombe, D.J., and Sen, P.K. 1977. The stereochemistry of beta-lactam formation in penicillin biosynthesis. *Eur. J. Biochem.* 75:133−147.

60. Bycroft, B.W., Wels, C.M., Corbett, K., Maloney, A.P., and Lowe, D.A. 1975. Biosynthesis of penicillin G from D- and L-[14C]- and [α-3H]- valine. *J. Chem. Soc., Chem. Commun.* 0:923−924.

61. Rychnovsky, S.D., Rogers, B.N., and Richardson, T.I. 1998. Configurational assignment of polyene macrolide antibiotics using the [13C] acetonide analysis. *Acc. Chem. Res.* 31:9−17.

62. Peng, J., Place, A.R., Yoshida, W., Anklin, C., and Hamann, M.T. 2010. Structure and absolute configuration of karlotoxin-2, an ichthyotoxin from the marine dinoflagellate karlodinium veneficum. *J. Am. Chem. Soc.* 132 (10):3277−3279.

63. Chini, M.G., Riccio, R., and Bifulco, G. 2015. Computational nmr methods in the stereochemical analysis of organic compounds: are proton or carbon NMR chemical shift data more discriminating?. *Eur J Org Chem.* 6:1320–1324.

64. Marell, D.J., Emond, S.J., Kulshrestha, A., and Hoye, T.R. 2014. Analysis of seven-membered lactones by computational NMR methods: proton NMR chemical shift data are more discriminating than carbon. *J. Org. Chem.* 79 (2):752–758.

65. Faulkner, D.J. 2002. Marine natural products. *Nat Prod Rep.* 19 (1):1–48.

66. Kock, M., Grube, A., Seiple, I.B., and Baran, P.S. 2007. The pursuit of palau'amine. *Angew. Chem., Int. Ed.* 46:6586−6594.

67. Meyer, S.W., and Köck, M. 2008. NMR studies of phakellins and isophakellins. *J. Nat. Prod.* 71 (9):1524–1529.

68. Gonnella, N.C., and Fandrick, K. Development of an accurate and efficient NMR chemical shift prediction procedure using B3LYP/cc-pVDZ: A powerful method for molecular structure elucidation. Wesleyan University, Invited Lecture, April 28, 2017.

69. Herzog, H.L., Gentles, M.J., Marshall, H., and Hershberg, E.B. 1961. Weak acid-catalyzed rearrangement of the dihydroxyacetone side chain in steroids. *J. Am. Chem. Soc.* 83 (19):4073–4076.

70. Grube, A., and Köck, M. 2007. Structural assignment of tetrabromostyloguanidine: does the relative configuration of the palau'amines need revision? *Angew. Chem., Int. Ed.* 46:2320–2324.

71. Seiple, I.B., Su, S., Young, I.S., Lewis, C.A., Yamaguchi, J., and Baran, P.S. 2010. Total synthesis of palau'amine. *Angew. Chem., Int. Ed.* 49 (6):1095–1098.

Glossary

NMR TERMS (2ND EDITION)

Acquisition Time: The time during which the signal is actually being recorded is called the acquisition time. This parameter set is based upon the spectral window setting and the number of digitized data points entered. It is important that the acquisition time be set long enough to prevent baseline distortions in the final spectrum.

α and β Spin States: The two possible states of a spin 1/2 nucleus in a magnetic field. By convention, if the magnetogyric ratio is positive, the α state will have lower energy (z magnetization aligned with B_0).

ADC (Analog-to-Digital Converter): Hardware component that translates a voltage from the detector into a binary number, commonly 16 bits long.

Atomic Orbital: An orbital described by a wave function for a single electron centered on a single atom.

B_0: Static magnetic field about whose direction the nuclear magnetic moment processes. Magnitude of the magnetic field is expressed in tesla or as the proton precession frequency (e.g., 11.7 T or 500 MHz).

B_1: The radio-frequency field that is applied to the nuclei that causes precession of the magnetization vector about an axis perpendicular to B_0.

B3LYP: Acronym for **B**ecke **3**-Parameter (Exchange), **L**ee, **Y**ang, and **P**arr (correlation; density functional theory). A widely used method for NMR chemical shift prediction.

Basis Set: A finite set of functions used to approximately express the molecular orbital wave function(s) of a system, normally atom centered, consisting of atomic orbitals differing in local angular momentum for each atom.

Boltzmann Average: The Maxwell–Boltzmann distribution is the classical distribution function for distribution of an amount of energy between identical but distinguishable particles. In statistical mechanics, Maxwell–Boltzmann statistics describes the average distribution of non-interacting material particles over various energy states in thermal equilibrium.

Boltzmann Distribution: In NMR, this may refer to the distribution of nuclear spins among their possible energy levels at thermal equilibrium. For a positive magnetogyric ratio, the excess population excess (polarization) is in the direction of B_0.

Broadband Decoupling: Irradiation of a group of nuclei with a wide distribution of chemical shifts by a radio-frequency field to effectively spread its spin decoupling effect evenly over all the nuclei. Often used to remove ^1H-^{13}C splittings from a ^{13}C spectrum.

cc-pVDZ: Dunning's correlation consistent basis set. This basis set has had redundant functions removed and has been rotated in order to increase computational efficiency.

Chemical Shift: A variation in the resonance frequency of a nuclear spin due to the chemical environment around the nucleus. Chemical shift is reported in parts per million (ppm).

CSA (Chemical Shift Anisotropy): The chemical shift effect that is dependent on the orientation of a molecule in the magnetic field. This effect is partially responsible for the very wide lines observed in solid state NMR. In solution, CSA is averaged out by molecular tumbling and sharp lines are observed; however, the modulation of the shielding can sometimes contribute to relaxation.

Chemical-Shift Reference: A chemical-shift reference is used to define the positions of the resonances in the spectrum in terms of parts-per-million, or ppm, of frequency. It is usually a compound such as tetramethyl silane (TMS) which can be directly added to the sample. The shift of this material is defined as $\delta=0$. For nuclei such as ^{1}H and ^{13}C, TMS is the generally accepted standard. However, for less common nuclei, there may be alternate standards. While modern spectrometers can perform frequency referencing operations automatically, highly precise chemical shifts require use of appropriate standards and manual frequency referencing.

Combined D*i*CE: This approach uses the product of probabilities of multiple nuclei for an individual isomer divided by the sum of the products of probabilities for each of the possible isomers.

Continuous-Flow NMR: Eluent containing a compound of interest that is sampled in "real-time" when flowing through the NMR detection coil.

Correlation Time (τ_c): A parameter related to the mean time during which a molecule maintains its spatial geometry. For an internuclear vector, τ_c is approximately equal to the average time for it to rotate through an angle of one radian.

COSY (COrrelation SpectroscopY): A 2D experiment used to identify nuclei that share a scalar (J) coupling. The presence of off-diagonal peaks (cross-peaks) in the spectrum directly correlates the coupled spins. Most often used to analyze coupling relationships between protons, but may be used to correlate any high-abundance homonuclear spins (e.g., ^{31}P, ^{19}F, and ^{11}B).

Coupling/Decoupling: Couplings, or interactions, between nuclei can be either ^{1}H–^{1}H or ^{13}C–^{1}H. In either case, they cause splitting or broadening of the resonances in the spectrum, complicating interpretation. Irradiation, at appropriate power levels, at the frequency of a particular nucleus, is called decoupling. The coupling interaction is thus removed, leading to a simpler spectrum.

Cross-Peaks: Off-diagonal peaks in a 2D spectrum that appear at the coordinates (usually chemical shifts) of the correlated nuclei.

CPCM (Conductor-like Polarizable Continuum Model): Conductor-like polarizable continuum model is an efficient way of accounting for solvent effects in quantum chemical calculations. The solvent is represented as a dielectric polarizable continuum and the solute is placed in a cavity of approximate molecular shape.

CryoProbe: Probe that relies on the principle that radio frequency electronics will generate a higher signal and less noise at lower temperatures. By reducing the temperature of the NMR coil and preamplifier, signal-to-noise ratio enhancements of greater than a factor of four can be achieved over conventional 5 mm probes. The coil assembly and preamplifier are cooled using cold helium gas in a closed-loop cooling system. Vacuum insulation in the cooling system allows the coil assembly to reach very low temperatures.

Deuteration: Deuteration refers to the replacement of some or all of the ^{1}H nuclei (protons) in a material by ^{2}H (deuterons). Most solvents used in NMR analysis are deuterated.

In ^{1}H NMR, this has the effect of making the solvent effectively invisible (except for very small residual ^{1}H resonances). In ^{13}C NMR, the presence of the deuterons alters the splitting pattern of the solvent peaks, making solvent peaks easily distinguishable. For all nuclei, the ^{2}H resonance from the solvent can serve as the lock signal, which is used to compensate for magnetic field drift, to optimize the magnetic field homogeneity (shimming), and sometimes to provide a frequency reference. Commonly used deuterated solvents include acetone-d$_6$, acetonitrile-d$_3$, chloroform-d, N,N,-dimethylsulfoxide-d$_6$, and deuterium oxide-d$_2$ (heavy water).

DFT: (**D**ensity **F**unctional **T**heory) ab initio electronic method from solid-state physics. Tries to find best approximate functional to calculate energy from electron density.

D*i*CE: (**D**iastereomeric *in-silico* **C**hiral **E**lucidation) probability theory algorithm for ^{1}H, ^{13}C, and ^{15}N NMR data that was developed to be compatible with chemical shift prediction performed at the B3LYP/cc-pVDZ level of theory. Calculations used a CPCM solvent model and geometry optimization of each conformer. Generalized linear scaling terms were developed for four different solvents providing improved accuracy in chemical shift prediction for ^{1}H and ^{13}C NMR spectra. The process was also developed for ^{15}N NMR in DMSO.

Dipole–Dipole Coupling: An interaction that is very large (often kilohertz) and depends upon the distance between nuclei and the angular relationship between the magnetic field and the internuclear vectors. This coupling is not seen for mobile molecules in solution because it is averaged to zero by tumbling of the molecule.

DI-NMR (**D**irect-**I**njection **NMR**): A simplified flow-NMR system where the mobile phase is removed along with any optional detectors. DI-NMR injects the sample directly into the NMR flow probe, and the whole injection process is driven by automation. In this case, the NMR flow probe may be viewed as a sample loop for the injector port of an automated injector.

DNP (**D**ynamic **N**uclear **P**olarization): A technique that attacks the basic NMR sensitivity problem in that it selectively increases the population of one nuclear spin state over the other. This is accomplished by polarizing the electron spins of a radical either through microwaves or optical pumping. This polarization is then transferred from the electrons of the radical to the nuclei of the analyte, and followed by a fast NMR acquisition before the nuclei have time to relax. Although DNP can achieve large signal enhancements, only one-dimensional experiments are possible due to the short lifetimes of the hyperpolarized state.

Doublet: A resonance that exhibits (scalar) coupling to one spin-1/2 nucleus appears as two peaks of equal intensity in the spectrum. The spacing between the peaks is called the **coupling constant** and is given in units of Hz.

DP4: A probability measure which helps to assign structure and stereochemistry by comparing experimental and calculated NMR spectra.

DP4+: A probability measure that differs from the original DP4 method by inclusion of unscaled data (combined with scaled probabilities) and the use of higher levels of theory in the NMR calculation procedure. In addition, combined ^{1}H and ^{13}C probabilities are employed.

Dynamic Range: The ratio of the largest to the smallest signal observable in a digitized spectrum. Determined by the number of bits in the output of the analog–digital converter.

ERETIC (**E**lectronic **RE**ference **T**o access **I**n vivo **C**oncentrations): A method that provides a reference signal, synthesized by an electronic device, which can be used for the determination of absolute concentrations.

FBDD (**F**ragment-**B**ased **D**rug **D**esign): This method screens for molecules with weak binding affinity for a particular target. The fragment approach is based upon the premise that small molecular scaffolds with weak binding affinity for a target may be synthetically modified using either structure-based approaches or via combinatorial or parallel synthesis to ultimately generate a potent biological inhibitor. In this area, NMR is vital because of its ability to identify weakly binding ligands which may be subsequently modified to increase binding affinity.

FIA-NMR (**F**low **I**njection **A**nalysis **NMR**): A technique that may be viewed as LC-NMR without a chromatography column.

FID (**F**ree **I**nduction **D**ecay): The magnetic resonance signal resulting from the decay of transverse magnetization.

Field Strength: The magnetic field strength, designated by B_0, reflects the strength of the instrument system's magnet. The field strength is most properly given in units of gauss (G) or tesla (T = 10^4G), although NMR spectroscopists commonly refer to the ^{1}H resonant frequency instead.

Fill Factor: The ratio of the sample volume being observed by the rf coil to the coil volume.

Flow Probe: A configuration that is designed to acquire NMR data on a sample introduced into a probe inside a magnet via capillary tubing. Optimization of the flow probe design introduces sample from the bottom of the magnet into the flow cell chamber and exiting from the opposite end.

Fourier Transform (FT): A mathematical technique capable of converting a time domain signal to a frequency domain signal.

GARP (**G**lobally optimized **A**lternating-phase **R**ectangular **P**ulses): A decoupling method that makes use of composite pulses. This approach uses fractional multipliers to describe a composite pulse more precisely obtaining improved inversion properties. The GARP pulse invokes a four-step $R\overline{R}\overline{R}\overline{R}$ supercycle to achieve a more uniform decoupling profile.

Gaussian Distribution: see Normal Distribution.

GIAO (**G**auge **I**ndependent **A**tomic **O**rbitals): Ditchfield's method for canceling out the arbitrariness of the choice of origin and form (gauge) of the vector potential used to introduce the magnetic field in the Hamiltonian when calculating chemical shielding and chemical shift tensor. An exponential term containing the vector potential is included with each atomic orbital.

Gradient: A variation of one quantity with respect to another. In the context of NMR, a magnetic field gradient is a variation in the magnetic field with respect to distance.

HETCOR (**HET**eronuclear **COR**relation Spectroscopy): A 2D NMR experiment where two different types of nuclei are correlated through single bond spin–spin couplings. The chemical shift of one nucleus, usually ^{13}C, is detected in the direct observe dimension, while the chemical shift of second nucleus usually ^{1}H is recorded in the indirect dimension.

HMBC (**H**eteronuclear **M**ultiple **B**ond **C**orrelation): A 2D experiment that correlates chemical shifts of two types of nuclei separated from each other with two or more chemical bonds. For ^{1}H,^{13}C–HMBC, the spectrum is frequently used for assigning quaternary and carbonyl carbons. In some cases, it can be used for assigning –OH protons, and protons bonded to other heteroatoms.

HMQC (Heteronuclear Multiple-Quantum Correlation): A 2D experiment used to correlate directly bonded carbon–proton nuclei. This experiment uses proton detection and can be acquired more rapidly than a 1D carbon spectrum. The correlations can be used to associate known proton assignments with their directly attached carbons. The 2D spectrum can also be used in the assignment of the proton spectrum by separating proton resonances in the ^{13}C dimension, thereby reducing proton multiplet overlap. It also provides a convenient way of identifying non-equivalent geminal protons which are sometimes difficult to distinguish unambiguously, since such protons will produce two correlations to the same carbon.

HSQC (Heteronuclear Single-Quantum Correlation): A 2D proton-detected heteronuclear shift correlation experiment which provides the same information as the closely related HMQC (e.g., one-bond H–X correlations). The principle advantage of HSQC is the slightly better resolution that can be obtained in the X-dimension where the resonances are broadened by homonuclear proton couplings in HMQC. For most routine applications, this difference is barely noticeable, but where crowding occurs in the X-dimension, the HSQC will provide better results (provided sufficiently high digital resolution is used).

INADEQUATE (Incredible Natural Abundance DoublE QUAntum Transfer Experiment): A 2D experiment that correlates carbons that are attached to each other. This experiment can help establish a carbon skeleton of the molecule. Since ^{13}C–^{13}C coupling is used, the probability of one ^{13}C is (0.01) 1% while two next to each other 0.01 × 0.01 = 0.0001 (0.01%) (~1 molecule in 10,000). Hence, the experiment requires a large amount instrument time.

Integration: An integrated intensity of a signal in a ^{1}H NMR spectrum gives a ratio for the number of hydrogens that give rise to the signal, thereby helping calculate the total number of hydrogens present in a sample.

Inversion Recovery Sequence: A pulse sequence producing signals which represent the longitudinal magnetization present after the application of a 180° inversion RF pulse. Used for measuring longitudinal relaxation (T_1).

J or Scalar Coupling: Internuclear couplings (prominent features of ^{1}H spectra), leading to splittings which provide information on the structure of the sample. Interpretation of couplings can be difficult when signal overlap is severe. The distance between any two peaks in a J-split is called the coupling constant, J_{AB}, where A and B indicate the coupled nuclei. The coupling constant, usually given in Hz, is independent of magnetic field. Problems with severe overlap of signals may be reduced by acquiring spectra at higher field to achieve greater peak dispersion.

Karplus Relationship: This relationship establishes a correlation between the dihedral angle between two protons separated by three bonds and the magnitude of the coupling constant. Dihedral angles between vicinal protons that are either near 0 or 180 degrees have large coupling between 9 and 13 Hz. Dihedral angles for protons that are near 90 degrees, however, give small couplings of 0–2 Hz.

Larmor Frequency: The resonance frequency of a spin in a magnetic field. The rate of precession of a set of spins in a magnetic field. The frequency which will cause a transition between the two-spin energy levels of a nucleus. If a system on which the ^{1}H nucleus resonates at 400 MHz is taken as a reference point, the NMR frequency for a nucleus ^{13}C can be calculated by multiplying its frequency factor (0.2514) by the ^{1}H frequency of the instrument system of interest. Thus, the ^{13}C nucleus will resonate at around 100 MHz on a system whose ^{1}H frequency is 400 MHz.

LC-NMR: Hyphenated technology: Liquid Chromatography interfaced with Nuclear Magnetic Resonance Spectroscopy.

Lock: During the course of an NMR experiment, the system is usually locked to the frequency of the ^{2}H signal from the solvent. By constantly making adjustments that ensure that this frequency does not drift, the entire system is kept stable. The lock signal also provides a means of monitoring the magnetic-field homogeneity, which is optimized by shimming.

Longitudinal Magnetization: The Z component of magnetization.

Loop Collection: An NMR-based collection mode where peaks of interest are "parked" in off-line sample loops awaiting transfer to an NMR flow probe.

Lossy Sample: A sample containing ions from salts or buffers. In "lossy" solutions, the rf power from the transmitter is absorbed by the electric fields created by the ions in the sample. This results in longer pulses and less efficient excitation of nuclei.

Magnetic Moment: Any NMR-active nucleus, when placed in an external magnetic field, begins to precess about the direction of the field. This rotating charge generates a small magnetic field of its own, called its magnetic moment.

Magnetogyric Ratio (γ): Inherent property of a nucleus; can be either positive (e.g., ^{1}H, ^{13}C) or negative (e.g., ^{15}N, ^{29}Si). For particular values of B_0 and the magnetic quantum number, it determines the energy level of the nucleus. This constant relates the magnetic field strength to the resonant frequency of the nucleus. It is designated by the Greek letter γ, and is sometimes called "gamma" by NMR spectroscopists. Note that is usually given in rad/sT, and that it can be either positive or negative. To find the resonant frequency, v_X, of nucleus X in Hz, multiply the field strength B_0 by $2\pi\gamma$:

$$v_X(Hz) = 2\pi\gamma B_0$$

Magnetic Susceptibility: (Latin: susceptibilis "receptiveness") is a dimensionless proportionality constant that indicates the degree of magnetization of a material in response to an applied magnetic field. Magnetic susceptibility may be viewed as a measure of how well a particular solution can accommodate magnetic field lines. Whenever there is a steep transition from one solution composition to another, the density of field lines will change at the interface. This condition will introduce magnetic field heterogeneity in the sample. Shimming to compensate for the heterogeneities is compromised due to the severity of the discontinuity across the sample; hence, this situation broadens the NMR resonances. In flow NMR, the heterogeneity of the sample may be attributed either to incomplete mixing of the contents in the flow cell, or to a steep ramp in solvent composition in the NMR flow cell.

Magnetic Susceptibility Fluids: Fluids used in the construction of microcoil probes where field gradients that vary over very short distances will produce a field profile that cannot be shimmed by macroscopic shim coils that are distant from the sample. Optimization of field uniformity in the sample can be accomplished by immersing the microcoil in solutions with the same susceptibility as a copper metal. Perfluorocarbons are used since these liquids have the same magnetic susceptibility as copper. Perfluorocarbons have the advantage that they introduce no proton background signal. They are commonly used in commercial fabrication of microcoil probes for capillary NMR.

Magnetization Vector (M): Resultant of the individual magnetic moment vectors for an ensemble of a particular type of nucleus. At equilibrium, "M" points in the direction of B_0 (positive γ).

Microcoil: RF coils in an NMR probe that optimize the sensitivity of an NMR spectrometer for analysis of microgram and submicrogram quantities of compounds. The microconfiguration enables the radio-frequency (RF) receiver coil to closely conform to the sample to ensure good detection sensitivity. A microcoil probe maximizes both the observe factor, which is the ratio of the sample volume being observed by the RF coil to the total sample volume required for analysis, and the fill factor. Microcoils produce considerably less background signal effects due to the 100-fold reduction in solvent volume for the mass detected.

Microcoil NMR Spectroscopy: Technology based on the increase of coil sensitivity for smaller coil diameters. Microcoil NMR probes deliver a remarkable mass-based sensitivity increase (8- to 12-fold) when compared with commonly used 5 mm NMR probes. Microcoil NMR probes are a well-established analytical tool for small-molecule liquid-state NMR spectroscopy. These probes have also become available for biomolecular NMR spectroscopy studies.

MicroCryoProbe: A MicroCryoProbe comprises sensitivity afforded with microcoil technology and that of cryo technology. The combination has resulted in unparalleled mass sensitivity (more than an order of magnitude mass sensitivity compared to the conventional 5 mm probe). The sensitivity is ideal for NMR analysis of samples with limited amounts of material (submicrogram quantities).

Microstrip (or Stripline) and Microslot Probe: Probes that use a thin metal strip positioned parallel to B_0 to produce a perpendicular B_1 magnetic field and detect the NMR signal. Such probes have an advantage that they may be easily manufactured using automated well-established lithographic methods. Sample volumes are approximately 5 nL.

Mixing Period: A part of a pulse sequence for correlation spectroscopy consisting of delay times that enable transfer of magnetization.

Multiplex Probes: Probes containing two or more sample chambers that enable simultaneous acquisition of NMR signals from multiple samples.

Natural Abundance: The natural nuclear spin prevalence of various atoms in the periodic table. For example, the predominant isotope of carbon is ^{12}C, with a natural abundance of nearly 99%; however, this has a spin of 0. The NMR-observable form is ^{13}C (I = 1/2), which is 1.1% abundant. This means that, during an NMR experiment, only about 1% of the carbon atoms in the sample are able to contribute to the signal. Hence, it will take much longer to acquire a ^{13}C spectrum than a ^{31}P spectrum (natural abundance of ^{31}P = 100%), given the same number of identical atoms in the samples.

Net Magnetization Vector: A vector representing the sum of the magnetization from a spin system.

NOE (Nuclear Overhauser Effect): A through-space phenomenon used in the study of 3D structure and conformation of molecules. It gives rise to changes in the intensities of NMR resonances of spins I when the spin population differences of neighboring spins S are altered from their equilibrium values (by saturation or population inversion). Proton–proton NOEs are the mostly widely used in structure elucidation. Since the effect has a (non-linear) distance dependence, only protons "close" in space (within 4–5 Å) give rise to such changes and the NOE is thus an extremely useful probe of spatial proximity. The NOE is a spin relaxation phenomenon and has very different behavior depending on molecular motion, and in particular molecular tumbling rates. Small molecules (<1000 Da) under typical solution conditions tumble rapidly and produce weak, positive proton NOEs that grow rather slowly, whereas large molecules (>3000 Da) tumble slowly in solution and so produce large, negative NOEs that grow quickly. Mid-size molecules (~1000–3000 Da) tumble at "intermediate" rates where the NOE crosses from the positive to the negative regime, and thus can have vanishingly small NOEs.

NOESY (**N**uclear **O**verhauser **E**ffect Correlation **S**pectroscop**Y**): Two-dimensional method that correlates pairs of nuclei that are related by the nuclear Overhauser effect.

Normal Distribution: A function that represents the distribution of many random variables as a symmetrical bell-shaped graph. A normal distribution is an arrangement of a data set in which most values cluster in the middle of the range and the rest taper off symmetrically toward either extreme.

One-Dimensional (1D) NMR: Experiment producing a spectrum with only one frequency axis.

Paramagnetism: Magnetic behavior of a substance containing unpaired electrons. When placed in a magnetic field, the induced magnetic field is parallel to the applied field.

Phase Correction (Phasing): Linear combination of the real and imaginary parts of a NMR spectrum to produce 1D spectral peaks with pure absorption mode shape. 2D spectral peaks may be absolute value (magnitude spectra) or phase dependent (signals in-phase or anti-phase with respect to the diagonal). Phasing may be performed either automatically by software or interactively by the operator.

Precession: When a magnetic moment is placed within an external magnetic field, it begins to oscillate slowly about the direction of the field. This motion is called precession. The frequency of the precession is determined by the resonant frequency of the nucleus in question at a particular field strength. Most considerations of the NMR phenomenon circumvent this motion by treating the system in a rotating frame of reference oscillating at the Larmor frequency.

Presaturation: Suppression of a signal from an interfering nucleus (e.g., a solvent peak) by selective irradiation to saturate it prior to application of a nonselective pulse to the whole system.

Probe: The probe is part of spectrometer hardware that holds the sample and some of the electronics necessary to deliver the radio-frequency (rf) pulses and to detect the sample's response. In instruments with superconducting magnets, the probe is a long cylinder (about 24″) which is inserted into the magnet from the base of the magnet. Probes may be capable of detecting one or more nuclei (most commonly ^1H and ^{13}C). Broadband, adjustable probes cover a wide range of frequencies hence enable detection of a wide range of nuclei.

PULCON (**PU**lse **L**ength–based **CON**centration determination): A method that correlates the absolute intensities of two spectra measured in different solution conditions.

Pulse Sequence: Modern NMR experiments are carried out in Fourier-Transform (FT) mode, in which the sample is exposed to one or more rf pulses, followed by recording of the resulting signal. By applying a suitable delay, the system is allowed to return to equilibrium, and the entire sequence is repeated. The level of complexity of pulse sequences varies as the number, phase, shape, and timing of the incoming pulses is manipulated to provide the desired information.

Pulse Fourier Transform NMR: Use of an rf pulse to rotate the magnetization vectors of all the nuclei to be observed. The time domain signal (the FID) is subsequently digitized and converted into the frequency spectrum by Fourier transformation providing a humanly interpretable spectrum.

Pulse Width: Time duration of the rf pulse which determines the tip angle of the magnetization vector.

Q-Factor: The sensitivity of a probe is commonly described in terms of the quality factor parameter, otherwise known as the Q-factor of the probe. The Q-factor is determined by the resonance frequency (ω), the inductance (L), and the resistance (R) of the entire resonant circuit.

qNMR (quantitative NMR): A process that refers to the use of NMR to determine the concentration of one or more chemical species in solution.

Quadruple Moment (Q): Nuclei whose spin I > 1/2 are said to be quadrupolar. Such nuclei possess a nonspherical electric field gradient around the nucleus. In practice, this usually means that NMR lines will be broader relative to nuclei with spin ½. The degree to which broadening occurs depends on the magnitude of Q.

Radiation Damping: An effect on intense signals in high field NMR. It appears during experiments with an intense proteo solvent peak (often water). It occurs when the rotating transverse magnetization of the sample is intense enough to induce a rotating electromotive force in the RF coil that is strong enough to act back on the sample. Hence, the amplitude and phase of the RF pulse is disturbed by feedback from the sample. Radiation damping causes peak broadening, peak asymmetry, phase shifting, and residual signals at 180 and 360 degrees in the flip cycle, all of which can interfere with spectral quality. The effect increases with increasing field strength and affects high-power pulses more than low-power pulses.

Residence Time (τ): In continuous-flow NMR, the residence time refers to the ratio of detection volume to flow rate. Residence time is proportional to the effective lifetime of particular spin states.

Resonance: An exchange of energy between two systems at a specific frequency. The tendency of a nucleus to oscillate at greater amplitude at a particular frequency.

RF Coil: An inductor–capacitor resonant circuit used to set up B_1 magnetic fields in the sample and to detect the rf signal from the sample.

RF Pulse: Short burst of rf energy which has a specific shape. When a burst, or pulse of rf, is sent into the NMR probe, the effect on the nuclei of the sample depends on the power level, duration, and direction of the pulse. In general, variations in the rf pulse are generated by changing the pulse length (or pulse width) and phase and not the pulse power.

ROESY (**R**otating-Frame **NOE** **S**pectroscop**Y**): The ROESY experiment is a 2D pulse sequence. It is similar to the NOESY in that it provides information concerning spatial distance between nuclei in a molecule. This technique is based on NOE in the rotating frame. This pulse sequence is almost identical to one used for TOCSY, hence, TOCSY artifacts may result. To avoid TOCSY artifacts, the power used to achieve spinlock should be reduced and the spinlock offset may be shifted to one end of the spectrum (instead of the center). Contrary to NOE that can experience phase differences in the cross-peaks due to molecular weight, the ROE is always positive. ROESY is particularly well suited for mid-sized molecules (1000–3000 Da) where conventional NOEs may be close to zero. Alternation in sign of the ROE effect in a ROESY spectrum allows one to distinguish effects due to chemical and conformational exchange.

Shimming: The field within the magnet may not be spatially homogeneous relative to the sample, because of slight changes in the position of the sample, concentration gradients, undissolved particles, or movement of objects around the spectrometer. Since the quality of the resulting spectrum depends very strongly on field homogeneity, small adjustments in the field need to be made. These can either be achieved automatically or manually, to compensate for as many of the irregularities as possible. Corrections made to improve field homogeneity are called shimming.

Signal Averaging: Repetition of an experiment "n" times to achieve an increase in signal-to-noise ratio of $n^{1/2}$.

Signal-to-Noise (S/N) Ratio: The ratio of peak signal intensity to the root-mean-square noise level. This value determines the minimum signal that can be detected. The signal-to-noise (S/N) ratio after n scans is given by: $S/N = n^{1/2}I_S/I_N$, where I_S and I_N are the intensities of the signal and noise, respectively. From the equation above, it can be seen that doubling the number of scans only increases S/N by a factor of 1.4 ($2^{1/2}$). To actually double S/N, one must quadruple the number of scans. A tenfold enhancement of S/N requires 100 n and a significant increase in experiment time results.

Solid-Phase Extraction in NMR: Peak trapping cartridges used to "concentrate" peaks as they elute off of a column for NMR detection.

Solvent Suppression: Reduction of an intense signal from the solvent by some means (e.g., a selective pulse or presaturation) to avoid dynamic range problems.

Spectral Density: This condition is defined as the concentration of fields at a given frequency of motion that affects the longitudinal relaxation of nuclei. For small molecules in solution, the tumbling motion is rapid and a larger population of different frequencies exists for a shorter period of time. Hence, the population of specific frequencies that induce relaxation is small. For small molecules, relaxation would not be significantly affected by field strength. The opposite is true for large slow-moving molecules where the concentration of fields induced by molecular tumbling is field dependent. In such cases, as the external field increases the population of nuclei tumbling at a rate needed to induce relaxation decreases. This results in a lower concentration of fields available to induce relaxation and T_1 becomes longer.

Spectral Resolution: Instrument performance parameter that determines minimum peak separation that can be distinguished. Usually specified in terms of line width at half-height called LW (1/2) or LW.

Spectrometer Frequency: Center frequency of a band of frequencies produced by the rf pulse generator. The spectrometer frequency defines the center of the NMR spectrum acquired (also the center of the spectral window).

Spin: A fundamental property of matter responsible for the NMR phenomenon.

Spin-Lattice Relaxation (T_1): After excitation by a pulse, the net magnetization regains its starting, or equilibrium position by interacting with the surroundings, or the lattice, in a process called spin-lattice relaxation. This exponential return is characterized by a time constant, T_1, which can be measured and related to molecular motions. T_1 must be considered when establishing experimental conditions for quantitative analyses; the pulse delay must be set to $5–10 \times T_1$ to insure that relative peak areas are correct.

Spin-Spin Relaxation (T_2): After excitation by a pulse, the net magnetization, which is the vector sum of the magnetic moments of many nuclei, begins to dephase, or spread out by interacting with other spins in a process called spin–spin relaxation. Because of intrinsic molecular differences or inhomogeneities in the magnetic field, the magnetic-moment components precess at slightly different rates. The rate at which they dephase is characterized by a time constant, T_2, which is reflected in the observed linewidths.

Stopped Flow in NMR: A process where the pump and chromatographic flow is stopped at a desired location (within the NMR flow cell) and data acquired.

Superconducting Magnet: A superconducting magnet is an electromagnet made from coils of superconducting wire. The basic construction of such superconducting magnets consists of a solenoid wound with superconducting wire, which comprises the fundamental design of every high-frequency NMR magnet. For superconducting magnets, the wire can conduct much larger electric currents than ordinary wire, hence, very strong magnetic fields can be created. The wire used is usually fabricated from a niobium-titanium alloy (NbTi) or a niobium-tin (Nb_3Sn) alloy. Superconducting magnets can achieve an order of magnitude stronger field than ordinary ferromagnetic-core electromagnets, which are limited to fields of around 2 T. The field is generally more stable, resulting in less noise in experimental measurement and its operation does not require high consumption of electrical power and cooling water as required for the electromagnets. They do require periodic replenishing of liquid nitrogen and liquid helium cryogens.

SUSHY (Spectral Unraveling by Space-selective Hadamard SpectroscopY): A technique that enables recording and deconvolution of NMR spectra of multiple samples loaded in multiple sample tubes in a conventional 5 mm solution NMR probe equipped with a tri-axis pulsed field gradient coil. The individual spectrum from each sample can be extracted by adding and subtracting data that are simultaneously obtained from all the tubes based on the principles of spatially resolved Hadamard spectroscopy.

t-Distribution: Student's t-distribution (or t-distribution) is any member of a family of continuous probability distributions that arises when estimating the mean of a normally distributed population in situations where the sample size is small and population standard deviation is unknown.

Time Domain: Condition where the independent variable of all functions is time (e.g., in the display of an FID).

Tip Angle: Upon experiencing an rf pulse, the magnetic moment associated with a nucleus (spin) aligned with the external magnetic field, B_0, along the +z axis, tips away from the z axis, down toward the x-y plane. The angle between this new position and the +z axis is called the tip angle or the pulse angle. The pulse width and tip angle are directly proportional. If a 5 μsec pulse moves the magnetic moment through 45°, a 10 μsec pulse will tip it 90°, onto the x-y plane. The pulse phase determines the direction in which the magnetic moment moves. If a 90° pulse is sent in on the $-y$-axis, this will place the magnetization along the +x-axis.

TINS (**T**arget **I**mmobilized **N**MR **S**creening): An NMR ligand screening method that reduces the amount of target required for the fragment-based approach to drug discovery. Binding is detected by comparing 1D ^{1}H NMR spectra of compound mixtures in the presence of a target protein immobilized on a solid support to a control sample. TINS can be used in competition mode, allowing rapid characterization of the ligand binding site. This technology provides a means of screening of targets that are difficult to produce or that are insoluble, such as membrane proteins.

Transmitter: A coil of wire (often the same as used by the receiver) and associated electronics that apply radio-frequency energy to the NMR sample.

TOCSY (**TO**tal **C**orrelation **S**pectroscop**Y**): A 2D NMR experiment (also called HOHAHA HOmonuclear HArtmann-HAhn) that permits assignment of a whole spin system of coupled nuclei. The TOCSY experiment requires a single mixing period and some knowledge of spin system topology and couplings. TOCSY generates phase-sensitive spectra without intensity differentiation of the relative coupling, giving better sensitivity for long range coupling interactions. The TOCSY experiment can be used to obtain coupling of units of spins characteristic of part of a molecule that has no scalar couplings to other spins in the molecules (e.g., sequences of polysaccharides, peptides, or proteins).

W_0 Transitions (Zero-Quantum Transitions): Energy-state transition involving a zero-quantum transition, where the α spin simultaneous flips to the β spin and the β spin simultaneous flips to the α spin.

W_1 Transitions (Single-Quantum Transitions): Energy-state transitions corresponding to T_1 relaxation of the spin involving a spin flip of only one of two spins (either I^1 or I^2).

W_2 Transitions (Double-Quantum Transitions): Energy-state transition corresponding to a net double-quantum transition involving a simultaneous spin flip of both spins.

WALTZ-16 Pulse Sequence: A commonly used decoupling sequence often applied in inverse heteronuclear experiments such as HMQC and HMBC. This pulse/phase cycle is known as "WALTZ16" because of the 123 basic building block of the sequence.

Zeeman Effect: The effect where energy levels become non-degenerate in a magnetic field. An essential condition for magnetic resonance spectroscopy.

CHROMATOGRAPHY TERMS

Analyte: A compound of interest to be analyzed by injection into and elution from an HPLC column.

Analytical Column: Typically, the main column used in the HPLC system to separate sample components. An HPLC column used for qualitative and quantitative analysis. A typical analytical column will be 50–250 mm × 4.6 mm but columns with smaller diameters (down to 0.05 mm i.d.) are also considered analytical columns. Columns can be constructed of stainless steel, glass, glass-lined stainless steel, PEEK, and other metallic and nonmetallic materials.

Anion Exchange: The ion-exchange procedure used for the separation of anions. Synthetic resins, bonded-phase silicas, and other metal oxides can be analyzed in this mode. A typical anion-exchange functional group is the tetraalkylammonium ion, which is a strong anion exchanger. An amino group on a bonded stationary phase is an example of a weak anion exchanger.

Back Pressure: A phrase used to describe the pressure required to force fluid at a determined flow rate along a system's flow path, typically expressed in psi, bar, or atm.

CE (**C**apillary **E**lectrophoresis): A special type of chromatographic analysis which employs the use of electricity to selectively separate charged species comprising a sample. It is a powerful technique that produces sharp bands with minimal tailing, with the main difficulty being the sensitivity threshold.

CEC (**C**apillary **E**lectro**C**hromatography): A type of chromatographic analysis which combines the benefits of capillary electrophoresis and HPLC. Typically, with standard HPLC, there is a pressure limitation with the equipment which limits the number of theoretical plates which can be employed. Because flow is driven electro-osmotically in CEC analyses, there is no theoretical plate threshold.

cITP (capillary **I**so**T**acho**P**horesis): A promising variant of microcoil CE. This separation technology separates and concentrates charged species based on their electrophoretic mobilities through the application of a high voltage across a capillary using a two-buffer system. This buffer system is composed of a leading electrolyte (LE) and a trailing electrolyte (TE). When hyphenated to capillary NMR spectroscopy, cITP allows focusing of sample components in the active volume of the capillary probe, affording active volumes on the order of tens of nanoliters and great gains in mass sensitivity.

Chromatogram: A plot stemming from a detector's signal output, typically versus time. It is identified often with a baseline offset by a series of peaks or bands.

Chromatography: A chemical separation technique based on the differential distribution of the constituents of a mixture between two phases, one of which moves relative to the other. Literally meaning "color movement," the separation technique occurs based upon the difference of interaction of sample components with two phases – a "stationary" phase (or non-moving) and a "mobile" phase (or moving).

Chromatographic Conditions: Those parameters that describe how an analysis was achieved, for the purpose of potential future duplication for verification purposes.

Chromatographic Methods: A record of the parameters used in a separation yielding a particular result. It allows another analyst following the method and conditions to reproduce the separation, and achieve the same results.

Column: A tube which contains the stationary phase. The stationary phase differentially interacts with the sample's constituent compounds as they are carried along in the mobile phase.

Column Chromatography: Any form of chromatography that uses a column to hold the stationary phase. Open-column chromatography, HPLC, and open-tubular capillary chromatography are all examples.

Column Packing: The particulate material packed inside the column, also called the stationary phase. This usually consists of silica- or polymer-based particles, often chemically bonded with a chemical function or group. For analytical work, 3μm or 5μm spherical particles are used. For semi-preparative, 10μm or larger spherical or irregular particles are favored.

Dead Volume: The volume outside of the column packing. The interstitial volume (intra-particle volume and inter-particle volume) plus extra column volume (i.e., injector detector, connecting tubing, and end fittings) which combine to create the dead volume.

Degassing: The process of removing dissolved gas from the mobile phase before or during use. Dissolved gas may come out of solution in the detector cell and cause baseline spikes and noise. Degassing is carried out by vacuum evacuation and purging with an inert gas such as nitrogen or helium.

Detector: An electronic device that quantitatively discerns the presence of the separated components as they elute.

Detector Types in HPLC:

- Monochromatic UV detector
- Variable UV/Vis detector
- Diode or photodiode array detector
- ELSD light scattering detector
- ESI-MSD
- ESI-MS^n
- Refractive index detector – (RID) detects all solutes
- Coulometric detector
- Amperometric detector
- Fluorescence detector – detects solutes with fluorophores with very high sensitivity
- Radioactive detector

Detector Sensitivity: The sensitivity setting is the line between normal background noise and a true peak. Perturbations from the baseline that fall below the sensitivity setting are considered noise and are filtered out. Setting the sensitivity too high can result in missing small peaks, while setting it too low may result in the capture of spurious peaks causing the software to attempt to integrate peaks out of the noise.

Diffusion: Dilution of a compound in the direction of the column axis depending on differences in concentration.

ELSD (**E**vaporative **L**ight-**S**cattering **D**etector): A type of detector that works by measuring the light scattered from the solid-solute particles remaining after nebulization and evaporation of the mobile phase. Because the detector's response is independent of the light-absorbing properties of molecules, it can reveal weakly chromophoric sample components that UV detectors miss and provide a more accurate profile of relative component abundance.

Eluent: The mobile phase used to perform a separation.

Flow rate: The volumetric rate of flow of mobile phase through an LC column. For a conventional HPLC column with a 4.6 mm i.d., typical flow rates are 1 to 2 mL/min.

GPC (**G**el **P**ermeation **C**hromatography): A type of size exclusion chromatography (SEC) that separates analytes on the basis of size. The technique is often used for the analysis of polymers.

Gradient Elution: A technique for decreasing separation time by increasing the mobile-phase strength over time during the chromatographic separation.

HILIC (**H**ydrophilic **I**nteraction **LI**quid **C**hromatography): Hydrophilic Interaction Chromatography a variation of normal-phase chromatography without the disadvantages of using solvents that are not miscible with water (also called "reverse reversed-phase" or "aqueous normal phase" chromatography). Stationary phase is a polar material such as silica, cyano, amino, diol, etc. The mobile phase is highly organic with a small amount of aqueous/polar solvent.

HPLC: **H**igh **P**erformance **L**iquid **C**hromatography is a vastly improved form of column chromatography. Instead of a solvent being allowed to drip through a column under gravity, it is forced through under high pressures of up to 400 atmospheres. This condition makes separation much more rapid.

HPLC-SPE-NMR: High-performance liquid chromatography-solid-phase extraction-nuclear magnetic resonance is a novel hyphenation technology that concentrates single chromatographic peaks to elution volumes matching those of NMR flow probes. The SPE unit facilitates the solvent exchange from the mobile phase of the optimized HPLC method to a deuterated NMR solvent.

Injector: A mechanism for accurately injecting a predetermined amount of sample into the mobile-phase stream. The injector can be a simple manual device, or a sophisticated autosampler, that permits automated injections of many different samples for unattended operation.

Ion-Exchange Chromatography: Mode of chromatography in which ionic substances are separated on cationic or anionic sites of the packing in aqueous mobile phases.

Isocratic: A constant composition mobile phase used in liquid chromatography.

Matrix: A medium in which the analyte is distributed.

Method Development: A process for optimizing the separation, including the sample pretreatment, to obtain a reproducible and robust separation. This process involves identifying and optimizing the stationary phase, eluent, and column temperature combination and provides a process to yield the desired separation.

Mobile Phase: The solvent that moves the solute through the column. In LC, the mobile phase interacts with both the solute and the stationary phase and, therefore, can have a powerful influence on the separation.

N (Number of Theoretical Plates): Theoretical plate numbers are an indirect measure of peak width for a peak at a specific retention time. Columns with high plate numbers are considered to be more efficient, meaning they have higher column efficiency than columns with a lower plate count. A column with a high number of theoretical plates will produce a narrower peak at a given retention time than a column with a lower N number.

Narrow-Bore Column: Columns of <0.5 mm i.d. used in HPLC.

Normal-Phase Chromatography: A mode of chromatography carried out with a polar stationary phase and a nonpolar mobile phase. (i.e., adsorption is on silica, aluminum oxide, or polar-bonded silica gels in a normal-phase system). It also refers to the use of polar bonded phases, such as CN or NH_2.

Organic Modifier: Water-miscible organic solvent added to an aqueous mobile phase to obtain separations in reversed-phase HPLC. Common organic modifiers include acetonitrile, methanol, isopropanol, and tetrahydrofuran.

Overload: The overload of a column in preparative chromatography is defined as the sample mass injected onto the column at which efficiency and resolution begins to be effected. Resolution is compromised as the injected sample size is increased further.

Particle Size: The average particle size of the packing in an LC column.

Peak: Elution profile of a compound. When the detector registers the presence of a compound, the normal baseline is affected. The result is a deflection from the baseline called a peak. Well-resolved peaks have Gaussian symmetry, touch the baseline, and do not interfere with other peaks.

Peak Identification: Peak identification is usually performed by comparing the sample chromatogram to a chromatogram of a standard solution separated under the same conditions. Peaks that appear at the same elution time as peaks in the standard are identified as the same component.

Peak Shape: The profile of a chromatographic peak. Theory assumes a Gaussian peak shape (perfectly symmetrical).

PDA (PhotoDiode Array): PDA detectors are UV/Vis detectors that record the absorbance of light at many different wavelengths simultaneously.

Polarity: Measure of lipophilic or hydrophylic properties of a stationary or mobile phase.

Preparative Chromatography: The process of using liquid chromatography to isolate a sufficient amount of material for other experimental or functional purposes. For pharmaceutical or biotechnological purifications, columns several feet in diameter can be used for multiple grams of material. For isolating just a few micrograms of a natural product or a trace impurity, an analytical column can be used. Both are preparative chromatographic approaches.

Resolution: Ability of a column to separate chromatographic peaks.

Retention Time: The time between injection and the appearance of the peak maximum. Delay time of separated compounds due to their interaction with the stationary phase.

Reversed-Phase Chromatography: The most frequently used mode in HPLC. Uses low-polarity packings such as octadecyl or octylsilane phases bonded to silica or neutral polymeric beads. The mobile phase usually is water or water-miscible organic solvents such as methanol or acetonitrile. Elution usually occurs based on the relative hydrophobicity or lipophilicity of the solutes. The more hydrophobic, the stronger the retention. The greater the water solubility of the analyte, the less it is retained. The technique has many variations in which various mobile-phase additives impart a different selectivity. For example, adding a buffer and a tetraalkylammonium salt to an anion analysis would allow ion-pairing to occur and generate separations that rival those of ion-exchange chromatography. More than 90% of HPLC analysts use reversed-phase chromatography.

Sample Preparation: All steps to prepare a sample before separation. Attention to solubility, temperature, and chemical stability is particularly important.

SEC (Size-Exclusion Chromatography): A separation mode employing control pore-size packing to achieve resolution of molecules based on size and shape.

SPE (Solid-Phase Extraction): A sample preparation technique that uses a solid-phase packing contained in a small plastic cartridge. The solid stationary phases are the same as HPLC packings but the principle is different from HPLC. It is sometimes referred to as digital chromatography. This process requires four steps: conditioning the sorbent, adding the sample, washing away the impurities, and eluting the sample in as small a volume as possible with a strong solvent.

Stationary Phase: The immobile phase involved in the chromatographic process.

SFC (Supercritical Fluid Chromatography): A form of normal phase chromatography that is used for the analysis and purification of low-to-moderate molecular weight, thermally labile molecules. It can also be used for the separation of chiral compounds. Principles are similar to those of high-performance liquid chromatography (HPLC); however, SFC typically utilizes carbon dioxide as the mobile phase. Therefore, the entire chromatographic flow path must be pressurized.

Tailing: Unsymmetrical peak formation in which the side of the peak away from the injection site returns very slowly to the baseline. Usually due to an unresolved equilibration and incomplete separation.

UV/Vis (UltraViolet/Visible Light): The tunable or variable wavelength UV/Vis detector is the most popular form of detector. For methods involving organic compounds in aqueous mobile phases, the UV/Vis detector takes advantage of compounds' varying absorptivities of ultraviolet and visible light.

Index